Modelling Molecular Structure and Reactivity
in Biological Systems

Modelling Molecular Structure and Reactivity in Biological Systems

Edited by

Kevin J Naidoo
Department of Chemistry, University of Cape Town, South Africa

John Brady
Department of Food Science, Cornell University, Ithaca, USA

Martin J Field
Laboratoire de Dynamique Moléculaire, Institut de Biologie Structurale, Grenoble, France

Jiali Gao
Univeristy of Minnesota, USA

Michael Hann
Structural and Biophysical Sciences, GlaxoSmithKline Medicines Research Centre, Stevenage, UK

RSCPublishing

The proceedings of the WATOC 2005, Modelling Structure and Reactivity meeting held in Cape Town, South Africa, 16-21 January 2005.

Special Publication No. 304

ISBN-10: 0-85404-668-2
ISBN-13: 978-0-85404-668-3

A catalogue record for this book is available from the British Library

Published by The Royal Society of Chemistry,
Thomas Graham House, Science Park, Milton Road,
Cambridge CB4 0WF, UK

Registered Charity Number 207890

For further information see our web site at www.rsc.org

Printed by Henry Lings, Dorchester, Dorset, UK

Preface

In 2005 more than 400 computational chemists and biophysicists met at the 7th triennial conference of the World Association of Theoretical and Computational Chemists (WATOC 2005) in Cape Town, South Africa. Half of the 120 invited papers describing the state of the art in computational chemistry had a special focus on applications in biology. While much is now known about the molecular functioning of proteins in cells compared with the previous ten years the elucidation of structure and corresponding function of just one class of biomolecules i.e., proteins is phenomenal. However only about one tenth of the proteins that the genome expresses has so far had its three dimensional structure determined and this figure become even smaller when the complexity of the multiple protein interactions that likely exist is taken into consideration. The complexity is further compounded by the incredible variety of mechanisms that proteins express as a result of their individual three dimensional structures. This is where computational chemistry can help to fill the gaps in our understanding and enable meaningful comments on the structure and reactivity of carbohydrates, lipids and proteins, here collectively referred to as biomolecules, to be made.

In this volume we present a thematic selection of papers lead by a review from Martin Quack followed by reports on cutting edge research in *Molecular Conformation and Electronic Structure of Biomolecules*. The state of the art in elucidating the tertiary structure of proteins and saccharides as pioneered in the laboratories of Valarie Dagget, Axel Brunger and John Brady forms part of a compendium of original papers introducing this book. This leads into an accompanying set of manuscripts laying out the landscape of modelling and current computational methods being developed to investigate *Chemical Reactivity in Biological Surroundings*. Here the spectrum of research from Ursula Rothlisberger's applications of Car-Parrinello quantum dynamics to the more pervasive QM/MD methods as reviewed by Jill Gready is covered.

Finally no text would be complete without an exposition of the coal face biological applications of computational methods which is currently being undertaken in the pharmaceutical industry. The leading industrial groups present their work and perspectives of the use of modelling *Toward Drug Discovery* with all of these contributions placed into context by Hugo Kubinyi's review of database driven methods. Computational chemistry is now a vital and integrated part of the drug discovery process both for helping to interpret experimental data. It is equally important and increasing viable for making predictions that help inform on the choices to be made in terms of which compounds to make next. As the challenges of making more cost effective drugs for the developed and developing worlds become ever more demanding, the computational chemist has a unique role to play in helping to understand what molecules we should and should not make for research purposes.

Kevin J. Naidoo, Martin Field and Michael Hann

Contents

Toward Drug Discovery

Molecular Conformation and Electronic Structure of Biomolecules

ELECTROWEAK QUANTUM CHEMISTRY AND THE DYNAMICS OF PARITY VIOLATION IN CHIRAL MOLECULES

Martin Quack

ETH Zürich, Laboratory for Physical Chemistry, Wolfgang-Pauli-Str. 10, CH-8093 Zürich
Switzerland

1 INTRODUCTION

In the introduction to his famous paper "Quantum Mechanics of Many Electron Systems" Paul Adrien Maurice Dirac wrote one of the most cited sentences in quantum chemistry [1]:

"The underlying physical laws for the mathematical theory of a large part of physics and the whole of chemistry are thus completely known and the difficulty is only that the exact application of these laws leads to equations much too complicated to be soluble. It therefore becomes desirable that approximate practical methods of applying quantum mechanics should be developed, which can lead to an explanation of complex atomic systems without too much computation".

It is remarkable that the second part of this statement, which forms a reasonable starting point for modern, approximate numerical quantum chemistry and computational chemistry is only rarely cited. The more frequently cited first sentence with the strong statement about understanding "the whole of chemistry" and the small restriction "the difficulty is only", which claims that the quantum physics of the first half of the 20th century contains all basic knowledge about chemistry, is the one that seems to be liked by many theoretical chemists and physicists. It turns out, however, that this statement is incorrect. There is at least one important part of chemistry, namely stereochemistry and molecular chirality, which can be understood properly only when including the parity violating weak nuclear force in our quantum chemical theory in the framework of what we have termed "electroweak quantum chemistry [2,3], completely and fundamentally unknown at the time of Dirac's statement [1].

Figure 1 summarizes the modern view of the origin of the fundamental interactions as publicized on the website of a large accelerator facility (CERN [4]) According to this view, the electromagnetic force, which is included in the "Dirac-like" ordinary quantum chemistry, leads to the Coulomb repulsion, say, between two electrons in a molecule by exchange of virtual photons. In the picture the two electrons exchanging photons are compared to the ladies on two boats throwing a ball. If we do not see the exchange of the ball, we will observe only the motion of the boats resulting from the transfer of momentum in throwing the ball, and we could interpret this as resulting from a repulsive "force" between the two ladies on the boats. Similarly, we interpret the motion of the electrons resulting from "throwing photons as field particles" as arising from the Coulomb law, which forms the basis of ordinary quantum chemistry. The Coulomb force with the 1/r potential energy law is of long range. The other forces arise similarly. The strong force with very short range (0.1 to 1 fm) mediated by the gluons is important in nuclear physics but has only indirect influence in chemistry by providing the structures of the nuclei, which enter as parameters in chemistry, but there is otherwise usually no need to retain the strong force explicitly in chemistry. The weak force, on the other hand, is mediated by the $W^{\pm}$ and Z^0 Bosons of very high mass (80 to 90 Daltons, of the order of the mass of a bromine nucleus!) and short lifetime (0.26 yoctoseconds = 0.26 x 10^{-24} s).

The Forces in Nature			
Type	**Intensity of Forces (Decreasing Order)**	**Binding Particle (Field Quantum)**	**Important in**
Strong Nuclear Force	~ 1	Gluons (no mass)	Atomic Nucleus
Electro-Magnetic Force	~ 10^{-3}	Photons (no mass)	Atoms and Molecules
Weak Nuclear Force	~ 10^{-5}	Bosons Z, W+, W-, (heavy)	Radioactive β-Decay, Chiral Molecules
Gravitation	~ 10^{-38}	Gravitons (?)	Sun and Planets etc.

The exchange of particles is responsible for the force

CERN AC_Z04_V25/B/1992

Figure 1   *Forces in the standard model of particle physics (SMPP) and important effects. This is taken from the CERN website ref. 4 , but the importance of the weak interaction for chiral molecules has been added here from our work (by permission after ref. 4).*

This force is thus very weak and of very short range (< 0.1 fm) and one might therefore think that similar to the even weaker gravitational force (mediated by the still hypothetical graviton of spin 2) it should not contribute significantly to the forces between the particles in molecules (nuclei and electrons). Indeed, the weak force, because of its short range, becomes effective in molecules, when the electrons penetrate the nucleus, and then it leads only to a very small perturbation on the molecular dynamics, which ordinarily might be neglected completely.

It turns out, however, that because of the different symmetry groups of the electromagnetic and the electroweak hamiltonians there arises a fundamentally important, new aspect in the dynamics of chiral molecules, which we therefore have added to the figure from CERN, where this was not originally included, in our Fig. 1. Indeed the electromagnetic hamiltonian commutes with the space inversion or parity operator $\hat{P}$

$$\hat{P}\hat{H} = \hat{H}\hat{P} \tag{1}$$

which leads to the consequence that in chiral molecules the delocalized energy eigenstates χ_+ and χ_- have a well defined parity and the localized handed states λ and ρ of chiral molecules have exactly the same energy by symmetry (see section 2 for details). Therefore one can also say that the reaction enthalpy $\Delta_R H_0^{\ominus}$ for the stereomutation reaction (2) between R and S enantiomers of a chiral molecule would be exactly zero by symmetry $\left(\Delta_R H_0^{\ominus} \equiv 0\right)$ a fact originally noted already by van't Hoff [5]

$$R \rightleftarrows S \tag{2}$$
$$\Delta_R H_0^{\ominus} = 0 \text{ (? van't Hoff)} \quad \text{or} \quad \Delta_R H_0^{\ominus} = \Delta_{pv} E \cdot N_A \text{ (today)}$$

Today, we know, that the electroweak hamiltonian does not commute with $\hat{P}$

$$\hat{P}\hat{H}_{ew} \neq \hat{H}_{ew}\hat{P} \tag{3}$$

and therefore parity is violated leading to a small but nonzero parity violating energy difference $\Delta_{pv}E$ between enantiomers and thus $\Delta_R H_0^{\ominus} \neq 0$ (for example about 10^{-11} J mol^{-1} for a molecule like CHFClBr [6]). We shall discuss in section 2 in more detail, under which circumstances such small effects lead to observable results dominating the quantum dynamics of chiral molecules.

This symmetry violation in chiral molecules is, indeed, the key concept, which leads to an interesting interaction between high energy physics and molecular physics and chemistry, indeed also biochemistry [7, 8]. It results in the following at first perhaps surprising three statements:

(1) The fundamentally new physics arising from the discovery of parity violation [9-12] and the consequent electroweak theory in the Standard Model of Particle Physics (SMPP) [13-17] leads to the prediction of fundamental new effects in the dynamics of chiral molecules and thus in the realm of chemistry .

(2) Molecular parity violation as encoded in Eqs. (2), (3) has possibly (but not necessarily) important consequences for the evolution of biomolecular homochirality in the evolution of life [7, 8, 18].

(3) Possible experiments on molecular parity violation open a new window to look at fundamental aspects of the standard model of high energy physics, and thus molecular physics might contribute to our understanding of the fundamental laws relevant to high energy physics. Indeed, going beyond parity violation and the standard model, molecular chirality may provide a new look at time reversal symmetry and its violation, in fact the nature of time [8].

It should thus be clear that electroweak quantum chemistry has interesting lessons to tell. A brief history of electroweak quantum chemistry is quickly told. After the discovery of parity violation in 1956/57 [9-12] it took about a decade until the possible consequences for chemistry and biology were pointed out by Yamagata in 1966.[18] While his numerical estimates were wrong by many orders of magnitude (as also a later estimate [19]) and even some of his qualitative reasonings were flawed (see [8]), the link of parity violation in high energy physics and the molecular physics of chirality was thus established and repeatedly discussed qualitatively in the 1970s.[20-27]

The first quantitative calculations on molecular parity violation were carried out following the work of Hegström, Rein and Sandars [28, 29] and Mason and Tranter [30-32] in the 1980s.[30-52] Some far reaching conclusions about consequences for biomolecular homochirality were drawn from some of these early calculations but we know now, that none of these early calculations prior to 1995 can be relied on (nor can one retain their conclusions), as they were wrong even by orders of magnitude.

Indeed, in 1995 we carefully reinvestigated the calculations of parity violating energies in molecules and discovered, surprising to many at the time, that an improved theoretical treatment leads to an increase of calculated parity violating energies by about two orders of magnitude in the benchmark molecules H_2O_2 and H_2S_2.[2, 3] This discovery triggered substantial further theoretical [6-8, 53-67] and experimental activity [68-79] and the numerical results were rather quickly confirmed in independent calculations from several research groups as summarized in table 1. While the earlier overoptimistic conclusions on the selection of biomolecular homochirality had to be revised [8], our work has led to a completely new and much more optimistic outlook on the possibility of doing successful spectroscopic experiments, which are now underway in our own group and others [68-77]. Although no successful experiment has as yet been reported, one may now expect such results in the relatively near future.

The outline of the remainder of this review is as follows:

We shall start in section 2 by a discussion of fundamental symmetries of physics leading us to the conceptual foundations of various types of symmetry breaking. We shall discuss the related opinions or "communities of belief" for the physical-chemical dynamics of chirality, for the selection of biomolecular homochirality and for irreversibility, which are conceptually closely related because of the relationship between parity symmetry P and time reversal symmetry T. We shall then introduce the foundations of electroweak quantum chemistry in section 3, allowing us to calculate $\Delta_{pv}E$ in chiral molecules. This will lead us to the concepts for current experiments in section 4. We shall conclude in section 5 with an outlook on the role of stereochemistry for future experiments on the foundations of the CPT symmetry and the ultimate nature of irreversibility. In part of this we follow an earlier review (in German [80], see also refs. 7, 8, 81-83. We draw also attention to recent reviews [84, 85] as well as to ref. 3 and an earlier review [86] with many further references (see also [87, 88]). We do not aim to be encyclopedic here but rather to provide a conceptual summary.

Table 1 *Parity-violating energy differences $\Delta_{pv}H_0^\circ = \Delta_t H_0^\circ(M) - \Delta_t H_0^\circ(P)$ for H_2O_2, H_2S_2 and Cl_2S_2 at the geometries of the P and M enantiomers indicated (close to but not exactly equilibrium geometries, chosen for intercomparison of different results). The geometry parameters used were: $r_{OO} = 149$ pm, $r_{OH} = 97$ pm, $\alpha_{OOH} = 100°$, $\tau_{HOOH} = 90°$ for H_2O_2, $r_{SS} = 205,5$ pm, $r_{SH} = 135.2$ pm, $\alpha_{SSH} = 92°$, $\tau_{HSSH} = 92°$ for H_2S_2, and $r_{SS} = 194.7$ pm, $r_{SCl} = 205.2$ pm, $\alpha_{SSCl} = 107.55°$, $\tau_{ClSSCl} = 85.12°$ for Cl_2S_2 (see also ref. 8 and references cited therein).*

Molecule	Method	References	$\Delta_{pv}H_0^\circ$ [10^{-12} J mol^{-1}]
H_2O_2	SDE-RHF 6-31G	32	−0.0036
	CIS-RHF 6-31G	2, 3, 54, 89	−0.60
	TDA 6-31G	56, 57	−0.84
	DHF	58	−0.44
	CASSF-LR/cc-pVTZ	55	−0.41
H_2S_2	SDE-RHF 6-31G	32	0.24
	TDA 4.31G	56, 57 (scaled 75%, ref. 54)	14.4
	CIS-RHF 6-31G	3, 54, 89	22.5
	MC-LR RPA/aug-cc-pVTZ	59	22.4
	DHF	58	33.5
Cl_2S_2	MC-LR RPA/aug-cc-pVTZ	60	−15.4

2 FUNDAMENTAL SYMMETRIES OF PHYSICS AND THE VIOLATION OF PARITY AND TIME REVERSAL SYMMETRY IN MOLECULES

2.1 Fundamental Symmetries in Molecular Physics

We shall address here the fundamental symmetries of physics and how they can be investigated by molecular physics, in particular molecular spectroscopy.[87, 88] The following symmetry operations leave a molecular hamiltonian invariant within the framework of traditional quantum chemical dynamics including only the electromagnetic interaction (see refs. 80, 81, 90, 91, for example).

(1) Translation in space
(2) Translation in time
(3) Rotation in space
(4) Inversion of all particle coordinates at the origin (parity operation P or E^*)
(5) "Time reversal" or the reversal of all particle momenta and spins (operation T for time reversal)
(6) Permutation of indices of identical particles (for instance nuclei and electrons).
(7) The replacement of all particles by their corresponding antiparticles (operation C for "charge conjugation", for instance replacing electrons by positrons and protons by antiprotons etc.).

These symmetry operations form the symmetry group of the molecular hamiltonian. It is well known following the early work of Emmy Noether that in connection to an exact symmetry we have a corresponding exact conservation law. For instance (1) leads to momentum conservation, (2) leads to energy conservation, (3) to angular momentum conservation and (4) to parity conservation, that is conservation of the quantum number parity which describes the symmetry (even or odd, positive or negative) of the wavefunction under reflection at the origin. Another interesting observation is that an exact symmetry and conservation law leads to a fundamentally non-observable property of nature (see the discussion in refs. 8, 80, 92). For instance P corresponds to the fundamentally non observable property of the left-handedness or right-handedness of space. That implies that it would be fundamentally impossible to say what is a left handed or right handed coordinate system or an R or S enantiomeric molecular structure, only the opposition of left and right would have a meaning. It would correspondingly be impossible to communicate by a coded message (without sending a chiral example) to a distant civilization that we are made of S (or L)-aminoacids and not of their R (or D) enantiomers. This impossibility is removed, if the exact symmetry is invalid, "broken de lege" as we shall see. We know from nuclear and high energy physics that P, C, and T are individually all violated, as is also the combination CP. Only the combined operation CPT remains an exact symmetry in the current "standard model".[87, 88, 90-97] Parity violation is in fact abundant in a variety of contexts in nuclear and atomic physics, [10-12, 98, 99] CP violation was originally observed in the K-meson decays only [100], but is now also found in the B-Meson system [97] and direct T-violation has been established in 1995 after it had already been inferred from the earlier CP violation experiments.[101]. We have speculated on several occasions that in principle all the discrete symmetries might be violated ultimately in molecular physics (see refs. 80-83, 87, 102, 103, where also some tests for symmetry violations are cited). We shall however, now usefully discuss first in more detail the

fundamental concepts of symmetry violations and symmetry breakings, starting out with
the easy to grasp concepts for parity violation.

2.2 Concepts of Symmetry breaking spontaneous, de facto, de lege in relation to molecular chirality

The three concepts of symmetry breaking should be carefully distinguished [7] and we
shall illustrate this with a chiral molecular example, H_2O_2 (Fig. 2). This molecule is non-
planar, bent, in its equilibrium geometry and an accurate full dimensional potential hypersurface has been formulated for its reaction dynamics.[104]

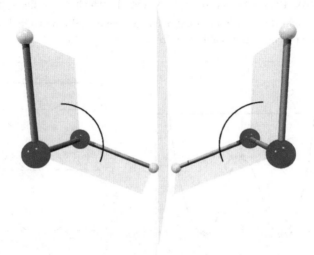

Figure 2   *The H_2O_2 molecule in its enantiomeric mirror image forms at the equilibrium
structure (by permission from ref. 80).* The angle between the two planes indicated is α.

In a simplified fashion the stereomutation reaction (2) interconverting the two enantiomers can be considered as a one dimensional torsional internal rotation about an angle α
indicated in the figure. One has barriers to this motion, a low one in the planar trans form

$$E_{trans} = 4.3 \text{ kJ mol}^{-1} (361 \text{ cm}^{-1}) \qquad (4)$$

and a much higher one in the planar cis structure

$$E_{cis} = 31.6 \text{ kJ mol}^{-1} \text{ (2645 cm}^{-1}\text{)} \qquad (5)$$

We thus represent the dynamics of stereomutation in a simplified manner as the motion of a reduced mass point in a one dimensional double minimum potential (Fig.3), even though the true dynamics correspond to a six-dimensional motion. Indeed, Fig. 3 corresponds to a standard textbook representation of this type of tunnelling stereomutation dynamics. In principle, with parity conservation (Eq. (1)), the eigenstates are delocalized states of positive (χ_+) and negative parity $(-\chi_-)$, which show the symmetry of the mechanical problem (Fig. 3). In classical dynamics there is also a symmetrical state at the maximum of the potential in unstable equilibrium (see Fig. 3A).

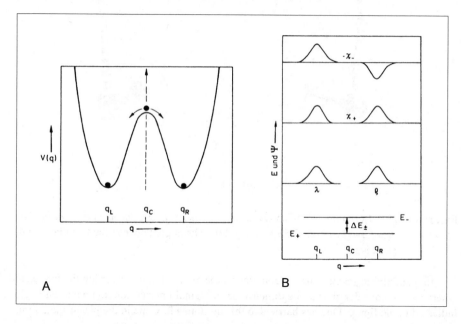

Figure 3 *The double minimum potential for illustrating symmetry breaking (by permission after ref. 105). Wavefunctions are shown at right.*

In practice, in chemistry and similar problems in physics, one finds states that are either left (λ, q_λ) or right (ρ, q_R), localized classically or quantum mechanically, which do not show the symmetries of the hamiltonian. The symmetry is broken *spontaneously* in classical dynamics, where at low energy only one (either left or right) solution to the dynamics is possible. The symmetrical state would be impossible at any energy below the barrier maximum. Spontaneous symmetry breaking is essentially a classical concept, which can be extended to quantum mechanical systems with an infinite number of degrees of freedom.[106-108]

In molecular quantum mechanics, the superposition principle guarantees that we have also the symmetrical state of positive parity

$$\chi_+ = \frac{1}{\sqrt{2}}(\lambda + \rho) \tag{6}$$

and the antisymmetrical state of negative parity

$$-\chi_- = \frac{1}{\sqrt{2}}(\lambda - \rho) \tag{7}$$

These states are delocalized, right and left structures simultaneously. They correspond to the eigenstates of both the Hamiltonian $\hat{H}$ and parity $\hat{P}$ operators in Eq. (1) for a parity conserving hamiltonian. According to Hund [109] one can generate from the eigenstates the initial condition of a localized state by superposition, either left

$$\lambda = \frac{1}{\sqrt{2}}(\chi_+ - \chi_-) \tag{8}$$

or right

$$\rho = \frac{1}{\sqrt{2}}(\chi_+ + \chi_-) \tag{9}$$

The symmetry is now broken *de facto* by the chosen initial condition and these states show a time evolution following the time dependent Schrödinger equation [110]

$$i\frac{h}{2\pi}\frac{\partial \Psi(q,t)}{\partial t} = \hat{H}\Psi(q,t) \tag{10}$$

with the solution

$$\Psi(q,t) = \sum_k c_k \varphi_k \exp(-2\pi i E_k t / h) \tag{11}$$

For just two levels with energies E_+ and E_- and $\Delta E_{\pm} = E_- - E_+$ (see Fig. 3) the evolution reduces to

$$\Psi(q,t) = \frac{1}{\sqrt{2}}\exp(-2\pi i E_+ t)\left[\chi_+ + \chi_- \exp(-2\pi i \Delta E_{\pm} t / h)\right] \tag{12}$$

The observable probability density

$$P(q,t) = \Psi(q,t)\Psi^*(q,t) = |\Psi|^2 = \frac{1}{2}\left[\chi_+ + \chi_- \exp\left(-2\pi i \Delta E_\pm t / h\right)\right]^2 \qquad (13)$$

shows a period of motion

$$\tau = \frac{h}{\Delta E_\pm} \qquad (14)$$

and a stereomutation tunnelling time from left to right (or the reverse) which is just half of this value

$$\tau_{\lambda \to \rho} = \frac{h}{2\Delta E_\pm} = \frac{1}{2c\Delta\tilde{\nu}} \qquad (15)$$

Table 2 shows stereomutation times for H_2O_2 for different degrees of excitation [111] in various modes. It should be made clear that the times shown in table 2 correspond to the full 6-dimensional wavepacket motion, not to the simple 1-dimensional textbook model [111, 112].

Table 2   *Tunneling splittings and stereomutation times in H_2O_2 (after refs. 80, 111, 112)*

$\tilde{\nu}/cm^{-1}$		Exp.	6D	RPH	$\tau_{\lambda \to \rho}(6D)$
0	ν_0	11.4	11.0	11.1	1.5 ps
3609	ν_1	8.2	7.4	7.4	2.0 ps
1396	ν_2	(2.4?)	6.1	5.0	2.7 ps
866	ν_3	12.0	11.1	10.8	1.5 ps
255	ν_4	116	118	120.0	0.14 ps
3610	ν_5	8.2	7.6	8.4	2.2 ps
1265	ν_6	20.5	20.8	21.8	0.8 ps

If we add now to the hamiltonian the electroweak parity violating part, this will add a tiny antisymmetrical effective potential as shown in Fig. 4 for the cases of H_2O_2 and H_2S_2. While these extra potentials should not naively be considered as Born-Oppenheimer like potentials, it is true that, without these extra potentials, the molecular Hamiltonian (beyond the Born-Oppenheimer approximation and exact to all orders) shows exact inversion symmetry (effectively about the planar structures at 180° and 0° in Fig. 4), whereas with the additional potentials the symmetry is broken *de lege* (from the Latin lex = law) as in the fundamental natural law now there appears an asymmetry (we prefer *de lege* in the nomenclature, as the alternative *de iure* would refer to the man made conventional law = ius in Latin). Obviously one of the minima will be slightly preferred and the explanation of the localized states as arising from a de lege symmetry breaking is fundamentally different from a situation with a de facto symmetry breaking. It turns out, that the distinction has a quantitative aspect, related to the relative size of the tunnelling splittings $\Delta E_\pm$ with the hypothetical symmetrical potential and the parity violating energy difference $\Delta_{pv}E$, which is about twice the parity violating potential at the left or right equilibrium geometry

$$\left|\Delta_{pv}E\right| \simeq \left|E_{pv}(R)-E_{pv}(L)\right| = 2\left|E_{pv}(R)\right| = 2\left|E_{pv}(L)\right| \tag{16}$$

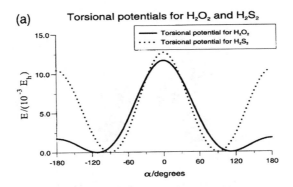

(a) Torsional potentials for H$_2$O$_2$ and H$_2$S$_2$

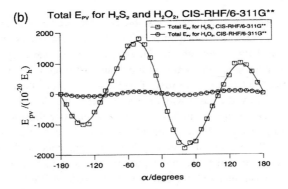

(b) Total E$_{pv}$ for H$_2$S$_2$ and H$_2$O$_2$, CIS-RHF/6-311G**

Figure 4 *Parity conserving and parity violating potentials for H$_2$O$_2$ and H$_2$S$_2$ (by permission from ref. 54)*

If one has

$$\Delta E_\pm \gg \Delta_{pv}E \tag{17}$$

then the "*de facto*" symmetry breaking provides the proper explanation for the chiral molecule dynamics whereas in the case with

$$\Delta_{pv}E \gg \Delta E_\pm \tag{18}$$

the *"de lege"* symmetry breaking prevails. Comparing the numerical results for H_2O_2 in Table 1 and Fig. 4 with the results in Table 2 it is obvious, that the de facto symmetry breaking is the proper explanation for H_2O_2 and parity violation is effectively irrelevant for the tunnelling dynamics of H_2O_2. However, H_2O_2 is also not a normal chiral molecule that has stable enantiomers. Indeed the lifetime of enantiomers even at low energy is only about one ps.

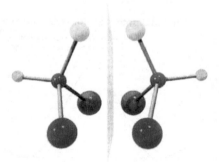

Figure 5   *The enantiomeric mirror image isomers of CHFClBr*

   If we consider as typical examples for stable chiral molecules the substituted methane derivatives shown in Fig. 5, the situation is quite different. The corresponding parity con-serving and parity violating potentials are shown in Fig. 6 schematically for chiral ground and some achiral excited electronic states. It turns out, that in molecules such as CHFClBr with barriers to stereomutation of the order of 100 to 400 kJ mol^{-1} the tunnelling would be completely suppressed by the parity violating asymmetry and thus the *de lege symmetry breaking* is the correct explanation in this case. This is only one illustration of the concep-tual subtleties arising in such a basic chemical question such as to the origin of molecular chirality. Indeed the three cases of symmetry breaking spontaneous, de facto, de lege are only three examples of several explanations put forward to interpret the physical chemical origin of chirality and summarized in our review of the topic in 1989.[7] Prior to this critical discussion, there existed essentially non interacting "communities of belief" for explaining chirality, each of which had a self consistent set of explanations. We shall summarize these here, as very similar situations arise in the discussion of biomolecular homochirality and irreversibility, which we shall discuss in turn thereafter.

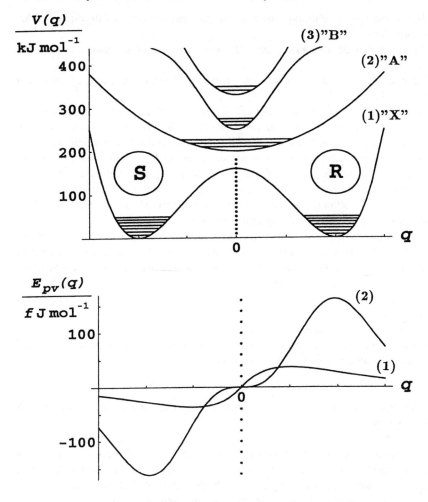

Figure 6 *Schematic representation of parity conserving and parity violating potentials for ground and excited states (by permission after ref. 3).*

2.3 Communities of belief and hypotheses for molecular chirality

As discussed in detail in ref. 7, we have in the scientific literature the following fundamental hypotheses for molecular chirality (or dissymmetry):

(1) The classical structural hypothesis according to van't Hoff [5, 113, 114] (a type of spontaneous symmetry breaking)

(2) The hypothesis of quantum mechanical de facto symmetry breaking following Hund 1927.[109]

(3) The hypothesis of chirality induced by the environment (collisions or other perturbations).[115, 116]

(4) The hypothesis of a superselection rule (another type of spontaneous symmetry breaking).[106-108]

(5) The hypothesis of dominant de lege symmetry breaking because of molecular parity violation.[7, 8, 18, 20, 21]

We shall not discuss details here but refer to refs. 7, 80 for further discussion. The short summary of the current theoretical status would be, that for molecules such as H_2O_2 hypothesis (2) is appropriate, whereas for CHFClBr and similar stable chiral molecules hypothesis (5) should apply. It is still not entirely clear, under which circumstances the hypotheses (3) and (4) might perhaps provide the most appropriate explanation. We have, however, pointed out, that the questions can, in principle, be decided by carefully designed experiments,[7, 80-82, 105] although the experiments are nontrivial and predict sometimes striking outcomes depending upon the hypothesis chosen [81].

Therefore there remain some open questions, although we think that some of the questions have been resolved today at least theoretically following the quantitative approaches described in section 3. It is of some interest that one can find today in larger audiences almost always a few proponents of every one of the "communities of belief", thus there remains cause for controversy, although the controversies are not usually made the subject of open debate.

We have taken some space here to describe this conceptually somewhat complex situation, because very similar situations arise for the questions of biomolecular homochirality and irreversibility, which we shall discuss now.

2.4 Communities of belief and hypotheses on the origin of biomolecular homochirality

The close relationship between chirality and the biochemistry of life was addressed at an early time by Pasteur, who found biomolecular homochirality (today expressed as the dominance of L-amino acids and D-sugars in the biopolymers of life). Since that time, the selection of observed enantiomers in living matter has been the subject of controversy. One has essentially the following main groups, which can be further subdivided in many variants.[80]

1. The stochastic "all or nothing" selection of one enantiomer (D or L) in an appropriate nonlinear biochemical selection mechanism [117-123] or by an abiotic mechanism (for instance crystallization).[124-126] According to this (de facto or chance selection) hypothesis only one enantiomer (either D or L) is selected in a given "evolution experiment", whereas in many different experiments both D and L appear with equal statistical probability.

2. There may be an external chiral influence inducing one type of chirality in a singular initial evolution step, which is later always propagated. There have been numerous different suggestions of this type starting already with the work of Pasteur and many later discussions (for instance life could have been formed under the influence of circularly polarized light or on the surface of an L-quartz mineral). We refer

to refs. 80, 127-131 for such hypotheses. These can be "*de facto*" but also "*de lege*", if one has the external influence, say, of parity violating β-radioactivity.

3. One might have abiotic selection of an enantiomer de lege in a low temperature phase transition under the influence of parity violation.[49-51, 132]

4. One might have preferred thermodynamic or kinetic selection in a mechanism that is asymmetric under the influence of parity violation de lege. In such a selection in many separate evolution experiments one would obtain preferentially one type (say L-amino acids).[18, 86, 133-135]

Roughly speaking all these can be classified in the two large classes of de facto ("chance") and de lege ("necessity") selection processes. There have been numerous heated debates but very little firm evidence – neither theoretical nor experimental. In our opinion, the question is completely open[8, 80] and not likely to be resolved in the near future. There have been claims of abiotic de lege selection in crystallization experiments[136]. It is not clear how plausible these are. In our opinion two necessary conditions would have to be met in order to provide adequate scientific experimental proof for any of these proposals.

(i) All possible counter-experiments to disprove the given hypothesis must be made in addition to the "positive" observation of a given outcome.

(ii) The detailed mechanism of the observed selection must be understood theoretically.

It will be difficult to meet these conditions, but it is true that, in principle, the question of which hypothesis applies could be resolved experimentally – it is not mere "philosophy". It might be, by the way, that under different quantitative conditions, different hypotheses might be applicable, somewhat similar to the discussion of molecular chirality in the previous section, but much more complex.

2.5 Communities of belief or hypotheses on irreversibility and the origin of the second law

After our discussion of the relatively simple concepts of symmetry breaking related to molecular chirality, we can now, by analogy, discuss the kind symmetry breaking of time reversal symmetry which ultimately is at the origin of irreversibility and the second law of thermodynamics[80-82]. Again we find several hypotheses that might be broadly classified as de facto (1,3) or de lege (2,4,5,6) symmetry breakings.

(1) Irreversibility as de facto symmetry breaking is the standard statistical interpretation of the origin of the second law of thermodynamics following Boltzmann[137-141]. There have been a number of detailed illustrations of this hypothesis[119, 142-147] and a special variant of this hypothesis uses information theory, which might also be classified as a separate hypothesis.

(2) One might rigorously stick to the second law as an observed empirical, macroscopic law.[148, 149]

(3) One can introduce irreversibility through interaction with the environment similar to hypothesis (3) on chirality. In connection with introducing a reduced density matrix description of the dynamics of a subsystem this is a common textbook trick to get the desired results on irreversibility.

(4) One might add irreversible damping in the mechanical laws.[150, 151] Sometimes a connection with the quantum mechanical measurement process [152] or with gravitation [153] is made.

(5) One possibility is also to invoke de lege violation of time reversal symmetry as observed in K-meson decay.[101]

(6) Finally, one might speculate about a possible CPT-symmetry violation as the most fundamental source of irreversibility, in a generalized sense.[80-83, 87, 102, 103]

Debates between proponents of these various hypothesis have sometimes been rather spiteful,[140, 154, 155] probably because of the somewhat metaphysical touch that the "arrow of time" has for all human beings, connected to the inevitable transitoriness of human existence. We can summarize again, that while various hypotheses can provide selfconsistent "explanations" of the observed irreversibility in chemistry and physics, some of the explanations contradict each other and the question which one applies remains open. One key question is, whether the second law of thermodynamics finds its origin in a de facto or de lege symmetry breaking.[80]

2.6   Violation of the Pauli Principle

For reasons of completeness we mention this point here, although it is much more speculative and there is really very little evidence in favour of a possible de lege symmetry breaking in this case (de facto symmetry breakings are common). We refer to [80] and the references cited there for further discussion.

2.7   Summary

We have shown, that the situation of our understanding of molecular chirality, biomolecular homochirality, and irreversibility has close parallels for all three cases, the understanding of molecular chirality being presumably most advanced.

In all cases one key question concerns the opposition of the hypotheses of de facto symmetry breaking versus de lege symmetry breaking. This question is not simply a qualitative one, but may depend on quantitative aspects of the problem or the relative magnitude of various parameters governing the dynamics. Its investigation is thus open to quantitative theory and experiment. As we have discussed in some detail for the case of molecular chirality and parity violation, the open questions concerning de facto and de lege symmetry breakings can be answered for a given molecule, in principle.

3   THE QUANTITATIVE THEORY FOR PARITY VIOLATION IN ELECTROWEAK QUANTUM CHEMISTRY

Early quantitative calculations were carried out by Hegström, Rein and Sandars 1980.[29] As we have pointed out in the introduction, there was a dramatic change in our quantitative understanding of parity violating potentials calculated from electroweak quantum chemistry about a decade ago.[2, 3] One starting point for this was our observation [2] that there were surprisingly large deviations of the older theoretical calculations from simple estimates following an equation proposed by Zel'dovich and coworkers

$$\frac{\Delta_{pv}E}{h} \simeq 10^4 \frac{Z^5}{100^5} \, \text{Hz} \tag{19}$$

This by itself was not such a strong argument, given the complexity of the problem and the many possibilities for compensation of contributions leading to lower values of parity violation than expected from simple estimates. Indeed, we could rationalize such compensations by analysing the calculated parity violating potentials in terms of a trace of a tensor [2, 3] under certain conditions, thus

$$E_{pv} = E_{pv}^{XX} + E_{pv}^{YY} + E_{pv}^{ZZ} \tag{20}$$

As the three components frequently differ in sign, this explains a certain lowering below the maximum possible values realized for the individual components. However, a more serious observation is related to the RHF wave functions used in the older calculation being really quite inappropriate. Indeed, the simplest improvement of using excited state CIS (configuration interaction singles) wavefunctions already introduced an increase of parity violating potentials by about 2 orders of magnitude,[2, 3, 54] a result later corroborated by our much improved MC-LR approach [55] and further confirmed independently by several other groups as well as in further calculations by our group.[56-67]

We shall provide here a brief outline of the new theory following [2, 3, 54, 55], in order to provide a basic understanding also of the limitations and omissions in current approaches (see also ref. 84 and further references cited therein).

3.1 Basic theory

In the framework of the standard model, the relevant parity violating interaction is mediated by the electrically neutral Z^0 Bosons. At molecular energies, which are much lower than the energy corresponding to mc^2 of the Z^0-Boson (91.19 GeV) the contribution of Z^0 becomes virtual [156, 157]. This leads at low energies to the hamiltonian density of the fully relativistic parity violating electron-neutron interaction of the following form (with $\hbar \equiv c \equiv 1$)

$$\hat{H}^{(e-n)}(x) = \frac{G_F}{2\sqrt{2}} g_A \left(1 - 4\sin^2 \Theta_w\right) j_\mu[\psi^{(el)}(x)] \tag{21}$$

$$\times j_{(ax)}^\mu[\psi^{(n)}(x)] + \frac{G_F}{2\sqrt{2}} j_\mu[\psi^{(n)}(x)]$$

$$\times j_{(ax)}^\mu[\psi^{(el)}(x)]$$

with $j_{(ax)}^\mu[\psi(x)]$ involving the familiar γ-matrices.[157]

$$j_{(ax)}^{\mu}[\psi(x)] \overset{def}{=} : \Psi^{+}(x)\gamma^{0}\gamma^{\mu}\gamma^{5}\Psi(x) \tag{22}$$

The γ^{5} matrix converts the 4vector $j^{\mu}[\psi(x)]$ into the axial vector $j_{(ax)}^{\mu}[\psi(x)]$. Similar expressions are obtained for the electron proton and the electron electron interaction. [3, 158]

In principle, as pointed out in [3] one can use these relativistic equations as a starting point for the theory: Relativistic theories of this type have been carried out at different levels of approximation for instance by the group of Barra, Robert and Wiesenfeld (Hückel-type)[38, 39, 42, 43] and four component relativistic theory by Schwerdtfeger and coworkers.[58, 64, 65] Omitting the two small components of the bispinors $\Psi''(x)$ and $\Psi^{el}(x)$ and thus converting from four component bispinors to two component spinors following [3, 98, 158, 159] one obtains

$$\hat{H}^{(e-n)}(x) \tag{23}$$

$$= \frac{G_{F}}{4\sqrt{2}\,\mu c}[-\psi^{\dagger(n)}(x)\psi^{(n)}(x)\{\psi^{\dagger(el)}(x)\sigma\left(P\psi^{(el)}(x)\right)+\left(P^{*}\psi^{\dagger(el)}(x)\right)\sigma\psi^{(el)}(x)\}$$

$$+ig_{A}\left(1-4\sin^{2}\Theta_{w}\right)P\left(\psi^{(el)}(x)\sigma\psi^{(el)}(x)\psi^{\dagger(n)}(x)\right)\sigma\psi^{(n)}(x)]$$

G_{F} is the Fermi constant, P the momentum operator, σ the doubled spin operator which has as components the familiar 2 x 2 Pauli matrices, x the spatial coordinate set and μ the reduced mass of the electron. The last term in Eq. (23) is taken to be small because of the factor $\left(1-4\sin^{2}\Theta_{w}\right)\simeq 0.08$ and because of the dependence on neutron (and similarly proton) spin with the tendency of spin compensation in nuclei. The form factor g_{A} (from the strong interaction of the neutron) can be taken as 1.25.[3]

Replacing finally the neutron density by a delta function because of the contact like nature of the very short range weak interaction

$$\psi^{\dagger(n)}(x)\psi^{(n)}(x) \simeq \delta^{3}\left(x-x^{(n)}\right) \tag{24}$$

one obtains a Hamilton operator for the electron neutron interaction

$$\hat{H}^{(e-n)} = -\frac{G_{F}}{4\mu c\sqrt{2}}\left(P\sigma\delta^{3}\left(x-x^{n}\right)+\delta^{3}\left(x-x^{n}\right)P\sigma\right) \tag{25}$$

For the electron proton interaction the hamiltonian is similarly

$$\hat{H}^{(e-p)} = -\frac{G_{F}}{4\mu c\sqrt{2}}\left(1-4\sin^{2}\Theta_{W}\right)\cdot\left(P\sigma\delta^{3}\left(x-x^{(p)}\right)+\delta^{3}\left(x-x^{(p)}\right)P\sigma\right) \tag{26}$$

Collecting the terms for neutrons and protons together and defining an electroweak charge Q_{a} of the nucleus a with charge number Z_{a} and neutron number N_{a}

$$Q_a = Z_a \left(1 - 4\sin^2 \Theta_W\right) - N_a \tag{27}$$

one gets an effective electron nucleus interaction

$$\hat{H}_a^{(e-nucleus)} = -\frac{G_F}{4\mu c\sqrt{2}} Q_a \left(\boldsymbol{P}\boldsymbol{\sigma}\delta^3 \left(\boldsymbol{x} - \boldsymbol{x}^{(nucleus)}\right) + \delta^3 \left(\boldsymbol{x} - \boldsymbol{x}^{(nucleus)}\right) \boldsymbol{P}\boldsymbol{\sigma} \right) \tag{28}$$

In addition to the electron nucleus interaction, one should consider the electron-electron interaction.

$$\hat{H}^{(e-e)} = \frac{G_F}{2\mu c\sqrt{2}} \left(1 - 4\sin^2 \Theta_W\right) \cdot \left\{ \delta^3 \left(\boldsymbol{x}^{(1)} - \boldsymbol{x}^{(2)}\right), \left(\boldsymbol{\sigma}^{(1)} - \boldsymbol{\sigma}^{(2)}\right) \cdot \left(\boldsymbol{P}^{(1)} - \boldsymbol{P}^{(2)}\right) \right\}_+ \tag{29}$$
$$+i \left[\delta^3 \left(\boldsymbol{x}^{(1)} - \boldsymbol{x}^{(2)}\right), \left(\boldsymbol{\sigma}^{(1)} \times \boldsymbol{\sigma}^{(2)}\right) \cdot \left(\boldsymbol{P}^{(1)} - \boldsymbol{P}^{(2)}\right) \right]_-$$

with obvious notation for the two electrons 1 and 2 in a pair $\{,\}_+$ for the anticommutator and $[,]_-$ for the commutator.

The electron-electron contribution to the effective parity violating potential is considered small,[3] below 1% of the other contributions, because of the small prefactor and the lack of a corresponding enhancement with Q_a as well as compensation of terms from different electron-electron pairs. Thus this term is usually neglected, although one must remember that it really consists of a sum over many electron pairs. Assembling all terms together and introducing the electron spin $\hat{s}$ (with dimension) together with linear momentum $\hat{p}$, the electron mass m_e and consistent SI units throughout one obtains finally for the electron-nucleus part of the hamiltonian

$$\hat{H}_{pv}^{(e-nucl)} = \frac{\pi G_F}{m_e hc\sqrt{2}} \sum_{i=1}^{n} \sum_{a=1}^{N} Q_a \left\{ \hat{s}_i \hat{p}_i \delta^3 \left(r_i - r_a\right) + \delta^3 \left(r_i - r_a\right) \hat{s}_i \hat{p}_i \right\} \tag{30}$$

The sums extend over n electrons and N nuclei. This operator can be evaluated in different ways. The simple perturbative sum over states expression in the Breit Pauli approximation for the spin orbit interaction reads for the parity violating potential

$$E_{pv} = 2\,\mathrm{Re} \left\{ \sum_n \frac{\left\langle \psi_0 \left| \hat{H}_{pv}^{e-nucleus} \right| \psi_n \right\rangle \left\langle \psi_n \left| \hat{H}_{SO} \right| \psi_0 \right\rangle}{E_0 - E_n} \right\} \tag{31}$$

The Breit Pauli spin orbit hamiltonian $\hat{H}_{SO}$ is as usual

$$\hat{H}_{SO} = \frac{\alpha^2}{2} \left[\sum_{i=1}^{n} \sum_{a=1}^{N} Z_a \frac{\hat{l}_{i,a} \hat{s}_i}{\left|\vec{r}_a - \vec{r}_i\right|^3} + \sum_{i=1}^{n} \sum_{j \neq i}^{n} \frac{\hat{l}_{i,j} \left(\hat{s}_i + \hat{s}_j\right)}{\left|\vec{r}_i - \vec{r}_j\right|^3} \right] \tag{32}$$

where $\hat{l}_{i,k}$ refers to the orbital angular momentum of electron i with respect to particle number k.

The sum over states expression (31) essentially mixes the electronic ground state singlet function with excited state triplets in order to obtain a parity violating energy expectation value for the true (mixed singlet-triplet) ground state (for a pure singlet this would vanish). However, the sum over states expression, when used explicitly, converges slowly for larger molecules. It is well known in the framework of propagator methods [160, 161] that the expression in Eq. (31) is equivalent to the expression from response theory in Eq. (33) [55]

$$E_{pv} = \left\langle\!\left\langle \hat{H}_{pv}; \hat{H}_{SO} \right\rangle\!\right\rangle_{\omega=0} = \left\langle\!\left\langle \hat{H}_{SO}; \hat{H}_{pv} \right\rangle\!\right\rangle_{\omega=0} \tag{33}$$

One can say, that the parity violating potential E_{pv} is the response of $\left\langle \psi_0 \left| \hat{H}_{pv} \right| \psi_0 \right\rangle$ to the static ($\omega = 0$) perturbation $\hat{H}_{SO}$ or vice versa. This multiconfiguration linear response approach (MCLR) was derived in [55], to which we refer for details. It shows much better convergence properties than when evaluating Eq. (31) directly.

We have given this brief summary of the theory developed in more detail in refs. [2, 3, 55] in order to show all the steps of the many successive approximations made. Each of these approximations can be removed, when the necessity arises. For instance, if one wishes to describe explicitly hyperfine structure components or NMR experiments one must not neglect the spin dependent terms and therefore one has to add to the operator of Eq. (30) a further operator given by Eq. (34) [55]:

$$\hat{H}_{pv2}^{(e-nucl)} = \frac{\pi G_F}{m_e hc\sqrt{2}} \sum_{i=1}^{n} \left[\sum_{a=1}^{N} (-\lambda_a)\left(1 - 4\sin^2\Theta_W\right)\left\{ \hat{p}_i \hat{l}_a, \delta^3\left(\vec{r}_i - \vec{r}_a\right)\right\}_+ \right. \tag{34}$$
$$\left. + \left(2i\lambda_a\right)\left(1 - 4\sin^2\Theta_W\right)\left(\hat{s}_i \times \hat{l}_a\right)\left[\hat{p}_i, \delta^3\left(\vec{r}_i - \vec{r}_a\right)\right] \right]$$

Also, sometimes the approximate "theoretical" value of the Weinberg parameter $\sin^2\Theta_W = 0.25$ is taken, which simplifies the expressions with $\left(1 - 4\sin^2\Theta_W\right) = 0.0$. But more generally the accurate experimental parameter will be used with $\left(1 - 4\sin^2\Theta_W\right) = 0.07$ which may further depend on the energy range considered. Furthermore, one might use the semirelativistic expressions using the Breit Pauli spin orbit operator (32) and the operators for parity violation in Eqs. (30) and (34). This should be an excellent approximation for nuclei with maximum charge number $Z_a = 20$ and acceptable up to $Z_a = 40$. However, for more highly charged nuclei one must return to the relativistic equation (21) and from there derive various approximate relativistic expressions for instance in the four component Dirac Fock framework [58] or within two component relativistic approximations.[162] On the other hand one might also use more approximate treatments such as density functional theory.[163]

One might also consider to investigating explicit "nonvirtual" couplings going beyond the use of Eq. (20) as starting point or one might include the electron-electron parity violating interaction in Eq. (29) in the calculations. Where one wishes to invest effort in removing some of the approximations used, depends upon one's intuition whether large improvements are to be expected. At present it seems unlikely that order of magnitude improvements will again be found in the future, although only experiment can give a definitive answer. We think that the currently largest chance for improvement resides in appropriate electronic wavefunctions that are highly accurate in particular near the nuclei and in further effects from molecular structure and motion to be discussed now.

3.2 Parity violating potential hypersurfaces and vibrational effects

Two qualitative aspects of the structure of parity violating potentials deserve mention. Firstly, similar to the parity conserving electronic potential, the parity violating potentials are a function of all 3N-6 internal nuclear degrees of freedom in the molecule. Thus the parity violating potentials E_{pv} defined by Eqs. (31) or (33) define a parity violating potential hypersurface

$$E_{pv} = V_{pv}\left(q_1, q_2, q_3 \cdots q_{3N-6}\right) \tag{35}$$

While isolated distortions or individual coordinate displacements have been considered for some time in such calculations (see Fig. 4 for instance) the true multidimensional aspects have been considered only more recently.[6, 61, 67, 164, 165]. The spectroscopically observable parity violating energy differences $\Delta_{pv}E$ have to be computed as appropriate expectation values of the parity violating potential in Eq. (35) for the multidimensional rovibrational state with anharmonically coupled vibrations. This leads to sizeable effects as was shown recently.[164] However, we know from our work in rovibrational spectroscopy and dynamics of polyatomic molecules [166] that this problem can be handled accurately for not too complex molecules [68-70, 81, 104, 111, 112] and a similar statement applies to the other important dynamical problem: tunnelling (section 3.3).

A second general aspect of the parity violating potential arises from the structure of the hamiltonian in Eq. (30). Because of the contact like interaction between electrons and nuclei, the parity violating potential can be written as a sum of contributions from the individual nuclei.

$$V_{pv}\left(q_1, q_2, q_3 \cdots q_{3N-6}\right) = \sum_{a=1}^{N} V_{pv}^a\left(q_1, q_2, q_3 \cdots q_{3N-6}\right) \tag{36}$$

Because of the approximate Z^5 scaling (see, however [165]) this allows for an easy analysis of calculations and also some rough estimates. Because the electronic wavefunction generally depends upon the coordinates $\left(q_1, q_2, q_3 \cdots q_{3N-6}\right)$ in a very complex fashion, there are, however, no really simple and generally accurate estimates to be expected. However, one can derive certain sets of rules for special cases (see [167] for example).

3.3   The interplay of tunnelling dynamics and parity violation

From our discussion in section 2.2 the important role of tunnelling dynamics for parity violation is obvious.

Table 3   *Tuning tunnelling and parity violation in a series of molecules (after ref. 168)*

Molecule	$\left|\Delta E_{pv}^{el}\right|\left(cm^{-1}\right)$	$\left|\Delta E_{\pm}\right|\left(cm^{-1}\right)$	Reference
H_2O_2	$4 \cdot 10^{-14}$	11	55, 104, 111, 169
D_2O_2	$4 \cdot 10^{-14}$	2	55, 111
T_2O_2	$4 \cdot 10^{-14}$	0.5	170
HSOH	$4 \cdot 10^{-13}$	$2 \cdot 10^{-3}$	170
DSOD	$4 \cdot 10^{-13}$	$1 \cdot 10^{-5}$	170
TSOT	$4 \cdot 10^{-13}$	$3 \cdot 10^{-7}$	170
$HClOH^+$	$8 \cdot 10^{-13}$	$2 \cdot 10^{-2}$	168
$DClOD^+$	$-^c$	$2 \cdot 10^{-4}$	168
$TClOT^+$	$-^c$	$7 \cdot 10^{-6}$	168
H_2S_2	$1 \cdot 10^{-12}$	$2 \cdot 10^{-6}$	59
D_2S_2	$1 \cdot 10^{-12}$	$5 \cdot 10^{-10}$	59
T_2S_2	$1 \cdot 10^{-12}$	$1 \cdot 10^{-12}$	59
Cl_2S_2	$1 \cdot 10^{-12}$	$\approx 10^{-76}$ a	60
H_2Se_2	$2 \cdot 10^{-10}$ d	$1 \cdot 10^{-6}$	171
D_2Se_2	$-^c$	$3 \cdot 10^{-10}$	171
T_2Se_2	$-^c$	$4 \cdot 10^{-13}$	171
H_2Te_2	$3 \cdot 10^{-9}$ b	$3 \cdot 10^{-8}$	168
D_2Te_2	$-^c$	$1 \cdot 10^{-12}$	168
T_2Te_2	$-^c$	$3 \cdot 10^{-16}$	168

a   Extrapolated value

b   Calculated in ref. 58 for the *P*-structure (r_{TeTe} = 284 pm, r_{HTe} = 164 pm, α_{HTeTe} = 92° and τ_{HTeTeH} = 90°) and the corresponding *M*-structure. An earlier, very approximate result by Wiesenfeld [43] should be cited as well, giving ΔE_{pv} = $8 \cdot 10^{-10}$ cm^{-1} for the following structure (r_{TeTe} = 271.2 pm, r_{HTe} = 165.8 pm, α_{HTeTe} = 90° and τ_{HTeTeH} = 90°)

c   Expected to be very similar to the corresponding hydrogen isotopomers.

d   Calculated in ref. 58 for the *P*-structure (r_{HSe} = 145 pm, α_{HSeSe} = 92° and τ_{HSeSeH} = 90°) and the corresponding *M*-structure.

Indeed, in spite of being a benchmark molecule for calculations on parity violation, H_2O_2 is not a useful molecule for its experimental observation in the electronic ground state because of the very large tunnelling splittings satisfying Eq. (17). Only if the condition (18) is satisfied can we speak of a directly measurable, observable parity violating energy difference $\Delta_{pv}E$ between enantiomers. It is thus necessary to at least calculate tunnelling with sufficient accuracy to guarantee the inequality (18). More generally accurate calculations for the multidimensional tunnelling dynamics are desirable. Methods for this have been developed in our group including light and heavy atom tunnelling [60, 111]. Table 3 provides a summary of results for simple H_2O_2 like molecules X–Y–Y´–X´. One can clearly see the transition from molecules like H_2O_2 where tunnelling dominates over parity violation (Eq. (17)) to molecules such as ClSSCl where parity violation dominates completely (Eq. (18)) with interesting intermediate cases. [168] While accurate full dimensional tunnelling calculations for larger polyatomic molecules are difficult, approximate treatments exist, [172] which at least would allow us to verify the validity of Eq. (18), if applicable.

3.4 Isotopic chirality and a new isotope effect

Recent investigations from our group have dealt with molecules that are chiral only by isotopic substitution, thus "isotopically chiral". [167, 173, 174]. Examples are the methanol isotopomers CHDTOH and molecules such as $PF^{35}Cl^{37}Cl$.

While such molecules have been considered and even synthesized for some time [7, 175] our work has provided insight into a fundamentally new isotope effect arising from the weak nuclear interaction.

Molecular isotope effects can arise from the following basic mechanisms

(1) Mass differences of the isotopes generating different rovibrational energy levels and also different translational motion, for instance in effusion. This is very common [166].

(2) Different spins of different isotopes again provide a very common origin for isotope effects including isotopic NMR spectra such as ^{13}C NMR. [176].

(3) Different symmetry properties of the molecular wave function for different isotopomers lead to different "Pauli principle" selection rules for spectroscopic levels and reaction dynamics. [90, 91]. This type of isotope effect would exist without difference of spin and mass for instance for certain pairs of nuclear isomers of almost the same mass. While predicted some time ago [90, 91] it is certainly less commonly studied.

(4) A new isotope effect arises from the different electroweak charge Q_a of different isotopes following Eq. (27). We have discussed that this leads to parity violating energy difference $\Delta_{pv}E$ in molecules like $PF^{35}Cl^{37}Cl$ which are expected to be on the order of about 10% of "normal" values and this is also borne out by quantitative calculations. [174]. This opens the route to the use of isotopically chiral molecules for experiments, as clearly the conditions for a measurable $\Delta_{pv}E$ will be frequently met.

Besides the potential application of this isotope effect (4) to measurements of molecular parity violation [177, 178] the addition of a new molecular isotope effect to the ones already previously known and applied in spectroscopy and reaction dynamics [179] is of quite general, fundamental interest.[174]

4 EXPERIMENTS ON THE DYNAMICS OF PARITY VIOLATION IN CHIRAL MOLECULES

4.1 Summary of experimental schemes

Three basic types of spectroscopic experiments have been proposed in the past. In the first proposal by Letokhov and coworkers in 1975/76 [21, 22] one tries to measure parity violating frequency shifts $\Delta\nu_{pv} = \nu_R - \nu_S$ in spectra of R- and S-enantiomers (or in the physicist notation R- and L-) molecules, as can be seen in the scheme of Fig. 7. Several experimental tests have been made according to this scheme in the infrared and microwave ranges [21-23, 68, 69] but in agreement with theoretical predictions for the relative shifts $\Delta\nu_{pv} \simeq 10^{-16}$ for the fairly typical case of CHFClBr [6] no positive results were obtained so far because the experimental uncertainties are on the order of 10^{-14} in the best cases. [71, 72] In Doppler limited experiments one has

$$\frac{\Delta\nu_D}{\nu} \cong 7 \cdot 10^{-7} \sqrt{\frac{T/K}{m/u}} \tag{37}$$

which defines the accuracy that can be achieved, if the maximum of the Doppler line is being determined at the level of 1% of the linewidth. Effective temperatures T can be reduced in molecular beam or supersonic jet experiments [69], as demonstrated by the Zürich group in IR and microwave spectra of CHFClBr. Doppler free experiments achieved an accuracy of 10^{-8} already decades ago [23], but not much better. Ultraprecise experiments on CHFClBr based on the first high resolution analysis of the IR spectrum of this molecule by the Zürich group are able to do much better today [71, 72], thus going far beyond the Doppler limit, but it remains true that this approach is unable, even in principle, to provide the value of the ground state energy difference $\Delta_{pv}E$ (see scheme in Fig. 7). Rather a difference of such differences $\left|\Delta_{pv}E^* - \Delta_{pv}E\right|$ is measured in the best case. Efforts along these lines continue in Paris in the infrared range and in Zürich in the microwave range.

A second approach proposes measuring time dependent optical activity in molecules, where $\Delta_{pv}E$ and $\Delta E_\pm$ are of the same order of magnitude [27]. Such experiments would provide parity violating couplings by reference to tunnel splittings $\Delta E_\pm$ and thus would have to rely on a precise analysis of both tunnelling and parity violation. In quantitative calculations we have identified a few candidates for such an approach (see table 3 and references cited therein), but to our knowledge there exists no current experimental effort along these lines.

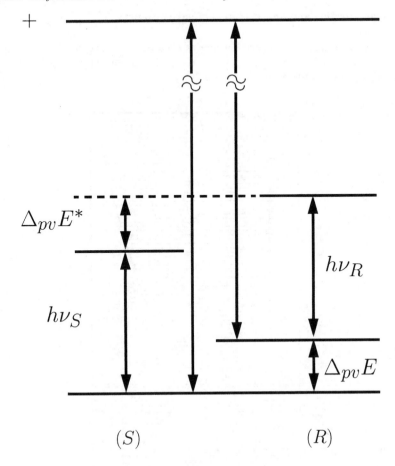

Figure 7 *Scheme for energy levels and experiments. One notes that*
$$\Delta\nu_{pv} = \nu_R - \nu_S = \left(\Delta_{pv}E^* - \Delta_{pv}E\right)/h \ \ in \ this \ scheme.$$

In the third approach we proposed to measure $\Delta_{pv}E$ directly as shown in Fig. 7, by using transitions to achiral excited states with well defined parity (labelled + in the scheme of Fig. 7). This scheme can be realized either as a combination difference measurement in the frequency domain [7] or as a time dependent measurement. [177] We shall discuss in more detail the latter, which is one of the main experimental efforts in Zürich.

4.2 The preparation and time evolution of parity isomers

Fig. 8 shows the basic experimental scheme as originally outlined in [7, 177]. A first laser selects an excited state of well defined parity on some electronically excited state (minus parity in the example of Fig. 8, see also Fig. 6).

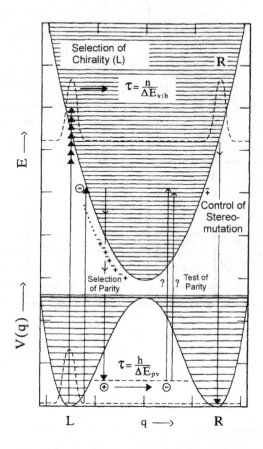

Figure 8 *Experimental scheme to measure time dependent parity violation following the proposal of ref 177 (by permission after ref 80).*

A second pulse selects a positive parity in the chiral electronic ground state because of the electric dipole selection rule $(-) \rightarrow (+)$ for parity. This state can be properly called a special achiral "parity isomer" of a "chiral" molecule, somewhat similar to the common nuclear spin isomers. The two complementary isomers (+ and – corresponding to the states χ_+ and $-\chi_-$ in Eqs. (6) and (7)) have different optical line spectra at high resolution because of the parity selection rule in electric dipole transitions. The parity isomers are, however, neither stable with respect to collisions nor intramolecularly, but show an intramolecular time evolution due to parity violation, the period of motion being

$$\tau_{pv} = h / \Delta_{pv}E \qquad (38)$$

and the initial time evolution for the concentration c_μ of the forbidden (minus) parity isomer would be (with the approximate equality holding for small times):

$$c_\mu = \sin^2\left(\pi t \Delta_{pv}E / h\right) \simeq \left(\pi\Delta_{pv}t / h\right)^2 \qquad (39)$$

Thus one has to measure a weak, slowly time dependent (quadratic) signal, for instance with absorption and subsequent ionization in the detection step. This scheme presents substantial challenges. But candidates for its future realization have been recently studied with the example of 1,3-difluoroallene. [178] As can be calculated from the fairly typical data in table 3, parity violating times are of the order of seconds. Thus initial time evolutions might be measured on a ms timescale.

5 OUTLOOK ON ANALYSIS OF FUTURE EXPERIMENTS AND ON POSSIBLE TESTS OF THE CPT THEOREM

We shall be a little more precise in conclusion of this review in our outlook on possible outcomes of possible theories and experiments of the type discussed here. Firstly, when we shall be able to measure and calculate parity violating energy differences $\Delta_{pv}E$ for relatively light molecules with very high precision, it should be possible to analyze the results of a fundamental precision experiment on the Standard Model (SMPP, see [8, 54]). In principle, this kind of thinking underlies the already existing experiments on parity violation in atomic spectra [98, 99]. However, these are limited to heavy atoms, which limits the precision of theoretical analyses (see ref. 99). The measurement of parity violation in light chiral molecules might overcome this limitation thus providing new information beyond the atomic experiments. It may be useful to discuss this point in slightly more detail. It is well known [180, 181] that the effective Weinberg parameter $\sin^2\Theta_w$ is a function of Q, the four momentum transfer between the interacting particles of effectively the energy range analysed by the experiments. At high energy the effective Weinberg parameter is well determined by the measurement of the properties of the Z-Boson ($\sin^2\Theta_w = 0.231$). [182] At lower energy, electron scattering leads to about $\sin^2\Theta_w = 0.24$, [183] whereas the lowest energy range is probed by spectroscopic measurements of atomic optical activity arising from parity violation. Here $\sin^2\Theta_w^{eff} \simeq 0.236$ has been reported with large uncertainty. [99, 184] In the analysis of the atomic experiments an important source of uncertainty arises from the theoretical calculations, because experimental results are only available on heavy atoms such as Cs, for which accurate calculations are difficult. [184] Difficulties arise form the complicated electronic structure in heavy atoms, but also from poorly known nuclear structure in heavy nuclei. On the other hand the close degeneracy of parity sublevels in chiral molecules leads to a maximum perturbation due to parity violation, which should become measurable for molecules involving only light atoms (up to Cl, say). Therefore very accurate calculations of both electronic structure and knowledge of nuclear structure should be available. The outlook is thus more promising for the analysis of parity violation in chiral molecules, in spite of the additional complications arising from the necessary averaging over the multidimensional vibrational wavefunction.

The sensitivity of molecular parity violation to the effective Weinberg parameter is obvious from Eqs. (27) and (30), where it enters through the weak charge Q_a of the nuclei. The use of isotopic chirality can be of interest here, because of compensations of contributions to the overall uncertainty in such molecules.

This is but one example of such an analysis of future experiments on molecular parity violation. Another possibility is to use the electroweak electron nucleus interaction in molecular parity violation to probe nuclear structure with respect to neutrons, which provide the dominant contribution, whereas the Coulomb interaction would provide a picture of the proton distribution in the nucleus. Of course in such future analyses, calculations would have to take nuclear structure into account explicitly.

Another obvious application of comparison of experiment and theory along those lines would be to test the theory for remaining fallacies.

Finally, at the most fundamental level, we have proposed to use this scheme to test for basic de lege violations of the CPT symmetry. [87, 102, 103] This can be done with high precision by finding different $\Delta_{pv}E$ in molecules and their enantiomers made of antimatter (ΔE_{pv}^{*} refers here to $\Delta_{pv}E$ of a molecule made of antimatter, with a different meaning to the * than in Fig. 7). This is shown in scheme of Fig. 9. [87, 103] It turns out that this leads to the most fundamental possible reinterpretation of the role of de lege time reversal symmetry violation for the second law. Indeed, even with our current knowledge of ordinary time reversal symmetry violation, it would be possible to reconstruct a time reversed state of a system made of antimatter and enantiomeric structures. Such a state would with CPT symmetry allow for apparent decrease of entropy (towards an "improbable "state) some time after time reversal (including C and P reversal), in agreement with a "de facto symmetry breaking" interpretation of the second law. However, if CPT symmetry violation were demonstrated, the "de lege violating" interpretation would become possible, subject to tests on its quantitative relevance under a given condition, very similar to the importance of parity violation for chiral molecules. This would be one of the most fundamental applications of the presents concepts, theories and experiments. The spectroscopic experiment on CPT symmetry violation would involve a measurement of spectra of chiral molecules (say L in Fig. 9) and their antimatter enantiomers R* in Figure 9. A difference in spectra of R* and L, say, in the infrared, would prove CPT violation. The sensitivity of this experiment is expected to be higher than in any other proposed experiment so far (see ref. 103 for a discussion.

Future experiments on CPT symmetry violation are a highly speculative subject and current wisdom assumes that CPT symmetry is valid. However, again the history of parity violation may teach us a lesson here. 150 years ago, Pasteur speculated about possible fundamental origins of chirality (called dissymmetry by him) in terms of a possible fundamental symmetry violation in "cosmic forces", which today we might call parity violation (see the citations in refs. 7 and 8). On the other hand during the century following Pasteur's discovery of molecular chirality, left-right symmetry was emphasized and even taken for granted, that is rigorous parity conservation in all aspects of fundamental dynamics. Prominent proponents include van't Hoff, Einstein, Hund and many others. [7, 8] The discovery of parity violation in β-decay in 1956/57 clearly changed our view on the space inversion symmetry or parity. But still, initially parity violation might have been considered as a very special, somewhat curious effect important only for some special phenomena such as β-decay (see fig. 1). However, a further theoretical understanding of the underlying symmetry violation gained over the past half century has led to the view that this symmetry violation is essential even to such basic theoretical structures such as the mere existence of particle masses, if these arise exclusively from the Higgs mechanism, as would be a common assumption in the standard model. Thus from this point of view, as expressed by M. Veltman,[185] parity violation is not some curiosity but parity *must* be violated if we want to understand the structure of matter as we do today! It is tempting to make a similarly strong but speculative statement at the other end of the structure of matter, that is biomolecular homochirality in the evolution of life. However, at this time such a statement would be premature. A better understanding of molecular chirality in relation to parity violation may provide some progress in this area as well.

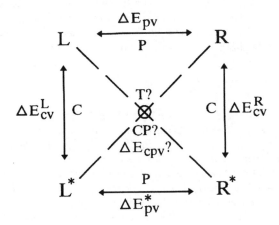

Figure 9 *Scheme for a CPT test with spectroscopy on chiral molecules (after ref. 102, 103).*

6 SUMMARY AND ABSTRACT

Parity (P), together with time reversal symmetry (T) and charge conjugation (C) constitute a fundamental set of discrete symmetries in physics. We review the current status of molecular parity violation in the framework of the fundamental symmetries of physics in general and in relation to intramolecular quantum dynamics. Work of the last decade in electroweak quantum chemistry, including the weak force of the standard model of particle physics in quantum calculations on chiral molecules, has resulted in an increase of the predicted parity violating energies by one to two orders of magnitude. This results in a new outlook on possible experiments which are discussed. We discuss furthermore the conceptual foundations of molecular symmetry breaking (spontaneous, de facto, de lege) in relation to molecular chirality, the evolution of biomolecular homochirality and irreversibility and the origin of the second law of thermodynamics. It is shown that there arise closely parallel situations in the lack of our current understanding of the true physical origins of all three phenomena. For molecular chirality some of the fundamental questions have been answered quantitatively by recent theory. The new theoretical approaches are summarized briefly, as well as some current results. Recent results on a new molecular isotope effect arising from parity violation are reviewed as well.

ACKNOWLEDGEMENT

I am greatly indebted to my coworkers (cited in the list of references) for their essential contributions, without which not much of the work in this report would exist. The parity violation projects have profited particularly from Sieghard Albert, Ayaz Bakasov, Robert Berger, Michael Gottselig, T. K. Ha, Luboš Horný, Achim Sieben, Jürgen Stohner, Greg Tschumper and Martin Willeke, but also others, listed more completely in [87]. Our work is supported financially by ETH Zürich and the Swiss National Science Foundation. I am very grateful to Ruth Schüpbach for substantial help in preparing a print ready manuscript.

References

1. P. A. M. Dirac, *Proc. R. Soc. London, A*, 1929, **123**, 714.
2. A. Bakasov, T. K. Ha, and M. Quack, *Ab initio calculation of molecular energies including parity violating interactions* in *Chemical Evolution, Physics of the Origin and Evolution of Life, Proc. of the 4th Trieste Conference (1995)*, ed. J. Chela-Flores and F. Raulin, Kluwer Academic Publishers, Dordrecht, 1996, p. 287.
3. A. Bakasov, T. K. Ha, and M. Quack, *J. Chem. Phys.*, 1998, **109**, 7263, (31 pages of supplementary material published as AIP Document, No PAPS JCP A6-109-303832 by American Institute of Physics, Physics Auxiliary Publication Service, 500 Sunnyside Blvd., Woodbury, N. Y. 11797-2999). Erratum: *J. Chem. Phys.* 1999, **110**, 6081.
4. CERN, Webseite http://doc.cern.ch//archive/electronic/cern/others/PHO/photo-di/9809005.jpeg, Bildnummer CERN AC_Z04_V25/B/1992
5. J. H. van't Hoff, *La chimie dans l'espace*, ed. B. M. Bazendijk, Rotterdam, 1887, reprinted in C. Bourgeois (ed). "Sur la dissymétrie moléculaire", Coll. Epistème Paris 1986.
6. M. Quack and J. Stohner, *Phys. Rev. Lett.*, 2000, **84**, 3807.
7. M. Quack, *Angew. Chem.-Int. Edit. Engl.*, 1989, **28**, 571, *Angew. Chem.* **1989**, *101*, 588-604.
8. M. Quack, *Angew. Chem., Int. Ed. Engl.*, 2002, **114**, 4618, Angew. Chem. **2002**, *114*, 4812-4825.
9. T. D. Lee and C. N. Yang, *Phys. Rev.*, 1956, **104**, 254.
10. C. S. Wu, E. Ambler, R. W. Hayward, D. D. Hoppes, and R. P. Hudson, *Phys. Rev.*, 1957, **105**, 1413.
11. R. L. Garwin, L. M. Lederman, and M. Weinrich, *Physical Review*, 1957, **105**, 1415.

12. J. I. Friedman and V. L. Telegdi, *Physical Review*, 1957, **105**, 1681.
13. S. L. Glashow, *Nucl. Phys.*, 1961, **22**, 579.
14. S. Weinberg, *Phys. Rev. Letters*, 1967, **19**, 1264.
15. A. Salam, Proceedings of the 8th Nobel Symposium, Almkvist und Wiksell, Stockholm, 1968, p. 367.
16. M. J. G. Veltman, *Rev. Mod. Phys.*, 2000, **72**, 341.
17. G. 't Hooft, *Rev. Mod. Phys.*, 2000, **72**, 333.
18. Y. Yamagata, *J. Theor. Biol.*, 1966, **11**, 495.
19. A. S. Garay and P. Hrasko, *J. Mol. Evol.*, 1975, **6**, 77.
20. D. W. Rein, *J. Mol. Evol.*, 1974, **4**, 15.
21. V. S. Letokhov, *Phys. Lett. A*, 1975, **53**, 275.
22. O. N. Kompanets, A. R. Kukudzhanov, V. S. Letokhov, and L. L. Gervits, *Opt. Commun.*, 1976, **19**, 414.
23. E. Arimondo, P. Glorieux, and T. Oka, *Opt. Commun.*, 1977, **23**, 369.
24. L. Keszthelyi, *Phys. Lett. A*, 1977, **64**, 287.
25. L. Keszthelyi, *Orig. Life Evol. Biosph.*, 1977, **8**, 299.
26. B. Y. Zeldovich, D. B. Saakyan, and Sobelman, II, *Jetp Lett.*, 1977, **25**, 94.
27. R. A. Harris and L. Stodolsky, *Phys. Lett. B*, 1978, **78**, 313.
28. D. W. Rein, R. A. Hegström, and P. G. H. Sandars, *Phys. Lett. A*, 1979, **71**, 499.
29. R. A. Hegström, D. W. Rein, and P. G. H. Sandars, *J. Chem. Phys.*, 1980, **73**, 2329.
30. S. F. Mason and G. E. Tranter, *J Chem Soc Chem Comm*, 1983, 117.
31. S. F. Mason and G. E. Tranter, *Chem. Phys. Lett.*, 1983, **94**, 34.
32. S. F. Mason and G. E. Tranter, *Mol. Phys.*, 1984, **53**, 1091.
33. S. F. Mason, *Nature*, 1984, **311**, 19.
34. G. E. Tranter, *Mol. Phys.*, 1985, **56**, 825.
35. G. E. Tranter, *Chem. Phys. Lett.*, 1985, **121**, 339.
36. G. E. Tranter, *Chem. Phys. Lett.*, 1985, **115**, 286.
37. G. E. Tranter, *Nature*, 1985, **318**, 172.
38. A. L. Barra, J. B. Robert, and L. Wiesenfeld, *Phys. Lett. A*, 1986, **115**, 443.
39. A. L. Barra, J. B. Robert, and L. Wiesenfeld, *Biosystems*, 1987, **20**, 57.
40. D. K. Kondepudi, *Biosystems*, 1987, **20**, 75.
41. A. J. Macdermott, G. E. Tranter, and S. B. Indoe, *Chem. Phys. Lett.*, 1987, **135**, 159.
42. A. L. Barra, J. B. Robert, and L. Wiesenfeld, *Europhys. Lett.*, 1988, **5**, 217.
43. L. Wiesenfeld, *Mol. Phys.*, 1988, **64**, 739.
44. P. Jungwirth, L. Skala, and R. Zahradnik, *Chem. Phys. Lett.*, 1989, **161**, 502.
45. A. J. Macdermott and G. E. Tranter, *Chem. Phys. Lett.*, 1989, **163**, 1.
46. A. J. Macdermott and G. E. Tranter, *Croat. Chem. Acta*, 1989, **62**, 165.
47. O. Kikuchi and H. Wang, *Bull. Chem. Soc. Jpn.*, 1990, **63**, 2751.
48. O. Kikuchi, H. Wang, T. Nakano, and K. Morihashi, *Theochem-J. Mol. Struct.*, 1990, **64**, 301.
49. J. Chela-Flores, *Chirality*, 1991, **3**, 389.
50. A. Salam, *J. Mol. Evol.*, 1991, **33**, 105.
51. A. Salam, *Phys. Lett. B*, 1992, **288**, 153.
52. O. Kikuchi and H. Kiyonaga, *Theochem-J. Mol. Struct.*, 1994, **118**, 271.
53. A. Bakasov, T. K. Ha, and M. Quack, *Chimia*, 1997, **51**, 559.
54. A. Bakasov and M. Quack, *Chem. Phys. Lett.*, 1999, **303**, 547.
55. R. Berger and M. Quack, *J. Chem. Phys.*, 2000, **112**, 3148, (cf. R. Berger and M. Quack, Proc. 37th IUPAC Congress Vol. 2, p. 518, Berlin, **1999**).
56. P. Lazzeretti and R. Zanasi, *Chem. Phys. Lett.*, 1997, **279**, 349.

57.   R. Zanasi and P. Lazzeretti, *Chem. Phys. Lett.*, 1998, **286**, 240.
58.   J. K. Laerdahl and P. Schwerdtfeger, *Phys. Rev. A*, 1999, **60**, 4439.
59.   M. Gottselig, D. Luckhaus, M. Quack, J. Stohner, and M. Willeke, *Helv. Chim. Acta*, 2001, **84**, 1846.
60.   R. Berger, M. Gottselig, M. Quack, and M. Willeke, *Angew. Chem.-Int. Edit.*, 2001, **40**, 4195, Angew. Chem. **2001**, *113*, 4342-4345.
61.   M. Quack and J. Stohner, *Z. Phys. Chemie*, 2000, **214**, 675, and prepublished summary in SASP 2000, Proc. 12th Symp. on Atomic and Surface Physics and related Topics, D. Bassi and P. Tosi eds., Folgaria, Trento (2000) pages PR-11, 1-4.
62.   R. Berger and M. Quack, *ChemPhysChem*, 2000, **1**, 57.
63.   R. Berger, M. Quack, and G. S. Tschumper, *Helv. Chim. Acta*, 2000, **83**, 1919.
64.   J. K. Laerdahl, P. Schwerdtfeger, and H. M. Quiney, *Phys. Rev. Lett.*, 2000, **84**, 3811.
65.   J. K. Laerdahl, R. Wesendrup, and P. Schwerdtfeger, *ChemPhysChem*, 2000, **1**, 60.
66.   A. C. Hennum, T. Helgaker, and W. Klopper, *Chem. Phys. Lett.*, 2002, **354**, 274.
67.   M. Quack and J. Stohner, *Chirality*, 2001, **13**, 745.
68.   A. Beil, D. Luckhaus, R. Marquardt, and M. Quack, *Faraday Discuss.*, 1994, **99**, 49.
69.   A. Bauder, A. Beil, D. Luckhaus, F. Müller, and M. Quack, *J. Chem. Phys.*, 1997, **106**, 7558.
70.   H. Hollenstein, D. Luckhaus, J. Pochert, M. Quack, and G. Seyfang, *Angew. Chem.-Int. Edit. Engl.*, 1997, **36**, 140, Angew. Chem. 109, 136-140 (1997).
71.   C. Daussy, *Premier test de très haute précision de violation de la parité dans le spectre de la molécule chirale CHFClBr, Thèse*, Université Paris 13, Villetaneuse (Paris Nord), 1999.
72.   C. Daussy, T. Marrel, A. Amy-Klein, C. T. Nguyen, C. J. Bordé, and C. Chardonnet, *Phys. Rev. Lett.*, 1999, **83**, 1554.
73.   J. Crassous and A. Collet, *Enantiomer*, 2000, **5**, 429.
74.   M. J. M. Pepper, I. Shavitt, P. v. R. Schleyer, M. N. Glukhovtsev, R. Janoschek, and M. Quack, *J. Comput. Chem.*, 1995, **16**, 207.
75.   M. Quack, *Faraday Discuss.*, 1994, 389.
76.   R. Berger, *Phys. Chem. Chem. Phys.*, 2003, **5**, 12.
77.   A. S. Lahamer, S. M. Mahurin, R. N. Compton, D. House, J. K. Laerdahl, M. Lein, and P. Schwerdtfeger, *Phys. Rev. Lett.*, 2000, **85**, 4470.
78.   H. Buschmann, R. Thede, and D. Heller, *Angew. Chem.-Int. Edit.*, 2000, **39**, 4033.
79.   P. Frank, W. Bonner, and R. N. Zare, in *Chemistry for the 21st century*, ed. E. Keinan and I. Schechter, Wiley-VCH, Weinheim, 2001, ch. 11, p. 175.
80.   M. Quack, *Nova Acta Leopoldina*, 1999, **81**, 137.
81.   M. Quack, *Molecular femtosecond quantum dynamics between less than yoctoseconds and more than days: Experiment and theory* in *Femtosecond Chemistry, Proc. Berlin Conf. Femtosecond Chemistry, Berlin (March 1993)*, ed. J. Manz and L. Woeste, Verlag Chemie, Weinheim, 1995, ch. 27, p. 781.
82.   M. Quack, *Time and Time Reversal Symmetry in Quantum Chemical Kinetics* in *Fundamental World of Quantum Chemistry. A Tribute to the Memory of Per-Olov Löwdin*, ed. E. J. Brändas and E. S. Kryachko, Kluwer Adacemic Publishers, Dordrecht, 2004, vol. 3, p. 423.
83.   M. Quack, *Zeit und Zeitumkehrsymmetrie in der molekularen Kinetik. Schriftliche Fassung des Vortrages am 7. Symposium der Deutschen Akademien der Wissenschaften, Berlin-Brandenburgische Akademie der Wissenschaften Berlin,*

Zeithorizonte in der Wissenschaften, 31.10. und 1.11.2002, ed. D. Simon, De Gruyter, Berlin, 2004, p. 125.

84. M. Quack and J. Stohner, *Chimia*, 2005, **59**, 530.

85. R. Berger, in *Relativistic Electronic Structure Theory*, ed. P. Schwerdtfeger, Elsevier, Amsterdam, 2004, vol. Part 2, ch. 4, p. 188.

86. R. Janoschek, *Theories of the Origin of Biomolecular Homochirality* in *Chirality*, ed. R. Janoschek, Springer-Verlag, Berlin, 1991, p. 18.

87. M. Quack, *Chimia*, 2003, **57**, 147.

88. M. Quack, *Chimia*, 2001, **55**, 753.

89. A. Bakasov, T. K. Ha, and M. Quack, *J. Chem. Phys.*, 1999, **110**, 6081.

90. M. Quack, *Mol. Phys.*, 1977, **34**, 477.

91. M. Quack, *Detailed symmetry selection rules for chemical reactions* in *Symmetries and properties of non-rigid molecules: A comprehensive survey. Studies in Physical and Theoretical Chemistry*, Elsevier Publishing Co., Amsterdam, 1983, vol. 23, p. 355.

92. T. D. Lee, *Symmetries, Asymmetries and the World of Particles*, University of Washington Press, Seattle, 1988.

93. G. Gabrielse, D. Phillips, W. Quint, H. Kalinowsky, G. Rouleau, and W. Jhe, *Phys. Rev. Lett.*, 1995, **74**, 3544.

94. C. Zimmermann and T. W. Hänsch, *Phys. Bl.*, 1993, **49**, 193.

95. M. Amoretti, C. Amsler, G. Bonomi, A. Bouchta, P. Bowe, C. Carraro, C. L. Cesar, M. Charlton, M. J. T. Collier, M. Doser, V. Filippini, K. S. Fine, A. Fontana, M. C. Fujiwara, R. Funakoshi, P. Genova, J. S. Hangst, R. S. Hayano, M. H. Holzscheiter, L. V. Jorgensen, V. Lagomarsino, R. Landua, D. Lindelof, E. L. Rizzini, M. Macri, N. Madsen, G. Manuzio, M. Marchesotti, P. Montagna, H. Pruys, C. Regenfus, P. Riedler, J. Rochet, A. Rotondi, G. Rouleau, G. Testera, A. Variola, T. L. Watson, and D. P. van der Werf, *Nature*, 2002, **419**, 456.

96. M. Kobayashi and T. Maskawa, *Prog. Theor. Phys.*, 1973, **49**, 652.

97. K. Abe, R. Abe, I. Adachi, B. S. Ahn, H. Aihara, M. Akatsu, G. Alimonti, K. Asai, M. Asai, Y. Asano, T. Aso, V. Aulchenko, T. Aushev, A. M. Bakich, E. Banas, S. Behari, P. K. Behara, D. Beiline, A. Bondar, A. Bozek, T. E. Browder, B. C. K. Casey, P. Chang, Y. Chao, K. F. Chen, B. G. Cheon, R. Chistov, S. K. Choi, Y. Choi, L. Y. Dong, J. Dragic, A. Drutskoy, S. Eidelman, V. Eiges, Y. Enari, R. Enomoto, C. W. Everton, F. Fang, H. Fujii, C. Fukunaga, M. Fukushima, N. Gabyshev, A. Garmash, T. J. Gershon, A. Gordon, K. Gotow, H. Guler, R. Guo, J. Haba, H. Hamasaki, K. Hanagaki, F. Handa, K. Hara, T. Hara, N. C. Hastings, H. Hayashii, M. Hazumi, E. M. Heenan, Y. Higasino, I. Higuchi, T. Higuchi, T. Hirai, H. Hirano, T. Hojo, T. Hokuue, Y. Hoshi, K. Hoshina, S. R. Hou, W. S. Hou, S. C. Hsu, H. C. Huang, Y. Igarashi, T. Iijima, H. Ikeda, K. Ikeda, K. Inami, A. Ishikawa, H. Ishino, R. Itoh, G. Iwai, H. Iwasaki, Y. Iwasaki, D. J. Jackson, P. Jalocha, H. K. Jang, M. Jones, R. Kagan, H. Kakuno, J. Kaneko, J. H. Kang, J. S. Kang, P. Kapusta, N. Katayama, H. Kawai, Y. Kawakami, N. Kawamura, T. Kawasaki, H. Kichimi, D. W. Kim, H. Kim, et al., *Phys. Rev. Lett.*, 2001, **87**, 091802 1 *Physical Review D*, 2002, **66**, 032007, *Physical Review D*, 2002, **66**, 71102.

98. M. A. Bouchiat and C. Bouchiat, *Journal De Physique*, 1974, **35**, 899.

99. S. C. Bennett and C. E. Wieman, *Phys. Rev. Lett.*, 1999, **82**, 2484.

100. J. H. Christenson, V. L. Fitch, J. W. Cronin, and R. Turlay, *Phys. Rev. Lett.*, 1964, **13**, 138.

101.   R. Adler, *et al (CPLEAR collaboration)*. in *Direct measurement of T violation in the neutral kaon system using tagged K^0, K^0 at LEAR. Proc. IV. Int. Symp. on Weak and Electromagnetic Interactions in Nuclei*, ed. H. Ejiri, T. Kishimoto, and T. Salo, World Scientific, Singapore, 1995, p. 53.

102.   M. Quack, *J. Mol. Struct.*, 1993, **292**, 171.

103.   M. Quack, *Chem. Phys. Lett.*, 1994, **231**, 421.

104.   B. Kuhn, T. R. Rizzo, D. Luckhaus, M. Quack, and M. A. Suhm, *J. Chem. Phys.*, 1999, **111**, 2565.

105.   M. Quack, *Die Symmetrie von Zeit und Raum und ihre Verletzung in molekularen Prozessen* in *Jahrbuch 1990-1992 der Akademie der Wissenschaften zu Berlin*, W. de Gruyter Verlag, Berlin, 1993, p. 467, (printed version of the 8th public academy lecture, Berlin 4.10.1990); there exists also a slightly changed English version: M. Quack, *The symmetries of time and space and their violation in chiral molecules and molecular processes*, pp. 172, in *Conceptual Tools for Understanding Nature. Proc. 2nd Int. Symp. of Science and Epistemology Seminar, Trieste April 1993*, edited by G. Costa, G. Calucci, M. Giorgi, (World Scientific Publ., Singapore, 1995).

106.   H. Primas, *Chemistry, Quantum Mechanics and Reductionism*, Springer, Berlin, 1981.

107.   P. Pfeiffer, in *Energy Storage and Redistribution in Molecules, Proc. of Two Workshops, Bielefeld, June 1980*, ed. J. Hinze, Plenum, New York, NY, 1983, p. 315.

108.   A. Amann, *J Math Chem*, 1991, **6**, 1.

109.   F. Hund, *Zeitschrift für Physik*, 1927, **43**, 788.

110.   E. Schrödinger, *Annalen der Physik IV. Folge*, 1926, **81**, 109.

111.   B. Fehrensen, D. Luckhaus, and M. Quack, *Chem. Phys. Lett.*, 1999, **300**, 312.

112.   D. Luckhaus and M. Quack, *Gas Phase Kinetics Studies* in *Encyclopedia of Chemical Physics and Physical Chemistry*, ed. J. H. Moore and N. Spencer, IOP publishing, Bristol, 2001, vol. 2 (Methods), ch. B. 2.5, p. 1871.

113.   R. S. Cahn, C. K. Ingold, and V. Prelog, *Experientia*, 1956, **12**, 81.

114.   R. S. Cahn, C. Ingold, and V. Prelog, *Angew. Chem.-Int. Edit.*, 1966, **5**, 385.

115.   M. Simonius, *Phys. Rev. Lett.*, 1978, **40**, 980.

116.   R. A. Harris and L. Stodolsky, *J. Chem. Phys.*, 1981, **74**, 2145.

117.   F. C. Frank, *Biochim. Biophys. Acta*, 1953, **11**, 459.

118.   M. Eigen, *Naturwissenschaften*, 1971, **58**, 465.

119.   M. Eigen and R. Winkler, *Das Spiel*, Piper, München, 1975.

120.   M. Eigen, *Nova Acta Leopoldina*, 1982, **52**, 3.

121.   M. Eigen, *Stufen zum Leben*, Piper, München, 1987.

122.   M. Bolli, R. Micura, and A. Eschenmoser, *Chemistry & Biology*, 1997, **4**, 309.

123.   J. S. Siegel, *Chirality*, 1998, **10**, 24.

124.   M. Calvin, *Chemical Evolution*, Oxford University Press, Oxford, 1969.

125.   G. Nicolis and I. Prigogine, *Proc. Natl. Acad. Sci. U. S. A.*, 1981, **78**, 659.

126.   W. A. Bonner, *Orig. Life Evol. Biosph.*, 1995, **25**, 175.

127.   P. R. Kavasmaneck and W. A. Bonner, *J. Am. Chem. Soc.*, 1977, **99**, 44.

128.   H. Kuhn and J. Waser, *Self organization of matter and the early evolution of life* in *Biophysics*, ed. W. Hoppe, W. Lohmann, H. Markl, and H. Ziegler, Springer, Berlin, 1983.

129.   W. Bonner, *Origins of Chiral Homogeneity in Nature* in *Topics in Stereochemistry*, ed. E. L. Eliel and S. H. Wilen, Wiley, New York, 1988, vol. 18, p. 1.

130.   F. Vester, T. L. V. Ulbricht, and H. Krauch, *Naturwissenschaften*, 1959, **46**, 68.

131. P. Kleindienst and G. H. Wagnière, *Chem. Phys. Lett.*, 1998, **288**, 89.
132. A. Salam, *On biological macromolecules and the phase transitions they bring about* in *Conceptual Tools for Understanding Nature. Proc. 2nd Intl. Symp. of Science and Epistemology Seminar, Trieste 1993*, ed. G. Costa, G. Calucci, and M. Giorgi, World Scientific Publ., Singapore, 1995, p. 209.
133. D. K. Kondepudi and G. W. Nelson, *Phys. Lett. A*, 1984, **106**, 203.
134. D. K. Kondepudi and G. W. Nelson, *Physica A*, 1984, **125**, 465.
135. D. K. Kondepudi and G. W. Nelson, *Nature*, 1985, **314**, 438.
136. A. Szabo-Nagy and L. Keszthelyi, *Proc. Natl. Acad. Sci. U. S. A.*, 1999, **96**, 4252.
137. L. Boltzmann, *Annalen der Physik und Chemie*, 1896, **57**, 773.
138. L. Boltzmann, *Annalen der Physik und Chemie*, 1897, **60**, 392.
139. L. Boltzmann, *Vorlesungen über Gastheorie*, Barth, Leipzig, 1896.
140. J. L. Lebowitz, *Rev. Mod. Phys.*, 1999, **71**, 346, *Phys. Today*, 1993, **46**, 32. (and subsequent letters from readers of Physics Today).
141. R. Peierls, *Surprises in Theoretical Physics*, University Press, Princeton, 1979.
142. P. Ehrenfest and T. Ehrenfest, *Phys. Z.*, 1907, **8**, 311.
143. P. Ehrenfest and T. Ehrenfest, *Encykl. Math. Wiss.*, 1911, **4**.
144. K. W. F. Kohlrausch and E. Schrödinger, *Phys. Z.*, 1926, **27**, 306.
145. J. Orban and Belleman.A, *Phys. Lett. A*, 1967, **A 24**, 620.
146. M. Quack, *Nuovo Cimento Soc. Ital. Fis. B*, 1981, **63B**, 358.
147. M. Quack, *Reaction dynamics and statistical mechanics of the preparation of highly excited states by intense infrared radiation* in *Advances in Chemical Physics*, ed. K. Lawley, I. Prigogine, and S. A. Rice, 1982, vol. 50, p. 395.
148. R. Clausius, *Poggendorfs Annalen der Physik und Chemie*, 1865, **125**, 353.
149. M. Planck, *Reversibilität und Irreversibilität. Vorlesung 1.* in *Acht Vorlesungen über Theoretische Physik.*, Hirzel, Leipzig, 1910.
150. I. Prigogine, *From Being to Becoming*, Freeman, San Francisco, 1980.
151. I. Prigogine and I. Stengers, *Dialog mit der Natur*, Piper, München, 1981.
152. P. Grigolini, *Quantum Mechanical Irreversibility and Measurement*, World Scientific Publ., Singapore, 1993.
153. S. Hawking and R. Penrose, *The Nature of Space and Time*, University Press, Princeton, 1996.
154. H. R. Pagels, *Phys Today*, 1985, **38**, 97.
155. I. Prigogine and I. Stengers, *Order out of Chaos*, Bantam Press, New York, 1984.
156. N. N. Bogoljubov and D. V. Shirkov, *Quantum fields*, Benjamin/Cummings, Reading, Massachusetts etc., 1983.
157. N. N. Bogoljubov and D. V. Shirkov, *Introduction to the theory of quantized fields*, Wiley, New York a.o., 1980.
158. A. Bakasov, T. K. Ha, and M. Quack, *Parity violating potentials for molecules and clusters* in *Proceedings of the Symposium on Atomic and Surface Physics and Related Topics, SASP 98, Going/Kitzbühel, Austria*, ed. A. Hansel and W. Lindinger, Institut für Ionenphysik der Universität Innsbruck, 1998, p. 4.
159. M. A. Bouchiat and C. Bouchiat, *Journal De Physique*, 1975, **36**, 493.
160. J. Linderberg and Y. Öhrn, *Propagators in Quantum Chemistry*, Academic Press, London, 1973.
161. O. Vahtras, H. Agren, P. Jorgensen, H. J. A. Jensen, T. Helgaker, and J. Olsen, *J. Chem. Phys.*, 1992, **96**, 2118.
162. R. Berger and C. van Wüllen, *J. Chem. Phys.*, 2005, **122**.
163. P. Schwerdtfeger, T. Saue, J. N. P. van Stralen, and L. Visscher, *Phys. Rev. A*, 2005, **71**, 012103 1

164.   M. Quack and J. Stohner, *J. Chem. Phys.*, 2003, **119**, 11228.
165.   A. Bakasov, R. Berger, T. K. Ha, and M. Quack, *Int. J. Quantum Chem.*, 2004, **99**, 393.
166.   M. Quack, *Annu. Rev. Phys. Chem.*, 1990, **41**, 839.
167.   R. Berger, M. Quack, A. Sieben, and M. Willeke, *Helv. Chim. Acta*, 2003, **86**, 4048.
168.   M. Gottselig, M. Quack, J. Stohner, and M. Willeke, *Int. J. Mass Spectrom.*, 2004, **233**, 373.
169.   W. B. Olson, R. H. Hunt, B. W. Young, A. G. Maki, and J. W. Brault, *J. Mol. Spectrosc.*, 1988, **127**, 12.
170.   M. Quack and M. Willeke, *Helv. Chim. Acta*, 2003, **86**, 1641, and to be published.
171.   M. Gottselig, M. Quack, and M. Willeke, *Isr. J. Chem.*, 2003, **43**, 353.
172.   B. Fehrensen, D. Luckhaus, and M. Quack, *Z. Phys. Chemie*, 1999, **209**, 1.
173.   R. Berger, M. Quack, A. Sieben, and M. Willeke, *Isotopic chirality and molecular parity violation* in *Proc. 18th Coll. High Resol. Mol. Spectroscopy*, Dijon, 2003, p. Paper F4.
174.   R. Berger, G. Laubender, M. Quack, A. Sieben, J. Stohner, and M. Willeke, *Angew. Chem.-Int. Edit.*, 2005, **44**, 3623.
175.   D. Arigoni, in *Topics in Stereochemistry*, ed. E. L. Eliel, 1969, vol. 221, p. 1213.
176.   R. R. Ernst, *Angew. Chem.-Int. Edit. Engl.*, 1992, **31**, 805, *Angew. Chem.* 1992,**104**, 817.
177.   M. Quack, *Chem. Phys. Lett.*, 1986, **132**, 147.
178.   M. Gottselig and M. Quack, *J. Chem. Phys.*, 2005, **123**, 84305 1
179.   M. Hippler and M. Quack, *Isotope Selective Infrared Spectroscopy and Intramolecular Dynamics* in *Isotope Effects in Chemistry and Biology, Part III, Isotope Effects in Chemical Dynamics*, ed. A. Kohen and H.-H. Limbach, Marcel Dekker Inc., New York, 2005, ch. 11, p. 305.
180.   B. Schwarzschild, *Phys Today*, 2005, **58**, 23.
181.   A. Czarnecki and W. J. Marciano, *International Journal of Modern Physics A*, 2000, **15**, 2365.
182.   T. Cvitas, J. Frey, B. Holmström, K. Kuchitsu, R. Marquardt, I. Mills, F. Pavese, M. Quack, J. Stohner, H. L. Strauss, M. Takami, and A. J. Thor, *Quantities, Units and Symbols in Physical Chemistry*, RSC, Oxford, 2006 (to be published).
183.   P. L. Anthony, R. G. Arnold, C. Arroyo, K. Bega, J. Biesiada, P. E. Bosted, G. Bower, J. Cahoon, R. Carr, G. D. Cates, J. P. Chen, E. Chudakov, M. Cooke, P. Decowski, A. Deur, W. Emam, R. Erickson, T. Fieguth, C. Field, J. Gao, M. Gary, K. Gustafsson, R. S. Hicks, R. Holmes, E. W. Hughes, T. B. Humensky, G. M. Jones, L. J. Kaufman, L. Keller, Y. G. Kolomensky, K. S. Kumar, P. LaViolette, D. Lhuillier, R. M. Lombard-Nelsen, Z. Marshall, P. Mastromarino, R. D. McKeown, R. Michaels, J. Niedziela, M. Olson, K. D. Paschke, G. A. Peterson, R. Pitthan, D. Relyea, S. E. Rock, O. Saxton, J. Singh, P. A. Souder, Z. M. Szalata, J. Turner, B. Tweedie, A. Vacheret, D. Walz, T. Weber, J. Weisend, M. Woods, and I. Younus, *Phys. Rev. Lett.*, 2005, **95**, 081601.
184.   V. M. Shabaev, K. Pachucki, Tupitsyn, II, and V. A. Yerokhin, *Phys. Rev. Lett.*, 2005, **94**, Art. No. 213002.
185.   M. Veltman in *Symmetry and Modern Physics,* eds. A. Goldhaber, R. Shrock, J. Smith, G. Sternman, P. van Nieuwenhuizen, W. Weisberger, World Scientific Publ., Singapore, 2003, p. 95.

CHARACTERIZATION OF PROTEIN FOLDING/UNFOLDING AT ATOMIC RESOLUTION

R. Day[1] and V. Daggett[1,2]

[1]Biomolecular Structure and Design Program, University of Washington, Seattle, WA 98195-7610
[2]Department of Medicinal Chemistry, University of Washington, Seattle, WA 98195-7610

1 INTRODUCTION

Protein folding and unfolding can be very complex reactions, with many stable states populated at different temperatures. For many proteins (1), the reaction appears relatively simple, with only two states, native and denatured, populated at equilibrium. In this simple case, folding and unfolding follow single-exponential kinetics, consistent with a single rate-limiting transition state (TS). Even in this case, however, the process is made more complex by many other conformations that are transiently populated along the folding/unfolding reaction coordinate. The large-scale chain reorganizations involved in folding and unfolding are necessarily continuous rearrangements, leading to a series of unstable conformations in passing from a conformationally homogenous native state to a heterogeneous denatured state.

Chymotrypsin inhibitor 2 (CI2) was the first protein whose folding and unfolding were experimentally determined to be two-state (2). Equilibrium measurements gave a sigmoidal curve characteristic of a two-state process, the rates of folding and unfolding are characterized by single exponentials, and their ratio agrees with the stability of the protein from equilibrium measurements. As the folding is two-state, the rates of folding and unfolding are dependent on the stability of a single transition state. Structure in this transition state can be inferred by considering the effects of mutations on the stability of the protein and on its rates of folding and unfolding (3). The change in the rate of folding is defined by the change in stability of the transitions state relative to the denatured state, $\Delta\Delta G_{\ddagger\text{-}D}$. The ratio of this change to the change in the overall stability is defined as a Φ_F-value. For appropriately chosen mutants, such as hydrophobic deletions in the protein core, this value may be interpreted in terms of structure. In the case where the mutated position is as structured in the TS as in the native state, the changes in stability will be similar and Φ_F will equal 1. In the case the position is unstructured in the TS, there will be no change in the relative stability of the TS, and Φ_F will equal 0. Intermediate values of Φ_F can indicate partial structure or heterogeneity in the TS. Φ_U-values may also be calculated by considering changes in the unfolding rate. If the transition state is the same for folding and unfolding, Φ_F will be equal to $1\text{-}\Phi_U$. The folding/unfolding TS for CI2 was probed by this protein engineering method (4-6), giving structural information about the partially folded transition state. It remained, however, to describe the rest of the transiently populated conformations along the folding/unfolding reaction coordinate and provide atomically detailed models of structures in the TS ensemble.

Molecular dynamics (MD) simulations provide a means of characterizing conformations along the unfolding reaction coordinate in atomic detail. In MD simulations, all of the atoms in the protein are modelled explicitly, as are solvent atoms in the simulations described here, and their interactions are described by a simple potential function. The potential function is used to determine the forces on all atoms, and the forces are used to calculate accelerations, which are then integrated over time to allow the system to change. These calculations must be carried out using a very short time step (typically ≤ 0.2 fs), so the timescales available to simulation are quite short relative to experimental timescales. The unfolding process can, however, be sped up by increasing temperature, such that the full unfolding can be simulated at atomic resolution. The native, transition, and denatured states from simulation can then be compared to experimental data, and the fidelity of the simulations and their ability to model reality can be determined. Assuming reasonable agreement, structures from the simulation provide atomically detailed models of transiently populated conformations along the folding/unfolding reaction coordinate. Here we compare simulations of CI2 at a variety of temperatures to experimental data and describe the regions of the unfolding pathway that are inaccessible to experiment.

2 METHOD AND RESULTS

2.1 The native state

A well defined native state is key to further studies of the unfolding of a protein. Without a thorough native state simulation, it is impossible to separate early unfolding events from fluctuations in the native structure. The native structure of CI2 has been defined by X-ray crystallography (7) and NMR spectroscopy (8). The native topology consists of an α-helix folded against a three stranded β-sheet. The N-terminus completes the main core by packing antiparallel to the helix and the third strand of the sheet (Figure 1). The active site loop connects the first two strands of the β-sheet.

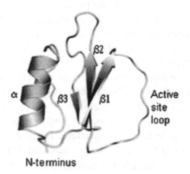

Figure 1 *CI2 native structure with secondary structure labels*

The native state has been characterized by a relatively long (50 ns) simulation starting from the X-ray crystal structure at 298 K (9). The average root mean square distance between α-carbons (Cα-RMSD) in structures in the simulation and the crystal structure is 1.7 Å. Most the movement occurs in the active site loop, in agreement with the crystallographic B-values and NMR relaxation data (10). When the loops are excluded from the calculation, the average $C\alpha$-RMSD to the crystal structure falls to 0.7 Å. Using a 5.5 Å cutoff for r^6 weighted average distances, 95% of the short range NOEs (sequence separation of 5 or fewer residues) and 92% of the long range NOEs (sequence separation greater than 5 residues) from Ludvigsen et al (1991) are satisfied in this simulation. A more detailed comparison between simulation and crystallographic and NMR data was performed using an earlier, shorter (5 ns) simulation of the crystal structure at 298 K (11). While this simulation did not maintain as low of a $C\alpha$-RMSD to the starting structure, the movement was again focused in the active site loop. This simulation was shown to be in good agreement with N-H order parameters, hydrogen exchange, and backbone coupling constants.

2.2 The transition state

For simple two state protein folding, only the native and denatured states are well populated and visible to experiment at equilibrium. The rate-limiting transition state can also be probed by experiment as perturbations to its structure are felt in the rates of folding and unfolding. As a protein unfolds in simulation, it passes through structures corresponding to the rate-limiting transition state. The challenge is in identifying those structures and verifying that they correspond to the experimental transition state.

2.2.1 Identification of the transition state. The transition state of folding/unfolding is naturally defined by the highest free-energy conformations on the folding/unfolding reaction coordinate. Unfortunately, free energies of individual conformations along an MD trajectory cannot be accurately calculated. A structural method for identifying the transition state has been developed instead (12, 13). A multidimensional scaling algorithm is used to determine a set of points in three dimensions whose pairwise distances approximate the pairwise $C\alpha$-RMSDs between conformations in the unfolding process. These points are connected according to their order in the simulation to provide a 3-dimensional map of the conformational space sampled in the simulation (Figure 2). Structures from the beginning of the simulation cluster tightly on this map. This cluster represents the native state, and we take the 5 ps window at the exit of this cluster to represent the transition state. When multiple simulations are mapped into the same space, this exit from the native cluster also corresponds to the point of divergence between simulations (14).

2.2.2 Verification of the transition state. The most complete experimental description of CI2's transition state ensemble comes in the form of Φ_F-values. 90 mutations were made at 41 positions spanning all regions of structure in CI2 (5, 6, 15). Three methods have been used to compare transition state structures from simulation to these Φ_F-values: calculation of Φ_{MD}-values, calculation of S-values, and free energy perturbation calculations. Φ_{MD}-values are calculated by modelling a given mutation onto the native and average transition state structures, and taking the ratio of the number of contacts lost on mutation in the transition state to the number lost in the native state (12, 13). This method is based on the observation that experimentally measured $\Delta\Delta G_{U\text{-}F}$ values are proportional to the number of methyl or methylene groups around the mutation site (15).

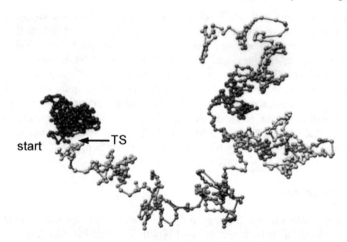

Figure 2   *3D clustering of 1 ns of simulation at 498 K. Each point represents a structure, with points connected by time. The distance between any two points approximates the Cα-RMSD between the two structures that they represent. The native cluster is colored dark grey with the location of the TS indicated by the arrow.*

Taking the ratio of transition state contact loss to native state contacts removes the proportionality constant and provides a reasonable estimate of Φ_F for core residues. Averaging over four simulations gave very good correlation (R=0.94) between Φ_{MD}-values and Φ_F-values for core residues. The free energy changes due to mutation can be modelled more explicitly using free-energy perturbation calculations. These calculations were carried out on folded, transition state, and unfolded conformations for 8 hydrophobic deletion point mutants and 2 double mutants in the core of CI2 (16). The correlation between Φ_F-values calculated by free energy perturbations and those from experiment was 0.85. S-values are a more general measure of the structure at a given position, and do not involve modelling in a specific mutation. S is defined as the product of $S_{3°}$, the ratio of number of native and non-native contacts in the TS to contacts in the native state, and $S_{2°}$, the fraction of TS time that a given residue spends in its native backbone conformation (17). S is much simpler to calculate than Φ_{MD} and can be calculated for all positions. The average S-values from the transition states identified in 100 simulations give a correlation coefficient of 0.74 with experimental Φ_F-values (14). The average S-values are mapped onto an example transition state structure in Figure 3.

In addition to comparing to experimental Φ_F-values, the transition state may be verified by making predictions based on the structures seen in simulation and by refolding simulations. Unfavorable interactions were identified in the C-terminal end of the α-helix and in the active site loop in the transition state from simulation (18). These interactions were removed and replaced with more favorable interactions by mutation, leading to a 40-fold increase in the rate of folding. Refolding simulations at 335 K of structures before, around, and after the transition state from one unfolding simulation were also performed (19). Starting structures taken from 10 to 30 ps before the transition state become significantly more native-like over the course of the simulations, whereas starting structures taken 5 to 30 ps after the TS become less native-like. Starting structures taken from within the 5 ps TS window become neither more nor less native-like on the timescale of the simulation.

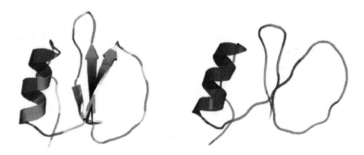

Figure 3 *The native structure of CI2 colored by experimental Φ_F-values (left) and an example TS structure from simulation colored by average S-values (right). In both cases, the values range from 0 (red) to 1 (blue).*

2.2.3 Structure of the transition state. The transition state maintains its native topology, but the core has expanded, the loops are more mobile and the β-sheet is disrupted (12, 13, 17). The native α-helix is generally maintained in the TS, as are its contacts to the hairpin connecting strands 2 and 3 in the β-sheet. The TS is more structurally heterogeneous than the native state (Figure 4). The average Cα-RMSD between transition state conformations from different simulations is 4.5 Å at 498 K (20).

Figure 4 *Overlay of TS structures from five independent 498 K MD simulations. The average Cα-RMSD between TS structures is 4.5 Å.*

The experimental Φ_F-values change with increasing temperature for some mutations (21). The average effect is slight, and different mutations at the same position change by different amounts, so it is not clear experimentally whether the change is due to a change in TS structure with temperature or changes in the effect of mutation with temperature. The structure of the transition state identified in simulation is not significantly changed by changing the temperature at which the simulation is performed (20). For several properties of the TS, including the total contacts, the total solvent accessible surface area, the number of main-chain hydrogen bonds, and the average S-value, there is no trend with temperature (Table 1). The number of native contacts maintained in the transition state, the radius of gyration, and the β-strand content do display trends with temperature in the direction of a more native-like transition state with increasing temperature, suggesting a slight Hammond effect (22, 23). These changes are all on the order of the expected error given the small number of simulations at lower temperatures, however. Thus, there appears to be no change in the global amount of structure, though a larger portion of that structure appears to be native-like at higher temperature. This may lead to differing energetic effects of mutation at different temperatures.

Table 1   *TS properties vs. temperature*

Temperature (K)	Total # of Contacts[a]	# Main-chain H-bonds[b]	SASA ($Å^2$)	Radius of gyration (Å)	# of Native Contacts[a,c]
Native State					
298[d]	210 ± 5[e]	21 ± 2	4538 ± 118	10.7 ± 0.1	187 ± 5
Transition State					
333[f]	172 ± 5	11 ± 1	5670 ± 179	12.0 ± 0.2	153 ± 6
373	173 ± 9	14 ± 2	5309 ± 283	12.0 ± 0.3	136 ± 10
398	172 ± 9	14 ± 2	5419 ± 283	11.9 ± 0.3	140 ± 10
448	179 ± 7	15 ± 2	5248 ± 231	11.8 ± 0.3	145 ± 8
473	173 ± 7	14 ± 1	5269 ± 231	11.7 ± 0.3	146 ± 8
498	170 ± 5	13 ± 1	5509 ± 163	11.8 ± 0.2	146 ± 6

[a]A contact is defined by a 5.4 Å distance between any heavy atom pair other than N-O pairs, which have a 4.6 Å distance cutoff.
[b]A hydrogen bond is defined by a 2.6 Å distance cutoff between donor and acceptor and a 35° angular cutoff.
[c]Native contacts are defined as those present in the crystal structure, 1YPC (7).
[d]298 K values are averaged over the last 40 ns of a 50 ns simulation.
[e]Error for the native state is based on the spread of values within the ensemble. Errors for the transition state are based on deviations within randomly chosen ensembles from a set of 100 simulations at 498 K.
[f]333 K simulations were solvated in 8 M urea.

The use of urea as a denaturant in simulation has a more pronounced effect on the structure of the transition state, though it does not significantly change the agreement between the calculated S-values and experimental Φ_F-values. The transition state in the presence of urea exposes more surface area to solvent and has fewer main-chain hydrogen bonds than the transition state in pure water. This is consistent with the combination of direct and indirect effects detailed in Bennion and Daggett (24).

2.3 The denatured state

While the denatured state is easily populated at equilibrium, it can be difficult to obtain experimental information about its structure as there is generally very little of it. In the classic view of the denatured state as a random coil, secondary structure is only transiently formed and long-range contacts are very short lived, making them hard to see in equilibrium experiments. The denatured state of CI2 from simulation appears to be well described as a random coil with minor amounts of residual structure, and experiments were performed to test this prediction (25). NMR relaxation times, correlation times, and backbone NOEs are all consistent with a random coil, and the denatured state ensemble from five simulations at 498 K reproduces them well. The only experimental indication of residual structure comes from chemical shift deviations, which suggest some residual structure at the C-terminal end of the α-helix and the C-terminal end of the first strand of the β-sheet. Some residual α-helix is observed in simulation, which could explain the chemical shift deviations. This structure is transient, however, and the average coupling constant calculated from the simulations is 7.0 Hz, compared to the experimental value of 7.2 Hz. The chemical shift deviations in the first strand may be explained by local, non-specific hydrophobic clustering in that region of the protein.

A more recent set of 100 simulations indicate that the earlier simulations considered in Kazmirski et al (25) were representative of the full unfolded ensemble (14). The average per residue properties of unfolded conformations are given in Figure 5. These properties appear to be well converged. Averaging over the last ns of simulation gives the same results as averaging over the last 10 ns of simulation. The average properties of sets of five simulations give correlation coefficients of at least 0.8 when compared to the average over all 100 simulations. This convergence is not due to complete sampling of the conformational space available to the denatured state. When the conformations in a single simulation are compared to those in the other 99, there is only a 25% chance that all of the conformations in the 100th simulation will be found in the other 99. In agreement with the earlier results, the most significant residual secondary structure is in the C-terminal end of the α-helix. The residues at the end of the first β-strand do not make more contacts on average than those in other parts of the protein, suggesting that the chemical shift deviations are due to more transient contacts.

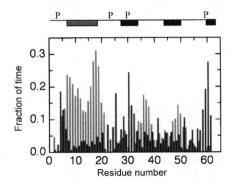 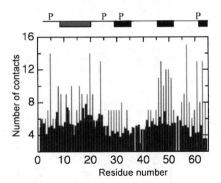

Figure 5 *Denatured state properties: A) Probability that a given residue will be in a three residue stretch of α-helix (grey) or β-strand (black) conformational space. α-helical conformational space is defined by $-100° < \phi < -30°, -80° < \psi < -5°$; and β-strand conformations by $-170° < \phi < -50°, 80° < \psi < -170°$. The native secondary structure and location of prolines is given above the graph. B) Heavy atom contacts, defined by a 4.6 Å distance cutoff for N-O pairs and a 5.4 Å distance cutoff for other heavy atom pairs, for each residue. The native contacts are given in light grey, and average denatured state contacts in black.*

2.4 The complete pathway

Given good agreement between simulation and experiment in the native, transition, and denatured states, it is reasonable to consider other conformations that are transiently populated in the unfolding process to be representative of the actual unfolding process. The unfolding process can be defined in terms of a pathway in which specific events in the unfolding are monitored. For CI2, an unfolding pathway was described as a five step process (9). First, the core is weakened and some movement is seen in the active site loop. This is followed by passage through the transition state as described above. After the transition state, the core opens up to solvent. This is followed by complete collapse of the active site loop as the β-sheet region is completely disrupted. The final step is passage into the fully unfolded state with loss of all native topology and structure in the α-helix. This order of events is well conserved across a range of temperatures from 373 K through 498 K (Figure 6). Temperature increases the rate of unfolding without changing the unfolding pathway.

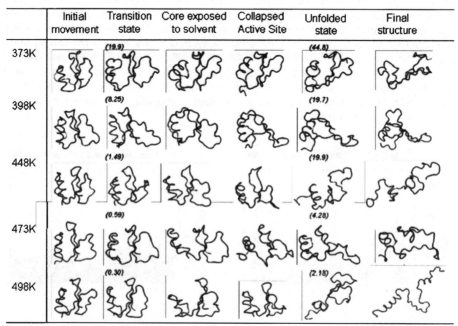

	Initial movement	Transition state	Core exposed to solvent	Collapsed Active Site	Unfolded state	Final structure
373K		(19.9)			(44.8)	
398K	(8.25)				(19.7)	
448K	(1.49)				(19.9)	
473K	(0.59)				(4.28)	
498K	(0.30)				(2.18)	

Figure 6   *Pathway structures vs. temperature. The times in ns at which the TS and unfolded state are reached are given in parenthesis above each structure.*

The unfolding pathway can also be thought of in terms of a most probable set of conformations on conformational landscape. A large set of unfolding simulations was used to populate a conformational landscape defined by the total number of contacts (tertiary structure) and number of residues in repeating α-helical or β-strand conformations (secondary structure) (14). The most probable path through this landscape involved concomitant loss of secondary and tertiary structure, with complete loss of secondary structure occurring after tertiary structure loss. When the trajectories of all individual simulations across this landscape are averaged, they closely mirror this average path (Figure 7). Although no individual simulation entirely follows the most probable path, the ensemble average is quite close to it.

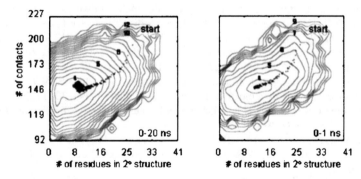

Figure 7   *Average pathway on a free energy landscape. A landscape based on the number of contacts and number of residues in α-helical or β-strand 2° structure was populated with structures found in the full 20 ns (left) or 1 ns (right) of 100 simulations at 498 K. Contact and 2° structure definitions are the same as those given in Figure 5. The contours give –ln(population).*

3 CONCLUSION

Simulations of CI2 are able to reproduce experimental observables in the native and denatured states, as well as the folding/unfolding transition state. It follows that the simulations capture the overall unfolding process and can be used to describe other transiently populated conformations in the unfolding pathway. Unfolding proceeds by concomitant loss of tertiary structure and secondary structure loss, with complete secondary structure loss, especially of the native α-helix, occurring after tertiary structure loss. Unfolding proceeds by an early loss of structure in the active site loop, followed by a loosening of the core packing and loss of well defined secondary structure in the β-sheet in the transition state. After the TS, the core becomes completely exposed to solvent. The active site loop then completely collapses as all structure in the β-sheet is lost. The last step in unfolding is the melting of the α-helix. While very high temperatures (498 K and above) were initially required to see full unfolding on MD timescales, advances in computational power have increased the timescales available to MD to the point where unfolding can now be simulated at experimental temperatures. These simulations show that the transition state and overall pathway of unfolding are unchanged by temperature.

Acknowledgements

Financial support was provided by the National Institutes of Health (GM50789). RD was also supported by a Molecular Biophysics Training Grant from the National Institutes of Health (2 T32 GM008268-17)

References

1   K.L. Maxwell, D. Wildes, A. Zarrine-Afsar, M.A. de los Rios, A.G. Brown, C.T. Friel, L. Hedberg, J.-C. Horng, D. Bona, E.J. Miller, A. Vallée-Bélisle, E.R.G. Main,

F. Bemporad, L. Qiu, K. Teilum, N.-D. Vu, A.M. Edwards, I. Ruczinski, F.M. Poulsen, B.B. Kragelund, S.W. Michnick, F. Chiti, Y. Bai, S.J. Hagen, L. Serrano, M. Oliveberg, D.P. Raleigh, P. Wittung-Stafshede, S.E. Radford, S.E. Jackson, T.R. Sosnick, S. Marqusee, A.R. Davidson and K.W. Plaxco, *Protein Sci.*, 2005, **14**, 602.

2   S.E. Jackson and A.R. Fersht, *Biochemistry*, 1991, **30**, 10428.

3   A.R. Fersht, A. Matouschek and L. Serrano, *J. Mol. Biol.*, 1992, **224**, 771.

4   S.E. Jackson, N. ElMasry and A.R. Fersht, *Biochemistry*, 1993, **32**, 11270.

5   D.E. Otzen, L.S. Itzhaki, N.F. ElMasry, S.E. Jackson and A.R. Fersht, *Proc. Natl. Acad. Sci. USA*, 1994, **91**, 10422.

6   L.S. Itzhaki, D.E. Otzen and A.R. Fersht, *J. Mol. Biol.*, 1995, **254**, 260.

7   Y. Harpaz, N. ElMasry, A.R. Fersht and K. Henrick, *Proc. Natl. Acad. Sci. USA*, 1994, **91**, 3.

8   S. Ludvigsen, H. Shen, M. Kjaer, J.C. Madsen, and F.M. Poulsen, *J. Mol. Biol.*, 1991, **222**, 621.

9   R. Day, B.J. Bennion, S. Ham, and V. Daggett, *J. Mol. Biol.*, 2002, **322**, 189.

10   G.L. Shaw, B. Davis, J. Keeler and A.R. Fersht, *Biochemistry*, 1995, **34**, 2225.

11   A. Li and V, Daggett, *Prot. Eng.*, 1995, **8**, 1117.

12   A. Li and V. Daggett, *Proc. Natl. Acad. Sci. USA*, 1994, **91**, 10430.

13   A. Li and V. Daggett, *J. Mol. Biol.*, 1996, **257**, 412.

14   R. Day and V. Daggett, *Proc. Natl. Acad. Sci. USA*, 2005, **102**, 13445-13450.

15   S.E. Jackson, M. Moracci, N. ElMasry, C.M. Johnson and A.R. Fersht, *Biochemistry*, 1993, **32**, 11259.

16   Y. Pan and V. Daggett, *Biochemistry*, 2001, **40**, 2723.

17   V. Daggett, A. Li, L.S. Itzhaki, D.E. Otzen and A.R. Fersht, *J. Mol. Biol.*, 1996, **257**, 430.

18   A.G. Ladurner, L.S. Itzhaki, V. Daggett and A.R. Fersht, *Proc. Natl. Acad. Sci. USA*. 1998, **95**, 8473.

19   D. De Jong, R. Riley, D.O.V. Alonso and V. Daggett, *J. Mol. Biol.*, 2002, **319**, 229.

20   R. Day and V. Daggett, *Protein Sci.*, 2005, **14**, 1242-1252.

21   M. Oliveberg, Y.-J. Tan, M. Silow and A.R. Fersht, *J. Mol. Biol.*, 1998, **277**, 933.

22   G.S. Hammond, *J. Am. Chem. Soc.*, 1955, **77**, 334.

23   A. Matouschek and A.R. Fersht, *Proc. Natl. Acad. Sci. USA*, 1993, **90**, 7814.

24   B.J. Bennion and V. Daggett, *Proc. Natl. Acad. Sci. USA*, 2003, **100**, 5142.

25   S.L. Kazmirski, K.-B. Wong, S.M.V. Freund, Y.-J. Tan, A.R. Fersht, and V. Daggett, *Proc. Natl. Acad. Sci. USA*, 2001, **98**, 4349.

THE ROLE OF ATTRACTIVE FORCES ON THE DEWETTING OF LARGE HYDROPHOBIC SOLUTES

Niharendu Choudhury[1] and B. Montgomery Pettitt[2]

Department of Chemistry, University of Houston, Houston, TX 77204-5003, USA

1. INTRODUCTION

The hydrophobic effect[1,2,3] between solutes in aqueous solutions plays a central role in our understanding of recognition and folding of proteins[4,5], and association of lipids[6,7]. This effect has recently been restudied and new mechanisms proposed for its origins.[8] Central to the debate is whether dewetting of the hydrophobic surfaces occurs prior to contact or at distances where water would not be allowed to intervene sterically.[8] Small hydrophobic solutes have been well studied.[9,10,11,12,13,14,15,16] The mechanism by which small solutes are accommodated into the natural cavities of water within the hydrogen bond network is well understood. Solvation of a large hydrophobic solute in water is thought to be associated with an energy cost due to partial disruption of hydrogen bond networks.[8] This mechanism leads to dewetting for specific cases and thus a particular mechanism for the hydrophobic effect. The question then is whether there are other energetic compensations which dictate a differing mechanism.

The energetic imbalance required for dewetting would be maximal when considering a purely repulsive model for nonpolar solutes in water, where incomplete hydrogen bonding might occur near such a repulsive solute with a large radius of curvature. Low dimensional networks of water might not be expected to have sufficient cohesion to be stable near such solutes. The imbalance in forces would cause an effective potential-cavity expulsion potential(CEP) to be generated. The effect of the CEP should increase[17] with increasing size of the solute due to an increasing interfacial region.

According to this picture,[8,18] for a large solute of nanometer size and above, the CEP causes water to be pushed away from the solute surface forming a thin vapor layer around it. When two such solutes come closer to each other, the fluctuations in the vapor-liquid interface between the individual solutes aids in growing the vapor layer and finally create a vacuous or dewetted region between the two solutes. Once the intersolute region is a vacuum, the solutes would then aggregate due to the solvent induced forces on them.

This theoretical picture of dewetting induced collapse [8,19,20,21,22] of large hydrophobic solutes has been supported by simulations[23,24,25,26] using purely repulsive or weakly attractive solute potentials. Theoretical works by Chandler and coworkers,[8,19,20,21] based on a mesoscopic square gradient theory for the liquid-vapor interface imply that hydration behavior interpolates between the traditional view of a small hydrophobic solute to this quite different picture for a large hydrophobic solute for such models.

[1]On leave from Theoretical Chemistry Division, Bhabha Atomic Research Centre, Mumbai 400 085, India
[2]Corresponding author. Fax: +1-713-743-2709, Email: pettitt@uh.edu

Recent simulation studies[27,28] on the behavior of water inside an atomistically modeled carbon nanotube (CNT) with realistic potentials have observed a one dimensional hydrogen-bonded chain of water molecules inside the hydrophobic nanotube. From the perspective of the theoretical arguments given above this would be unexpected. Earlier simulation studies by Rossky and coworkers[29,30] did not find any water density depression near an infinite repulsive wall.

Differing behavior in other systems has been seen. Further studies on the behavior of water near hydrophobic materials from the recent experimental literature,[31,32] indicate that water may wet a graphite surface and therefore may act as a lubricant between two layers of graphite like other gases. A number of recent experimental investigations[33,34,35,36] indicating contrasting results on the wetting/dewetting of large hydrophobic surfaces by water have made the study of hydrophobic hydration more interesting[37]. Although some recent computational studies[38,39,40] have attempted to elucidate the microscopic mechanism behind the strong attractive interaction between two large plates as observed in some surface force measurements,[41,42,43,44] it is clear that the fairly long ranged attraction, which may extend over several thousands of angstroms[43] to a few hundred angstroms[45] are computationally not feasible at present.

The role of the solute details in governing system behavior is thus apparent from both the theoretical and experimental sides. The role of simulation artifacts like boundary conditions have been explored by Patey and coworkers[46]. That group also explored the role of attractions in simple fluids near plates and found a strong dependence[47]. Observations from Hummer and coworkers[27,48,49] and from the recent studies by Berne and coworkers[25,50] lead one to question the role of the small attractive component of the usual van der Waals interaction, which is often thought[10,51] to have only a minor influence in determining the liquid structure and hence on the hydration phenomena. Although effects of attractive solute-solvent interactions on the hydration water structure around spherical solutes have been studied,[9,10,17,21] conclusions from such studies still show contradictions.

In a series of recent studies[52,53] we have demonstrated how the features of hydrophobic hydration of large solutes dramatically changes with the addition of normal attractive dispersion interaction to its usual purely repulsive counterpart. Here, we review that work and we demonstrate the importance of attractive solute-solvent interaction on the hydration behavior of large hydrophobic solutes and the mechanism of the hydrophobic effect.

2. METHODS

We consider here the results from our recent isothermal isobaric (NPT) simulation studies on the solvation behavior of nanoscopic solutes in water. Water was represented by the standard SPC/E[54] model and each of the hydrophobic solutes was modeled as a graphene sheet with carbon-carbon bond lengths of 1.4 Å. The solute was kept fixed and rigid during simulation. The carbon atoms of the solute were modeled as uncharged particles interacting with a Lennard-Jones (LJ) potential with diameter $\sigma_{CC} = 3.4$ Å. The well depth of the LJ potential was varied from $\varepsilon_{CC} = 0.086$ kcal mol^{-1} (or 0.3598 kJ mol^{-1}), corresponding to the sp^2 carbon atom of the AMBER 96 force field,[27,55] to a purely repulsive potential as obtained by the Weeks-Chandler-Andersen (WCA) truncation of the above LJ potential.

The potential of mean force between two solutes in water was calculated from thermodynamic perturbation theory.[56] Details of the simulation methods and other computational procedures have been given elsewhere.[52,53]

3. RESULTS

Potential of Mean Force and Hydration Structure of Nanoscopic Nonpolar Solutes

Let us first discuss the importance of considering solute attraction, which may be small at the atomic scale, on the large solute correlation by comparing the PMFs between two large plates in water with and without attractive dispersion interactions. We specifically compare the PMF between two 60 atom plates with the usual LJ interaction having parameters from the AMBER force field (type I) to that of geometrically equivalent but

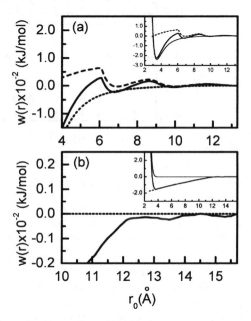

Figure 1. *Solute-solute potential contribution (dotted line) and solvent contribution (dashed line) to the potential of mean force (solid line) for (a) type I solute-water system and (b) for type II solute-water system. The inset in each figure shows the same over the entire range of r_0.*

purely repulsive analogues (type II) of the above LJ solutes. The potential of mean force, $w(r)$, as a function of the separation between the two large parallel plates of type I, is shown in Fig.1(a) and for type II in (b).Although at large separations, the undulating nature of the PMF corresponding to the solvent separated states with varying numbers of intervening water layers, has been observed in both the cases, at shorter separations a remarkable difference between the two PMFs can be seen.

From the decomposition of the PMFs into direct and solvent induced contributions, it was found that the difference in the mechanism of contact pair formation in these two types of solute is striking. For the repulsive solute-solvent system (type II) we find a large, purely solvent induced stabilization near contact, the direct contribution being zero. This result is consistent with earlier results[24] on the PMF of purely repulsive ellipsoidal solutes and thus eliminates several doubts raised in the literature about the specific water potential[25] and the effect[48] of curvature of the solute used in that study on their results. For the attractive solute plates (type I), however, we find a small solvent stabilization near contact with the overall PMF dominated by the solute-solute attractive potential. Beyond the contact minimum we also find a solvent separated minimum at around an intersolute distance of 6.8 Å, separated by a barrier between the contact and first solvent separated states. For repulsive solutes there is no barrier or solvent separated state around this distance.

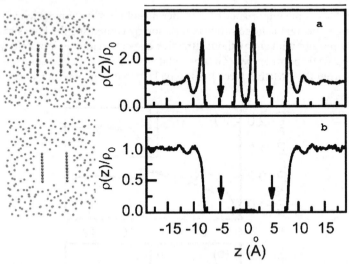

Figure 2. *Solute-solvent distributions. (a) Plot of a configuration for the system with full attractions. Red circles are the positions of the water oxygen atoms and blue circles are the carbon atoms on the solute plates for a slab through the sample. Right: Plot of the normalized single particle density $\rho(z)/\rho_0$ as a function of z, the distance perpendicular to the plates. Two arrows on the x-axis have been drawn to indicate the position (z-coordinate) of the two plates. (b) same as (a) but without attractions.*

The influence of attractive interaction on both solute-solute and solute-water correlation with simple small spherical solutes has been studied some time ago by Pratt and Chandler[10]. Although using previously used models for solute-water and solute-solute interactions, they found only a weak effect of attraction on the solute-water correlation, with a realistic methane potential, they found a 40 % change in the solute-solvent correlation. Thus, our observation that the solute correlation and PMF are entirely different for the cases of a purely repulsive solute and a LJ solute, is not surprising if we note that the size of the solute considered here is much larger than those used in that study[10].

For the type I system well outside of contact, Fig. 1(a) shows shallow minima in the PMF at separations (r_0) of 6.8 Å, 10 Å and 13 Å. In order to visualize the solvent structure near the solute plates corresponding to the minimum in the PMF, we have shown in Fig. 2(a) the coordinates (single configuration) and the number density profile corresponding to the minimum at $r_0 = 10$ Å that corresponds to two intervening water layers between the two solutes. The system without attractions (type II) is shown in 2(b). The lack of attractive forces expels the layer of water molecules from the intersolute region.

This shows the importance of taking into account solute attraction in determining the solute-solute correlation and hence predicting the dewetting induced collapse of large hydrophobic particles. Our results are consistent with the observation of a stable water chain inside a CNT[27] and nano cavity[49] as well as the results of Ashbaugh et al.[17] that found enhanced hydration (wetting) of a large single methane cluster in water with the full LJ interactions.

In order to test the stability of the configuration corresponding to the barrier in the PMF at $r_0 = 6.2$ Å that contains a highly strained water layer in the inter solute region, we have calculated (not shown) water occupancy in the intersolute region at this separation as a function of time, for a total simulation time of 2.5 nano seconds. It reveals that the number of water molecules confined between the two plates fluctuates around 14 for most of the time, indicating that the configuration is stable. Although once in that period a large fluctuation made the confined region empty for around 30ps, water molecules again entered the region and within 100ps, the number of water molecules continued to fluctuate around 14, where the average remained for rest of the time.

In order to investigate whether the solute-solvent configuration corresponding to the barrier observed in the solvent induced PMF at $r_0 = 6.2$ is metastable, we also performed a 1.2 ns simulation at this r_0 starting from a dry initial state where the inter plate region is initially empty. This test showed that an empty inter solute cavity is filled by around 14 water molecules within 120ps and thereafter the average number of water molecules remains the same. It is important to mention at this point that there is no first solvent separated state and hence barrier connecting that with the contact pair state in case of purely repulsive solutes (compare PMF of solute I with that of solute II). For the purely repulsive case, the PMF is monotonically decreasing with decreasing intersolute distance below 12 Å. This is due to the fact that all the water molecules are expelled from the intersolute region (see Fig.2(b)) even when the available space from geometric consideration can easily accommodate one to two layers of water. In fact the essentially linear decrease in the PMF (ΔG) with the decrease in the area ΔA of the liquid vacuum interface of the cavity makes it possible to calculate the liquid-vapor surface tension γ_{lv} from the relation $\Delta G = \gamma_{lv} \Delta A$. The liquid-vapor surface tension γ_{lv} obtained[52] in this way for SPC/E water is about 71.7 mJ/m^2. This is in good agreement with that calculated recently[20] from a simulation study of the liquid-vapor interface of the same water model. This indicates that the bubble has been formed in the intersolute region of repulsive plates with a center to center separation of around 10 Å and below and the free energy cost of maintaining the interface strongly resembles that of a free, planar water-vapor interface[52].

In contrast, for the attractive solute, not only is the loss of hydrogen bonds less, there is compensation from the solute-solvent attractive interaction as well. Therefore, in this case we do not observe any dewetting as long as there is steric space available for physically accommodating water in between the two solutes.

Effect of solute size on hydration

In order to investigate the effect of solute sizes, we have performed several MD simulation runs in the NPT ensemble with single flat planar solutes of various sizes immersed in water with the same two types of solute-solvent interaction: one with normal attractions and a purely repulsive potential obtained using the WCA approximation[51]. We show the results for the density profiles for three different solute sizes with the repulsive interactions in Fig.3 and with LJ interactions in Fig.4. The nature of density profiles for the repulsive solutes is distinctly different from that of their LJ counter parts. The LJ solutes are significantly wetted as indicated by the sharp well defined density peaks with a contact value or peak height considerably above the bulk density. Moreover, no significant change is observed in this feature with a change

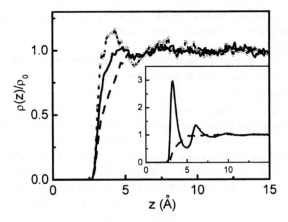

Figure 3. *Plot of the normalized single particle density* $\rho(z)/\rho_0$ *of water oxygen as a function of the distance perpendicular to the solute plate with solute atoms interacting with repulsive potential (type II) for three different solute sizes (half of the box is shown). The smallest one with 28 atoms shown by an open circle with a line, 60 atoms by a solid line and 178 atoms by a dashed line. In the inset the density profile of water around the largest solute with a purely repulsive interaction (dashed line) is compared with that including the LJ interaction (solid line).*

in solute size. In the case of the repulsive solute, however, the water density profile is drastically changed from its LJ analogue, with no strong layering around the solute (see the comparison in the inset of Figure 3). Most importantly, we observe a significant change in the water density profile when we change the solute size. The contact peak not only decreases with increasing sizes of the solutes, it rises from zero more slowly over a wide range as well (see the increase in the position of the first peak with increasing solute size).

The water density profile in Figure 3 around the largest repulsive solute plate considered here (21 Å × 21 Å) has been interpreted as showing a vapor-liquid like interface in the vicinity of the solute.[8,20,25] However if one compares (See inset of Fig 3) this density profile with the one for the same solute size but with LJ interactions, the difference is apparent.

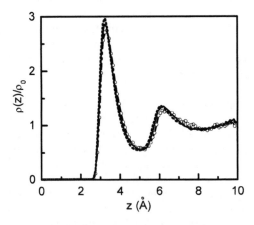

Figure 4. *Same as Fig.3 but the potential of the solute atoms modeled with full LJ interaction.*

This difference in size dependence of the water accumulation around large nonpolar solutes with and without attraction can be well understood in terms of the CEP[48]) that arises due to the imbalance in the attractive forces in water near a large repulsive surface. An increase in solute size causes the CEP to increase[17] due to the imbalance in the attractive interactions and thus water accumulation near the solute decreases with the increase in solute size. However, when an attractive dispersion interaction of the solute is taken into account, the cumulative attractive interaction between the solute and the solvent compensates for this CEP and thus we observe a significant hydration of the nanoscopic solutes without much solute size effect.

4. CONCLUSION

The results from a series of investigations[52,53] from our laboratory demonstrate the importance of the weak dispersion interactions naturally present in essentially all nonpolar substances to study hydration behavior of such solutes in general and dewetting in the intersolute region in particular. Our results indicate that the dewetting induced collapse of large hydrophobic solutes as predicted by many recent investigations[8,19,22,24,25] based on either purely repulsive or weakly attractive solute potential may not apply to hydrophobic solutes like protein interfaces which have significant polarity and dispersion interactions. These results should be considered when interpreting the mechanism of aggregation phenomena observed in many areas of chemistry and biology.[50]

Acknowledgments
BMP thanks L. Pratt and A.D.J. Haymet for many stimulating discussions about the nature of interfaces and the hydrophobic effect. We thank G.C. Lynch and K.-Y. Wong for many valuable technical discussions. We gratefully acknowledge NIH, the R.A. Welch foundation, and TiiMES, funded by NASA Cooperative Agreement No. NCC-1-02038 for partial financial support of this work. The computations were performed in part using the NSF meta center facilities and the Molecular Science Computing Facility in the W.R. Wiley Environmental Molecular Sciences Laboratory, a national scientific user facility sponsored by DOE's Office of Biological and Environmental Research and located at Pacific Northwest National Laboratory, operated for DOE by Battelle.

References

[1] W. Kauzmann, *Adv. Protein Chem.*, 1959, , 1.
[2] Tanford, C. *The Hydrophobic Effect: Formation of Micelles and Biological Membranes*; John Weiley: New York, 1973.
[3] L. R. Pratt and A. Pohorille, *Chem. Rev.*, 2002, , 2671.
[4] C. L. Brooks, J. N. Onuchic and D. J. Wales, *Science*, 2001, , 612.
[5] J. Shea and C. L. Brooks, *Annu. Rev. Phys. Chem.*, 2001, , 499.
[6] G. Hummer, S. Garde, A. E. Garcia and L. R. Pratt, *Chem. Phys.*, 2000, , 349-370.
[7] L. Lu and M. L. Berkowitz, *J. Am. Chem. Soc.*, 2004, , 10254.
[8] K. Lum, D. Chandler and J. D. Weeks, *J. Phys. Chem. B*, 1999, , 4570.
[9] L. R. Pratt and D. Chandler, *J. Chem. Phys.*, 1977, , 3683.
[10] L. R. Pratt and D. Chandler, *J. Chem. Phys.*, 1980, , 3434.
[11] C. Pangali, M. Rao and B. J. Berne, *J. Chem. Phys.*, 1979, , 2975.
[12] D. E. Smith and A. D. J. Haymet, *J. Chem. Phys.*, 1993, , 6445.
[13] G. Hummer, S. Garde, A. E. Garcia, A. Pohorille and L. R. Pratt, *Proc. Natl. Acad.*

Sci. U.S.A., 1996, , 8951.

[14] N. T. Southall and K. A. Dill, *Biophys. Chem.*, 2002, , 295.

[15] J.S. Perkyns and B.M. Pettitt, *J. Phys. Chem.*, 1996, , 1323-1329.

[16] N. M. Cann and G. N. Patey, *J. Chem. Phys.*, 1997, , 8165.

[17] H. S. Ashbaugh and M. E. Paulaitis *J. Am. Chem. Soc.*, 2001, , 10721-10728.

[18] F. H. Stillinger, *J. Solution Chem.*, 1973, , 141.

[19] D. M. Huang and D. Chandler, *Proc. Natl. Acad. Sci. U.S.A.*, 2000, , 8324.

[20] D. M. Huang, P. L. Geissler and D. Chandler, *J. Phys. Chem. B*, 2001, , 6704.

[21] D. M. Huang and D. Chandler, *J. Phys. Chem. B*, 2002, , 2047.

[22] P. R. ten Wolde and D. Chandler, *Proc. Natl. Acad. Sci. U.S.A.*, 2002, , 6539.

[23] A. Wallqvist and B. J. Berne, *J. Phys. Chem.*, 1995, , 2885.

[24] A. Wallqvist and B. J. Berne, *J. Phys. Chem.*, 1995, , 2893.

[25] X. Huang, C. J. Margulis, and B. J. Berne, *Proc. Natl. Acad. Sci. U.S.A.*, 2003, , 11953.

[26] A. Walqvist, E. Gallicchio and R.M. Levy, *J. Phys. Chem.*, 2001, , 6745-6753.

[27] G. Hummer, J. C. Rasaiah and J. P. Noworyta, *Nature*, 2001, , 188.

[28] M. S. P. Sansom and P. C. Biggin, *Nature*, 2001, , 156.

[29] C. Y. Lee, J. A. McCammon and P. J. Rossky, *J. Chem. Phys.*, 1984, , 4448.

[30] S. H. Lee and P. J. Rossky, *J. Chem. Phys.*, 1994, , 3334.

[31] M. E. Schrader, *J. Phys. Chem.*, 1980, , 2774.

[32] M. Luna, J. Colchero and A. M. Baro, *J. Phys. Chem. B*, 1999, , 9576.

[33] R. Steitz, T. Gutberlet, T. Hauss, B. Klosgen, R. Krastev, S. Schemmel, A. C. Simonsen, G. H. Findenegg, *Langmuir*, 2003, , 2409.

[34] T. R. Jensen, M. O. Jensen, N. Reitzel, K. Balashev, G. H. Peters, K. Kjaer and T. Bjornholm, *Phys. Rev. Lett.*, 2003, , 086101.

[35] D. Schwendel, T. Hayashi, R. Dahint, A. Pertsin, M. Grunze, R. Steitz and F. Schreiber, *Langmuir*, 2003, , 2284.

[36] V. Yaminsky and S. Ohnishi, *Langmuir*, 2003, , 1970.

[37] P. Ball, *Nature*, 2003, , 25.

[38] J. Forsman, B. Jonsson and C. E. Woodward, *J. Phys. Chem.*, 1996, , 15005.

[39] D. Bratko, R. A. Curtis, H. W. Blanch and J. M. Prausnitz, *J. Chem. Phys.*, 2001, , 3873.

[40] T. Hayashi, A. J. Pertsin and M. Grunze, *J. Chem. Phys.*, 2002, , 6171.

[41] P. M. Claesson and H. K. Christenson, *Science*, 1988, , 390.

[42] H. K. Christenson, In *Modern Approaches to wettability: Theory and Applications*; M. E. Schrader, G. Loeb, Eds.; Plenum: New York, 1992.

[43] J. L. Parker, P. M. Claesson and P. Attard, *J. Phys. Chem.*, 1994, , 8464.

[44] R. M. Pashley, P. M. McGuiggan, B. W. Ninham and D. F. Evans, *Science*, 1985, , 1088.

[45] J. Wood and R. Sharma, *Langmuir*, 1995, , 4797.

[46] J.C. Shelley and G.N. Patey, *Molecular Physics* , 1996, , 385.

[47] D. R. Berard, Phil Attard, and G. N. Patey, *J. Chem. Phys.*, 1993, , 7286.

[48] G. Hummer and S. Garde, *Phys. Rev. Lett.*, 1998, , 4193.

[49] S. Vaitheeswaran, H. Yin, J. C. Rasaiah and G. Hummer, *Proc. Natl. Acad. Sci. USA*, 2004, , 17002-17005.

[50] R. Zhou, X. Huang, C. J. Margulis and B. J. Berne, *Science*, 2004, , 1605.

[51] D. Chandler, J. D. Weeks and H. C. Andersen, *Science*, 1983, , 787; J. D. Weeks, D. Chandler and H. C. Andersen, *J. Chem. Phys.*, 1971, , 5237.

[52] N. Choudhury, and B. M. Pettitt, *J. Am. Chem. Soc.*, 2004, , 3556.

[53] N. Choudhury, and B. M. Pettitt, *Mol. Sim.*, 2004, , 457.

[54] H. J. C. Berendsen, J. R. Grigera and T. P. Straatsma, *J. Phys. Chem. B*, 1987, , 6269.

[55] D. W. Cornell, W. D. Cornell, P. Cieplak, C. I. Bayly, I. R. Gould, K. M. Merz, D. M. Ferguson, D. C. Spellmeyer, T. Fox, J. W. Caldwell and P. A. Kollman, *J. Am. Chem. Soc.*, 1995, , 5179.

[56] R. W. Zwanzig, *J. Chem. Phys.*, 1954, , 1420.

STRUCTURE AND MECHANISM OF THE ATPASE VCP/P97: COMPUTATIONAL CHALLENGES FOR STRUCTURE DETERMINATION AT LOW RESOLUTION

A.T. Brunger[1] and B. DeLaBarre[1]

[1]Howard Hughes Medical Institute and Departments of Molecular and Cellular Physiology, Neurology and Neurological Sciences, and Stanford Synchrotron Radiation Laboratory, J.H. Clark Center E300C, 318 Campus Drive, Stanford CA 94305, USA

1 INTRODUCTION

It is often assumed that crystal structures of biological macromolecules have to be obtained at sufficiently high resolution (typically 1-3.5 Å) in order to obtain biological insights about the system that is studied. However, there are many questions that can be addressed at minimum Bragg spacings lower than 3.5 Å, such as overall assembly and topology of macromolecular complexes, large-scale motions, and conformational changes. Such low resolution structural studies have been typically restricted to special cases, such as model-free display of electron density maps, approximate modeling of such maps, rigid body refinement, or all-atom refinement in the presence of high non-crystallographic symmetry. Examples of these types of studies include (with increasing resolution, this is not meant to be an all inclusive list): detection of large motions of the ribosome by interpretation of electron density maps of crystals that diffracted to 9 Å resolution [1]; studies of the reverse transcriptase from HIV in complex with a target RNA pseudoknot at 4.75 Å resolution [2]; mechanistic studies of electron density maps at 4.4 Å resolution of the E. coli F1 ATPase [3]; the structure of heptameric protective antigen bound to an anthrax toxin receptor at 4.3 Å resolution with 14-fold NCS averaging [4]; study of the interactions between human CD4 two-domain receptor fragments and MHC class II molecules at 4.3 Å resolution [5]; the crystal structure of core of the AP-1 complex at 4 Å resolution with six-fold NCS averaging [6]; and a crystal structure of the dimeric HIV-1 capsid protein at 3.7 Å resolution[7].

There is clearly a need for development of more general methods to analyze low resolution structures. A first step in this direction has been the recent demonstration that robust all-atom refinement is possible even for structures solved in the range from 3.8 to 4.7 Å [8-13]. This review summarizes the techniques that were used to achieve structure solution in these cases and discusses the challenges ahead.

2 METHOD AND RESULTS

2.1 Crystallographic phase problem

The problem of low resolution crystallography is directly related to the well-known phase problem [14], that is, a monochromatic diffraction experiment for a single crystal typically only produces information about the amplitudes of the molecular structure factors. However, both phases and amplitudes are required in order to calculate the electron density by Fourier transformation. Phase information thus has to be obtained by computational methods (direct methods), experimental techniques (single or multiple isomorphous replacement, single or multi-wavelength anomalous dispersion), or phasing by homologous search models (molecular replacement) [15]. The phase problem becomes more pronounced at lower resolution ranges. In the absence of experimental phase information, model refinement progressively diverges as the resolution limit is lowered. For example, for a simulated test case with artificially truncated diffraction data at 4 Å resolution, the resulting coordinate error is 0.6Å, while at 4.7 Å resolution, the coordinate error is 1.25 Å (Figure 1) [16]. Due to the artificial nature of the truncation of the diffraction data used in this study, the coordinate error is expected to be somewhat larger for actual diffraction data obtained at low resolution.

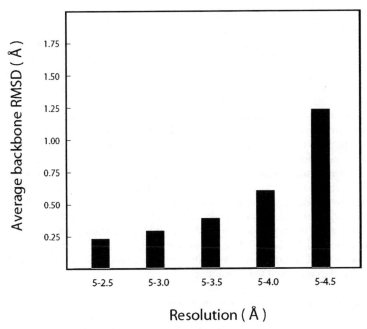

Resolution (Å)

Figure 1   *Radius of convergence of torsion-angle simulated annealing [18]. The test case consisted of an initial model of α-amylase inhibitor built using experimental phase information from multiple isomorphous replacement diffraction data [19]. The initial model was refined against artificially truncated sets of the original diffraction data [19] for the specified resolution ranges. Averages over ten torsion-angle simulated annealing refinements are shown. The average root-mean-square-difference (RMSD) refers to the refined high resolution structure of amylase inhibitor [19].*

It is instructive to compare the accuracy of low resolution crystal structures with that of solution Nuclear Magnetic Resonance (NMR)-derived structures [21]. Typically, the accuracy of NMR structures ranges from 0.5 to 1.0 Å as determined by the root-mean-square differences (RMSD) between independently solved structures of the same protein by NMR and X-ray crystallography (these RMSDs are generally within the precision of the corresponding NMR structures) [22]. Thus, the accuracy of a crystal structure solved at, for example, 4 Å resolution should be comparable to that of a typical solution NMR structure.

However, this comparison is somewhat misleading since nuclear Overhauser effect (NOE)-derived distance information is specific to particular pairs of protons that can be readily assigned by multi-dimensional heteronuclear techniques whereas a low-resolution electron density map of a macromolecule without heavy atom or anomalous scatterers has few or no pronounced markers that could aide in the assignment of molecular fragments.

2.2   Experimental phase information by MAD

The technique of multi-wavelength anomalous dispersion (MAD) [23] can be used to overcome some of the shortcomings of low resolution crystal structures. First, the identification of heavy atom or selenium positions leads to markers that guide tracing of the macromolecule(s). Second, the high accuracy of the experimental MAD phase probability distributions allows them to be used directly in the refinement process, making refinement robust, even below 3.5 Å resolution. For example, the structure of the complete p97/valosin containing protein (VCP) · ADP ·AlF$_x$ hexamer, an ATPase involved in endoplasmic reticulum associated degradation, was solved at 4.4 Å resolution with three independent protomers in the asymmetric unit [10, 11]. The structure solution process involved MAD phasing, and positional (torsion angle simulated annealing and minimization) and group B-factor refinement with the "MLHL" target function that incorporates experimental phase information [24]. B-factor sharpening [8] of phase-combined electron density maps allowed us to identify many side chains in addition to the main chain which was visible throughout each of the protomers. B factor sharpening also enabled identification of the bound nucleotide. Individual atoms in the ligand could not be seen clearly, but this is often the case in structures solved at 'traditional' resolutions of 2 -3 Å. The quality of the electron density around the ligand was sufficient to differentiate ADP from the ADP · AlF$_x$ species. Prior to this study, the use of B-factor sharpening had been shown to be useful, e.g. for the crystal structure of *E. Coli* MscS, a voltage-modulated and mechanosensitive channel solved at 3.9 Å resolution by MAD phasing [8]. Seven-fold crystallographic averaging combined with B-factor sharpening produced electron density maps with unusually high quality at that resolution.

2.3   Iterative improvement of the heavy atom model

As mentioned above, incorporation of experimental phase information into maximum likelihood refinement [24] is essential in order to make refinement robust at low resolution. It has also been observed that it is important to iteratively improve the heavy atom (or selenium) model and experimental phase probability distribution by using the refined model phases as a prior phase probability distribution in order to identify additional scatterers or to improve the parameters of the heavy atom scatterers [11]. During this iterative process, the resulting experimental phase probability distributions improve as assessed by map quality, figures of merit, and free *R* value of the MLHL-refined complete model. An alternative approach may consist of incorporating a multivariate likelihood function into refinement [25] although this new method yet has to be tested at low resolution. This new target function simultaneously refines both heavy atom and all-atom model parameters.

2.4 Refinement at low resolution

The refinement of p97/VCP · ADP ·AlF$_x$. made use of the standard Crystallography & NMR System protocols [26] except that the bulk solvent model had to be modified and secondary structure restraints added to stabilize the α-helices during simulated annealing refinement. MAD phasing was essential to obtain robust refinement, while phasing by single anomalous dispersion (SAD) was insufficient (unpublished). Interestingly, B-factor sharpening only performed well when phase-combined maps were used; B-factor sharpened maps obtained by molecular replacement with a partial model were not interpretable. After the p97/VCP · ADP ·AlF$_x$. structure was solved and published crystals of p97/VCP complexed with ADP and AMP-PNP were obtained that diffracted to 4.2 and 3.5 Å, respectively. Structures were solved and refined with a similar protocol as that for the ADP ·AlF$_x$ complex except that no secondary structure restraints were required [11].

These higher resolution structures revealed many more side chain positions and essentially confirmed the chain trace of the p97/VCP · ADP ·AlF$_x$ model that was originally solved at 4.7 Å. Interestingly, the D2 domain was largely disordered in the AMP-PNP complex. In contrast, the electron density maps for the complexes that diffracted to lower resolution showed well-defined electron density for the D2 domains. Thus, overall resolution is not necessarily correlated with quality of electron density maps.

The structure of a fully glycosylated SIV gp120 envelope glycoprotein in an unliganded conformation at 4.0 Å resolution [9, 27] used a strategy similar to that used for p97/VCP, but it employed another important tool: multi-crystal averaging. Molecular replacement failed due to large conformational differences to the liganded form of gp120, so heavy atom phasing had to be used. A combination of phase combination, B-factor sharpening, density modification, multi-crystal averaging, model building, and heavily restrained refinement was used to solve the structure. Secondary structure information (backbone torsion angle restraints) was used to restrain the structure. Initially, the MLHL target and later an amplitude-based maximum likelihood target was used. Bulk solvent parameters were modified as described for the structure solution of p97/VCP [10] and B-factor sharpening was successfully employed.

Two other recent examples of crystal structures refined at low resolution further illustrate the usefulness of such structures. First, the structure of the tetrahymena ribozyme was solved at 3.8 Å resolution [13]. Four independent monomers were present, but they showed significant conformational differences, so no averaging was possible. Combinations of MAD and SAD datasets were used to solve the structure. MLHL refinement and B-factor sharpening was used, although bulk solvent refinement failed for unknown reasons. Second, the inactive state of the *E. Coli* DNA polymerase clamp loader complex was solved in complex with ATPγS (at 3.5 Å resolution) and with ADP (at 4.1 Å resolution) [12]. MAD phasing was used for the ATPγS complex, while the ADP complex was solved by molecular replacement with the ATPγS form. B-factor sharpening was unsuccessful for the ADP crystal structure, presumably due to unavailability of experimental phase information.

2.5 Model building at low resolution

Structure solution and refinement at 4 Å is challenging since pattern recognition by a skilled crystallographer is essential to interpret noisy low resolution electron density maps;

with current technology essentially no automation can be used (Figure 2). It is essential to to obtain markers provided by peaks in anomalous Fourier difference maps that correspond to selenomethionine positions (red peak in Figure 2). In the case of p97/VCP, 16 methionines were roughly equally distributed through the primary sequence of the protomer of a molecular weight of 89.5 kDa. This in combination with the availability of high resolution structures for the N-D1 fragment, and judicious homology modeling of the D2 domain allowed the creation of a complete model of p97/VCP. α-helices and β-sheets were generally well resolved (Figures 2 and 3), with the latter becoming better defined at 4.25 Å resolution (Figure 3). The structures of p97/VCP in three representative nucleotide states refined to reasonable free R values and acceptable geometry [11].

A   B

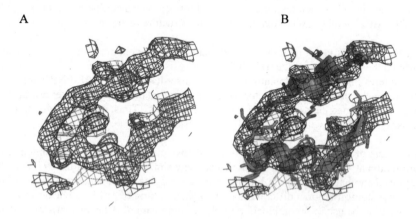

Figure 2 *Illustration of model building at low resolution assisted by the knowledge of selenium positions. (A) and (B): shown is a B-sharpenedσ_A weighted 2Fo-Fc phase combined map of residues 540:565 of the ATPase p97/VCP · APD at 4.25 Å resolution [11], contoured at 1.4 σ (magneta). Superimposed is an anomalous Fourier difference map at 3 σ (red) for the diffraction data that was collected at the anomalous peak wavelength for selenium. (B) The refined model is shown in green. Note that several side chains are visible, including the selenomethionine at residue 550.*

2.6   Motion and conformational changes of p97/VCP

The structures of p97/VCP that were solved in the range from 3.5 to 4.4 Å resolution [10, 11, 28] allowed a description of the motions and conformational transitions during the ATP hydrolysis cycle. The largest motions occurred at two stages during the hydrolysis cycle: after, but not upon, nucleotide binding and then following nucleotide release [11]. The motions primarily occurred in the D2 domain, the D1 α-helical domain, and the N-terminal domain, relative to the relatively stationary and invariant D1α/β domain. In addition to the motions, we observed a transition from a rigid state to a flexible state upon loss of the γ-phosphate, and a further increase in flexibility within the D2 domains upon nucleotide release. The domains within each protomer of the hexameric p97/VCP deviated from strict six fold symmetry, with the more flexible ADP state exhibiting larger asymmetry compared to the relatively rigid ADP·AlF$_3$ state, suggesting a mechanism of action in which hydrolysis and conformational changes move about the hexamer in a processive fashion [11].

A

B

C

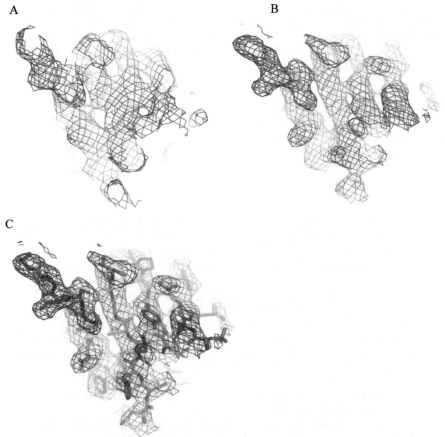

Figure 3 *Illustration of the improvement of the definition of the β-sheet formed by residues 515:520, 575:580, 616:623, and 639:645 of p97/VCP upon increasing resolution. Shown are B-sharpened σ_A weighted 2Fo-Fc phase-combined maps of the ATPase p97/VCP at 4.7 (A) and 4.25 Å (B) resolution [10, 11], contoured at 1.4 σ (red). In (C) the model (green) is superimposed on the 4.25 Å electron density map.*

3   CONCLUSION

As crystallographers study larger and larger macromolecular complexes, the intrinsic flexibility and high solvent content of some of these complexes will preclude structure solution at high resolution. Yet, to understand biological function it is important to study assemblies that are as close as possible to their physiological counterparts. Thus, the techniques discussed in this review will play an increasingly important role in biological crystallography. In addition, new methods need to be developed - to name a few: aides to interpret noisy low resolution maps, estimators of individual atomic coordinate errors, and statistically correct combination of structural information from a variety of sources (e.g., secondary structure restraints and homology models). The deposition of low resolution structures will also require special attention; perhaps electron density maps should be deposited in addition to coordinates and structure factors, so one could judge the accuracy of the model by direct inspection of the maps. These techniques should also be useful for cryo-electron microscopy studies that move closer to the low resolution limits that were discussed here [29, 30].

Acknowledgements
We would like to thank Paul Adams and Piet Gros for useful discussions and critical reading of this review.

References

[1]   A. Vila-Sanjurjo, W. K. Ridgeway, V. Seymaner, W. Zhang, S. Santoso, K. Yu, and J. H. Cate, *Proc Natl Acad Sci U S A*, 2003, **100**, 8682.

[2]   J. Jaeger, T. Restle, and T. A. Steitz, *Embo J*, 1998, **17**, 4535.

[3]   A. C. Hausrath, R. A. Capaldi, and B. W. Matthews, *J Biol Chem*, 2001, **276**, 47227.

[4]   D. B. Lacy, D. J. Wigelsworth, R. A. Melnyk, S. C. Harrison, and R. J. Collier, *Proc Natl Acad Sci U S A*, 2004, **101**, 13147.

[5]   J. H. Wang, R. Meijers, Y. Xiong, J. H. Liu, T. Sakihama, R. Zhang, A. Joachimiak, and E. L. Reinherz, *Proc Natl Acad Sci U S A*, 2001, **98**, 10799.

[6]   E. E. Heldwein, E. Macia, J. Wang, H. L. Yin, T. Kirchhausen, and S. C. Harrison, *Proc Natl Acad Sci U S A*, 2004, **101**, 14108.

[7]   C. Momany, L. C. Kovari, A. J. Prongay, W. Keller, R. K. Gitti, B. M. Lee, A. E. Gorbalenya, L. Tong, J. McClure, L. S. Ehrlich, M. F. Summers, C. Carter, and M. G. Rossmann, *Nat Struct Biol*, 1996, **3**, 763.

[8]   R. B. Bass, P. Strop, M. Barclay, and D. C. Rees, *Science*, 2002, **298**, 1582.

[9]   B. Chen, E. Vogan, H. Gong, J. Skehel, D. Wiley, and S. Harrison, *Structure*, 2005, **13**, 197.

[10]   B. DeLaBarre and A. T. Brunger, *Nat Struct Biol*, 2003, **10**, 856.

[11]   B. DeLaBarre and A. T. Brunger, *J Mol Biol*, 2005, 2005, **347**, 437.

[12]   S. L. Kazmirski, M. Podobnik, T. F. Weitze, M. O'Donnell, and J. Kuriyan, *Proc Natl Acad Sci U S A*, 2004, **101**, 16750.

[13]   F. Guo, A. R. Gooding, and T. R. Cech, *Mol Cell*, 2004, **16**, 351.

[14]   H. Hauptman, *Physics Today*, 1989, **42**, 24.

[15]   Drenth, 'Principles of Protein X-ray Crystallography', Springer, 1999.

[16]   A. T. Brunger and L. Rice, *Methods in Enzym*, 1997, **277**, 243.

[17]   A. T. Brunger, A. Krukowski, and J. W. Erickson, *Acta Crystallogr A*, 1990, **46**, 585.

[18]   L. M. Rice and A. T. Brunger, *Proteins*, 1994, **19**, 277.

[19]   J. W. Pflugrath, G. Wiegand, R. Huber, and L. Vertesy, *J Mol Biol*, 1986, **189**, 383.

[20]   A. T. Brunger, *Nature*, 1992, **355**, 472.

[21]   A. T. Brunger, *Nat Struct Biol*, 1997, **4**, 862.

[22]   A. Gronenborn and G. M. Clore, *Crit. Rev. Biochem. Mol. Biol.*, 1995, **30**, 351.

[23]   W. Hendrickson, *Science*, 1991, **254**, 51.

24   N. S. Pannu, G. N. Murshudov, E. J. Dodson, and R. J. Read, *Acta Crystallogr D Biol Crystallogr*, 1998, **54**, 1285.

25   P. Skubak, G. N. Murshudov, and N. S. Pannu, *Acta Crystallogr D Biol Crystallogr*, 2004, **60**, 2196.

26   A. T. Brunger, P. D. Adams, G. M. Clore, W. L. DeLano, P. Gros, R. W. Grosse-Kunstleve, J. S. Jiang, J. Kuszewski, M. Nilges, N. S. Pannu, R. J. Read, L. M. Rice, T. Simonson, and G. L. Warren, *Acta Crystallogr D Biol Crystallogr*, 1998, **54**, 905.

27   B. Chen, E. Vogan, H. Gong, J. Skehel, and S. C. Harrison, *Nature*, 2005, **433**, 834.

28   T. Huyton, V. E. Pye, L. C. Briggs, T. C. Flynn, F. Beuron, H. Kondo, J. Ma, X. Zhang, and P. S. Freemont, *J Struct Biol*, 2003, **144**, 337.

29   S. J. Ludtke, D. H. Chen, J. L. Song, D. T. Chuang, and W. Chiu, *Structure (Camb)*, 2004, **12**, 1129.

30   A. Fotin, Y. Cheng, P. Sliz, N. Grigorieff, S. C. Harrison, T. Kirchhausen, and T. Walz, *Nature*, 2004, **432**, 573.

THEORETICAL ANALYSIS OF MECHANOCHEMICAL COUPLING IN THE BIOMOLECULAR MOTOR MYOSIN

Q. Cui

Department of Chemistry and Theoretical Chemistry Institute, University of Wisconsin, Madison, 1101 University Avenue, Madison, WI 53706

1 Introduction

One of the most remarkable aspects of life processes concerns the efficient transformation of energy from one form to the other.[1] This is beautifully illustrated by biomolecular motors[2, 3, 4], which are "nanomachines" that convert the chemical free energy in the form of ATP binding and/or hydrolysis into mechanical work (unidirectional motion) with high efficiency; the precise efficiency of different motors varies and has been estimated to be about 60% for several myosin and kinesin systems [3] and nearly 100% for F_1-ATPase. [5]. How biomolecular motors achieve such efficient energy transduction has been a fascinating topic in the biophysics and biochemistry community for more than thirty years, with most notably early work from T. L. Hill and co-workers [6, 7] in terms of kinetic and thermodynamic analysis. In recent years, striking progress has been made in the biomolecular motor field. X-ray and cryo-EM structures became available for several motor systems, [8, 9] which provided snapshots of their conformations at different stages of the functional cycle; innovative single-molecule experiments have provided astonishing insights regarding the stepsize [10], movement pattern [11, 12] and stoichiometry of ATP consumption [10, 13] in several motor systems; on the larger scale, progress in genomics and cellular biology has revealed new families of motor proteins with intriguing biological functions [14]. The availability of those new information at different spatial and time scales has provided a unique opportunity for theoretical and computational analysis, which is now able to establish coherent mechanistic pictures of bioenergy transduction at a far more detailed level [15] compared to previous phenomenological studies [6, 7].

The specific motor system we focus on is myosin II, which is also known as the conventional myosin and is involved in muscle contraction[2, 3]. In contrast to a few "high-profile" members in the myosin superfamily, e.g., myosin V [16] and VI [17], the conventional myosin is not processive and has a very low duty ratio.[3] Nevertheless, myosin II [18] is uniquely suited for theoretical analysis due to the availability of multiple x-ray structures at high resolution and the rich experimental background. In the following, we briefly summarize our recent studies [19, 20, 21] on the mechanochemical coupling in myosin II and then comment on future studies that are required to make our understanding more complete and quantitative.

2   Theoretical and computational analysis

As discussed in more details in a recent article [23], although force (work) generation is accomplished when myosin is attached to actin, properties of the *detached states* of myosin also make crucial contributions to the efficiency of energy transduction. Two key points are illustrated in Fig.1: (i). The coupling between the ATP hydrolysis activity and the large-scale lever-arm swing has to be tight; otherwise, kinetic pathways that involve futile ATP hydrolysis may be followed, which compromises the efficiency of the motor. (ii). Processes in the detached states, such as the lever-arm swing and ATP hydrolysis in the motor domain, should involve small free energy changes, such that a large fraction of the total free energy drop (the hydrolysis free energy of ATP *in solution*) is reserved for the force generation step(s); a combination of large drops and increases in the free energy for non-force-generating steps would also produce a small cumulative free energy change, but large uphill transitions would compromise the overall rate (flux) of the cycle. The major goal of our research is to understand how these thermodynamic and kinetic constraints are implemented in biomolecular motors in terms of their sequence, structure and dynamical properties. In the following, we briefly summarize findings regarding those issues based on our recent theoretical analysis [19, 20, 21].

Two classes of x-ray structures have been determined for truncated myosin II from *D. discoideum*, which were proposed to correspond to the pre-hydrolysis (Mg·ATP bound) [24] and hydrolysis (Mg·ADP· VO_4^- bound) [25] ki-

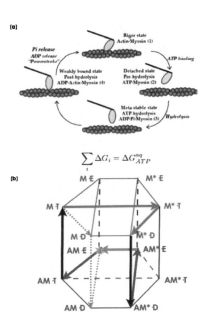

Figure 1: (a) The Lymn-Taylor kinetic pathway[22] for the functional cycle of conventional myosin; note that the sum of free energy changes along the pathway is rigorously equivalent to the hydrolysis free energy of ATP in solution (ΔG_{ATP}^{aq}). (b) A more complete, *minimal* kinetic model for the myosin-actin-ATP system, which includes two conformational states of the myosin (M), two binding states of myosin to actin (A) and 3 states for ATP in myosin: empty (E), ATP bound (T) and ADP/Pi bound (D). The Lymn-Taylor kinetic pathway in (a) is indicated as bold lines; a kinetic pathway involves futile ATP hydrolysis, which needs to be avoided for high efficiency, is shown as dotted lines.

netic states in the functional cycle (Fig.1). The key differences between the two structures involve the large-scale rotation of the C-terminal converter region (Fig.2a) and more subtle changes in the Switch I/II region in the ATP binding site (Fig.3a); the former corresponds to the lever-arm swing, and the latter controls whether the active-site is "open" or "closed". The major challenge is to understand the coupling between these events of different physicochemical nature and length-scales; the first step, however, is to determine the causal relationship between ATP hydrolysis, "open-closed" transitions in the active-site and the large-scale conformational changes in the converter. We proceed by first analyzing active-site properties that directly regulate the ATP hydrolysis energetics, then probing the intrinsic structural flexibility of the motor domain, and finally discussing the causality between the active-site activities and the converter rotation.

2.1   Regulation of ATP hydrolysis [19]

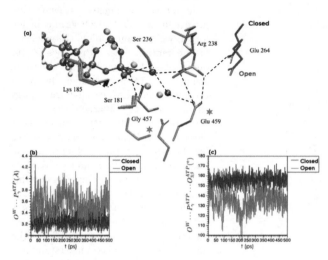

Since the orientation of water molecules has been proposed to be important to the hydrolysis activity, we first carried out *classical* molecular dynamics (MD) simulations of the myosin active-site in the two conformational states; ATP was the ligand in both simulations. In the "closed" active-site, the water molecules resolved in the x-ray structure [25] were found to be anchored near the γ phosphate of ATP due to the extensive hydrogen bonding network facilitated by the *closed* Switch I and Switch II motifs (Fig.2a); the distance between the lytic water and γ phosphate fluctuated mildly around 3.2 Å, and the critical O-P-O angle (Fig.2b,c) fluctuated around 155°, reflecting an ideal geometry for an efficient nucleophilic attack. In the "open" active-site, by contrast, the salt-bridge be-

Figure 2:   Protein configuration and water dynamics in the myosin active-site. (a) Superposition of averaged structures from MD simulations for the open (green) and closed (colored by atom type) conformations of the active-site; ATP was the ligand in both simulations (only that in the closed active-site was shown for clarity). The yellow and purple spheres indicate the position of water oxygen in the open and closed simulations, respectively. The major difference is that water molecules are positioned to carry out an efficient nucleophilic attack due to the extensive hydrogen-bonding network provided by the salt-bridge between Switch I (Arg 238) and Switch II (Glu 459) motifs, while this is not the case in the open active-site where the salt-bridge is broken. (b), (c) Instantaneous values for the critical distance and angle between the lytic water and γ phosphate of ATP during the MD simulations, which complement the average structure in (a) for illustrating the different behaviors of water in the two conformations of the active-site.

tween Switch I and Switch II was broken; as a result, the water molecules in the active-site were found highly mobile; the shortest distance between water and γ phosphate fluctuated wildly between 3.2 and 4.2 Å, and the critical O-P-O angle was often found in the range of 120-130 °. Clearly, the water structure is modulated by the conformation of the Switch I and II motifs; without the critical salt-bridge, the water molecules are typically far from the ideal orientation for an efficient nucleophilic attack.

Although the above observation is interesting and of mechanistic importance, it is likely that a small energy cost is associated with orienting water molecules into favorable positions, even in the "open" active-site. Thus, it is important to explicitly study the hydrolysis energetics as a function of active-site conformations, which requires QM/MM calculations. Due to the lack of robust semi-empirical QM methods for phosphate chemistry, calculations were carried out using the HF/3-21+G/MM and B3LYP/6-31+G(d,p)/MM methods for geometry optimization and single-point energy calculations, respectively; the cost of the calculations allowed only minimum energy path analysis so far. The QM region included a large portion of ATP, the Mg^{++} and all its ligands, the lytic water as well as the sidechain of the conserved Ser 236. Only the associative mechanism has been studied, for which two possibilities were examined where the Ser 236 acts either as a spectator or is explicitly involved as a proton relay group. The most important result from the QM/MM study is the striking difference between the hydrolysis activity in the open and closed active-site conformations. In the open active-site, the hydrolysis is very unfavorable from both kinetic and thermodynamic points of view. The activation barrier is more than 25 kcal/mol higher than that in the closed active-site and approaches the value in solution; the hydrolysis product state is about 17 kcal/mol higher in energy than the ATP state, in contrast to the nearly thermal-neutral situation found for the closed active-site. These differences are significantly larger than the intrinsic error expected for the QM/MM protocol used in Ref. [19] and therefore clearly illustrated that hydrolysis is not possible in the open active-site. This is also supported by alchemy free energy simulations (Li and Cui, unpublished) using classical force field [26], which did include protein thermal fluctuations; the result showed that hydrolysis is more favorable (thermodynamically) in the closed active-site by at least 9.6 kcal/mol, which is in qualitative agreement with the QM/MM result of 17 kcal/mol.

Perturbation analysis of the QM/MM energetics further revealed a few residues that interacted differently with the reactive moiety in the two active-site conformations. Arg 238, for example, swung closer to the ATP once the salt-bridge interaction with Glu 459 was broken in the open active-site conformation; this allowed Arg 238 to stabilize the ATP state more than the hydrolysis product state, thus contributing to the unfavorable hydrolysis in the open active-site. Therefore, our calculations strongly support the key role of the salt-bridge between the Switch I and Switch II motifs, which include not only controlling the water structure in the active-site but also directly impacting the hydrolysis energetics through electrostatic interactions. Similar regulatory mechanism may apply to other nucleotide dependent motor proteins, such as Ca^{++}-ATPase, F_1-ATPase and kinesin. For myosin, the importance of this salt-bridge was also shown by mutation studies [27, 28], which found that mutating either residue (Arg 238, Glu 459) to Ala abolished the ATPase activity; however, a curious result was that reversing the positions of these two residues also abolished the hydrolysis activity [28], which deserves further analysis.

2.2 Structural flexibility of myosin [20]

As mentioned above, the lever-arm swing should be a process that involves a small free energy change. Indeed, fluorescence [29] and ESR measurements[30] observed dynamical disorder of the motor domain, which indicates intrinsic structural flexibilities of myosin in the detached states. To better characterize the structural flexibility and reveal its implication in the myosin function, we performed normal mode analysis (NMA) [20], which is a simple but well-suited tool for such a purpose.

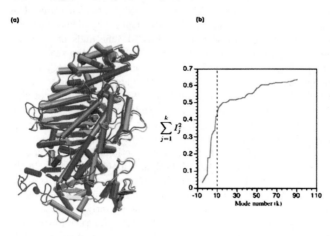

Figure 3: Conformational transition in myosin motor domain and low-frequency modes. (a) The superposition of the two x-ray structures of myosin II motor domain when the first 650 amino acids were used for best fit; the green and blue indicate the pre-hydrolysis and (post) hydrolysis structures, respectively, and the ATP molecule in the former is shown in van der Waals representation. Clearly, most of the changes occur in the C-terminal converter region, although more subtle changes are also present (e.g., active-site, see Fig.2a, and the relay helix). (b) Cumulative square of involvement coefficients (see text) as a function of normal mode indices. The strong correlation between the conformational change and low-frequency modes is evident: using only the ten lowest normal modes can describe nearly 50% of the conformational difference between the two structures shown in (a).

The most striking result from the NMA is that there is a significant degree of correlation between the characters of the low-frequency modes and the conformational transition observed in the x-ray structures. Useful variables for quantifying the correlation are involvement coefficients ($I_k = \mathbf{L}_k \cdot \frac{\Delta \mathbf{X}}{|\Delta \mathbf{X}|}$) and their cumulative squares up to a specific mode ($\sum_{i=1}^{k} I_i^2$). As shown in Fig.3b, nearly 50% of the conformational transitions observed in the x-ray structures of the pre-hydrolysis and hydrolysis conformations (Fig.3a) can be captured by only 10 lowest frequency modes. A similar behavior was observed for several other molecular motor systems [31, 32, 33], which strongly indicates that these systems are intrinsically flexible and nature has conveniently taken advantage of these floppy degrees of freedom for implementing important functions (also see below).

In addition to *rationalizing* conformational transitions in terms of structural flexibility (or plasticity), predictions can be attempted based on NMA regarding functionally important residues. The idea is that hinge residues in the low-frequency modes that strongly correlate with functional transitions (i.e., modes with large I_k) are the key motifs for maintaining motional coupling between distant regions. For myosin, for example, hinge residues in the two normal modes with the largest I_k values were identified [34] to be important for the coupling between the active-site and the converter domain. Among those, several residues (e.g., Phe 482 [35], Gly 680, Phe 692 [36]) were *independently* demonstrated by mutation studies to be crucial to the motility of myosin. Such encouraging agreement supports the value of NMA in structural analysis of other motor systems, and also highlights the importance of including sequence information in dynamical analysis.

2.3 Coupling between active-site and converter [21]

Although the NMA has revealed strong correlation between structural flexibility of myosin and the expected conformational transitions, the results do *not* contain sufficient information about the magnitude of the large-scale motion or how it is coupled to the hydrolysis and open-closed transition in the active-site; as discussed above, revealing the causal relationship among domain motion, hydrolysis and active-site structural rearrangement is a key step in understanding the mechanochemical coupling. Despite rapid progress in the experimental techniques, it remains difficult to resolve this issue due to the different length-scales involved in the process. Indeed, studies with different experimental techniques seem to reach conflicting conclusions [29, 37, 38].

Based on the theoretical analysis so far, and taking into account information from various experiments, we proposed a coupling pathway for myosin in the detached states, which is illustrated in Fig.4a. Starting with the ATP-bound, pre-hydrolysis, x-ray structure [24], the NMA clearly indicated that the motor domain has intrinsic flexibility, especially in the converter region. This suggests that significant motion in the converter domain persists prior to any major conformational changes in the active-site and ATP hydrolysis, which seems to be consistent with the fluorescence quenching experiment [29] that indicated dynamical disorder (change in the environment of Trp 501) in the absence of ATP hydrolysis. Next, largely independent of the converter motion, we *hypothesize* that the active-site may close upon thermal fluctuation, which, according to the QM/MM analysis, turns on the ATP hydrolysis activity. The insignificant role of the converter swing in the activation of ATP hydrolysis is consistent with the mutation studies mentioned above concerning hinge residues [35, 36], which observed sharply decreased motility but almost normal ATPase activity. Finally, once ATP is hydrolyzed, we propose that the active-site is strongly stabilized to remain closed, which in turn stabilizes the full swing of the converter.

The idea that full-scale rotation of the converter is coupled to the closure of the active-site, which in turn is stabilized by the hydrolysis product, rationalizes the tight coupling between hydrolysis and lever-arm swing as required for the high-efficiency of energy transduction; it is also consistent with observations from bulk FRET experiment of Spudich [37] and electric birefringence experiment of Highsmith [38]. The detailed mechanisms for the converter/active-site coupling and hydrolysis/active-site stabilization, however, remain to be illustrated. Concerning the latter, recall the QM/MM results that hydrolysis is highly unfavorable when the active-site is open but nearly thermal neutral when the active-site is closed; if we further *assume* that the open-closed transition in the active-site is facile with ATP bound, the fact that free energy is a state function immediately implies that the closed active-site is strongly favored after ATP is hydrolyzed (Fig.4b). It is interesting to note that closure of the active-site in fact occurred in the early stage of targeted molecular dynamics (TMD) simulations [39, 21]; since TMD simulations based on Cartesian-RMSD are

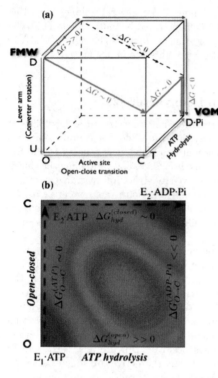

Figure 4: (a) Possible coupling schemes for the three events of different nature and length-scales for myosin in the detached states ($M \cdot T \rightarrow M^* \cdot D$): large-scale rotation of the converter domain, open-closed transitions in the active-site and ATP hydrolysis. Three possibilities are shown following different sequences of these events; the red and blue arrows indicate two extremes often referred to in the literature and the thick green arrows indicate a scheme proposed here based on theoretical analysis and available experimental data (see text). (b) Coupling between ATP hydrolysis and open-closed transition in the active-site; since hydrolysis was found in QM/MM simulations to be highly endothermic and nearly thermoneutral in the open and closed active-site, respectively, if we assume that the open-closed transition is facial with ATP bound, then the active-site is strongly stabilized to adopt the closed conformation once ATP is hydrolyzed.

usually argued to artificially favor large-scale conformational changes in the early stage of the simulation (because they are most effective in decreasing Cartesian-RMSD), the fact that relatively subtle rearrangements in the active-site were observed in the early stage strongly implies that the closure of the active-site is indeed a facile process.

3　Conclusion and future perspectives

Bio-energy transduction processes are fascinating because they involve physical and chemical changes at multiple length and time scales. For the same reason, they are difficult to study using a single technique; theoretical and computational analyses provide a framework for combining information from different sources at variable resolutions to construct a coherent mechanistic picture[4, 15]. Here I described some findings from our group (till Jan. 2005, WATOC) regarding the mechanochemical coupling in a representative biomolecular motor, myosin. Guided by the phenomenological ideas in earlier studies [6], we identified important mechanistic questions concerning the properties of myosin in the *detached states*, which we argue [23] are important in determining the efficiency of the motor. Up to date, QM/MM analysis [19] has underlined, in energetics and mechanistic terms, the importance of the Switch I and Switch II motifs in the hydrolysis activity. Normal mode analysis [20], on the other hand, revealed intrinsic structural flexibility that strongly correlates with the large-scale conformational transitions observed in the x-ray structures. Based on these results and available experimental observations, we proposed a coupling scheme for ATP hydrolysis, the open-closed transitions in the active-site and the large-scale rotation in the C-terminal converter domain.

Much work is needed to further validate the proposed coupling scheme and fill in additional details concerning the mechanism of the coupling (i.e., crucial residues and interactions). This involves more quantitative QM/MM studies of pathways and free energies associated with the ATP hydrolysis, which requires the development of efficient and accurate QM/MM methods as well as novel ways of computing reliable free energies for chemical reactions in complex systems. Moreover, the coupling between the active-site closure and converter rotation remains to be understood, which requires more robust methods for determining pathways and energetics associated with large-scale conformational transitions. Finally, the relationship between hydrolysis, converter rotation and the actin binding interface needs to be better clarified, because it is possible that actin-binding further ensures a tight coupling between hydrolysis activity and orientation of the converter; progress towards a high-resolution structure of the myosin-actin complex is extremely exciting in this regard.

Acknowledgments

The research was generously supported in part by various funding agencies (Research Corporation, American Chemical Society, Alfred. P. Sloan Foundation and the National Institutes of Health). Q.C. acknowledges discussions with Prof. I. Rayment, S. Gilbert on myosin related issues.

References

[1] Nicholls, D. G.; Ferguson, S. J. *Bioenergetics 3;* Academic Press, 2002.

[2] Schliwa, M. *Molecular Motors;* Wiley-VCH, 2003.

[3] Howard, J. *Mechanics of Motor Proteins and the Cytoskeleton;* Sinauer Associates, Inc., 2001.

[4] Bustamante, C.; Keller, D.; Oster, G. *Acc. Chem. Res.* **2001**, *34*, 412.

[5] Yasuda, R.; Noji, H.; Kinosita, K.; Yoshida, M. *Cell* **1998**, *93*, 1117–1124.

[6] Hill, T. L. *Free energy transduction in Biology;* Academic Press, 1977.

[7] Hill, T. L.; Eisenberg, E. *Q. Rev. Biophys.* **1981**, *14*, 463–511.

[8] Houdusse, A.; Sweeney, H. L. *Curr. Opin. Struct. Biol.* **2001**, *11*, 182–194.

[9] Holmes, K. C.; Angert, I.; Kull, F. J.; Jahn, W.; Schroder, R. R. *Nature* **2003**, *425*, 423–427.

[10] Schnitzer, M. J.; Block, S. M. *Nature* **1997**, *388*, 386–390.

[11] Yildiz, A.; Forkey, J. N.; McKinney, S. A.; Ha, T.; Goldman, Y. E.; Selvin, P. R. *Science* **2003**, *300*, 2061–2065.

[12] Yasuda, R.; Noji, H.; Yoshida, M.; Kinosita, K.; Itoh, H. *Nature* **2001**, *410*, 898–904.

[13] Sako, Y.; Yanagida, T. *Nat. Cell Biol. Suppl. S.* **2003**, pages SS1–SS5.

[14] Schliwa, M.; Woehlke, G. *Nature* **2003**, *422*, 759–765.

[15] Karplus, M.; Gao, Y. Q. *Curr. Opin. Struct. Biol.* **2004**, *14*, 250.

[16] Vale, R. D. *J. Cell. Biol.* **2003**, *163*, 445–450.

[17] Buss, F.; Spudich, G.; Kendrick, J. *Ann. Rev. Cell Dev. Biol.* **2004**, *20*, 649–676.

[18] Holmes, K. C.; Geeves, M. A. *Annu. Rev. Biochem.* **1999**, *68*, 687–728.

[19] Li, G.; Cui, Q. *J. Phys. Chem. B* **2004**, *108*, 3342.

[20] Li, G.; Cui, Q. *Biophys. J.* **2004**, *86*, 743–763.

[21] Yu, H.; Ma, L.; Yang, Y.; Cui, Q. **2006**, *To be submitted.*

[22] Lymn, R. W.; Taylor, E. W. *Biochem.* **1971**, *10*, 4617–4624.

[23] Cui, Q. *Theor. Chem. Acc.* **2006**, *In Press.*

[24] Bauer, C. B.; Holden, H. M.; Thoden, J. B.; Smith, R.; Rayment, I. *J. Biol. Chem.* **2000**, *275*, 38494–38499.

[25] Smith, C. A.; Rayment, I. *Biochem.* **1996**, *35*, 5404–5417.

[26] Yang, W.; Gao, Y. Q.; Cui, Q.; Ma, J.; Karplus, M. *Proc. Natl. Acad. Sci. USA* **2003**, *100*, 874–879.

[27] Sasaki, N.; Shimada, T.; Sutoh, K. *J. Biol. Chem.* **1998**, *273*, 20334–20340.

[28] Onishi, H.; Ohki, T.; Mochizuki, N.; Morales, M. F. *Proc. Natl. Acad. Sci. USA* **2002**, *99*, 15339–15344.

[29] Malnasi-Csizmadia, A.; Pearson, D. S.; Kovacs, M.; Woolley, R. J.; Geeves, M. A.; Bagshaw, C. R. *Biochemistry* **2001**, *40*, 12727–12737.

[30] Thomas, D. D.; Ramachandran, S.; Roopnarine, O.; Hayden, D. W.; Ostap, E. M. *Biophys. J.* **1995**, *68*, S135–S141.

[31] Li, G.; Cui, Q. *Biophys. J.* **2002**, *83*, 2457–2474.

[32] Cui, Q.; Li, G.; Ma, J.; Karplus, M. *J. Mol. Biol.* **2004**, *340*, 345–372.

[33] Ma, J.; Karplus, M. *Proc. Nat. Acad. Sci.* **1998**, *95*, 8502–8507.

[34] Haywards, S.; Kitao, A.; Berendsen, H. J. C. *Proteins: Struct. Funct. Gen.* **1997**, *27*, 425.

[35] Ito, K.; Uyeda, Q. P.; Suzuki, Y.; Sutoh, K.; Yamamoto, K. *J. Biol. Chem.* **2003**, *278*, 31049–31057.

[36] Sasaki, N.; Ohkura, R.; Sutoh, K. *Biochem.* **2003**, *42*, 90–95.

[37] Shih, W. M.; Gryczynski, Z.; Lakowicz, J. R.; Spudich, J. A. *Cell* **2000**, *102*, 683–694.

[38] Highsmith, S.; Polosukhina, K.; Eden, D. *Biochemistry* **2000**, *39*, 12330–12335.

[39] Schlitter, J.; Engels, M.; Kruger, P.; Jacoby, E.; Wollmer, A. *Mol. Sim.* **1993**, *10*, 291.

MOLECULAR DYNAMICS AND NEUTRON DIFFRACTION STUDIES OF THE STRUCTURING OF WATER BY CARBOHYDRATES AND OTHER SOLUTES

J.W. Brady[1], P.E. Mason[1], G.W. Neilson[2], J.E. Enderby[2], M.-L. Saboungi[3], K. Ueda[4], and K.J. Naidoo[5]

[1]Department of Food Science, Cornell University, Ithaca, NY 14853, USA
[2]H.H. Wills Physics Laboratory, University of Bristol, Bristol BS8 1TL, UK
[3]Centre de Recherche sur la Matière Divisée, 1 bis rue de la Férollerie, 45071 Orléans, FRANCE
[4]Department of Material Science, Yokohama National University, Yokohama 240-8501, JAPAN
[5]Department of Chemistry, University of Cape Town, Rondebosch 7701 SOUTH AFRICA

1 INTRODUCTION

Solvent structuring by solutes has long been known to significantly affect the properties of solutions, helping to determine, for example, the folded conformations of globular proteins and the organization of lipids into micelles and bilayers. In disaccharides and oligosaccharides, hydration can affect the conformational equilibrium, and even for monosaccharides, the degree of solvent structuring can affect the solubility and affinity of solutes as protein ligands. Molecular mechanics calculations, including molecular dynamics (MD) simulations, allow solvent structuring to be modeled directly and analyzed in specific detail. The reliability of these results remains unproven, however, since the results of simulations can in principle depend on the various approximations employed in the calculations, such as the molecular force fields, water models, and treatment of long-range forces.

Unfortunately, such solvent structuring is quite difficult to probe experimentally, which has hampered efforts to validate the results of modelling calculations by comparison with experimental data. The most useful experimental probe of liquid structure has been neutron diffraction. Neutron scattering experiments have been extremely powerful in revealing the water structuring around simple monatomic ions[1-3], noble gases[4,5], and molecular species with quasi-spherical symmetry such as tetramethylammonium[6,7]. The measured scattering is a direct function of the structure of the solution, even if it is difficult to interpret for complex molecules.

The combination of neutron diffraction experiments with MD simulations, however, could be particularly powerful, since the results of an MD simulation can be used to interpret the complex scattering data produced by even modest-sized molecular solutes. When coupled with the technique of neutron diffraction with isotopic substitution (NDIS), such an approach in principle could be used to extract information about the atomic structuring around a single atom in a molecular solute if the experiments were performed on solutes specifically isotope-labelled at a single position. This approach to studying the solvation of various molecular solutes will be discussed in detail below.

2 METHOD AND RESULTS

2.1 MD Simulations of Solvent Structuring by Carbohydrates

With the increasing awareness of the role which carbohydrates play in molecular recognition, and the exploitation of this role in the design of pharmaceuticals and vaccines, a renewed interest has developed in the conformations of sugars and how they are affected by solvation. In some cases solvent interactions can have significant influences on carbohydrate conformations[8,9]. A number of molecular dynamics (MD) simulations studies, in various laboratories and using a range of force fields and models, have demonstrated the highly complex nature of the solvent structuring imposed by asymmetric multifunctional solutes like the carbohydrates. Figure 1 illustrates an example of such solvent structuring, for the monosaccharide β-D-xylopyranose[10,11]. As can been seen from this figure, first hydration shell water molecules around such a solute occupy specific positions as a result of the geometric constraints of their hydrogen bonds to the solute. In sugars in particular, the hydration requirements of the multiple functions groups of the solute overlap and interfere, forcing these water molecules into the best compromise satisfying these constraints. In general, such structuring is a sensitive function of the solute molecular architecture, and varies with the stereochemistry and conformation. In the case of disaccharides, such structuring can sometimes affect the disaccharide conformation, since unlike the case of the monosaccharide rings, the two rings of a disaccharide can change their relative orientations through changes in the glycosidic torsional angles ϕ and ψ [12].

Figure 2 illustrates an example of how specifically-structured water molecules can affect the conformation of a disaccharide. The conformational potential of mean force for

Figure 1 *Left: β-D-xylopyranose, indicating the H4 proton which was labelled by an H/D substitution in diffraction experiments. Right: Contours of solvent density around a molecule of: β-D-xylopyranose, as calculated from MD simulations.*

the α-(1→4)-linked disaccharide of D-xylose in vacuum resembles that of the related maltose disaccharide, with a global minimum at (0,-40°), but in aqueous solution, a new minimum at (-37°,52°), which is not even a low-energy structure on the vacuum surface, is actually the global energy minimum, as the result of a bridging water molecule which makes a hydrogen bond to hydroxyl groups on both rings[12]. A similar situation was found in studies of the disaccharide neocarrabiose[8,9], where two experimental conformations, that of the crystal structure and for the related polysaccharide ι-carrageenan, are not minima on the vacuum energy map. However, in aqueous solution, a similar bridging hydrogen bonded water molecule destabilizes the vacuum global minimum energy conformation and causes a shift to the experimental conformations. Such arrangements are specific to particular molecular topologies, however; the conformation of the related carrabiose molecule, a model for the other carrageenan repeat unit, is unaffected by solvation[13].

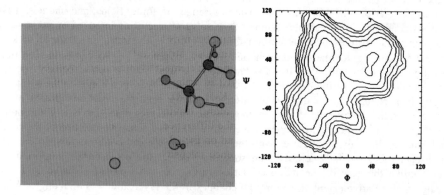

Figure 2   *Right: The Ramachandran conformational free energy map for the α-(1→4)-linked disaccharide of D-xylose (left) in aqueous solution, as calculated from MD simulations. Contours are at 2 kcal/mol intervals. Left: An MD snapshot illustrating a bridging .water molecule hydrogen bonded to hydroxyl groups on both monosaccharide rings. Such bridging favours the illustrated conformation, which corresponds to the broad well in the upper left in the energy map.*

2.2 Neutron Diffraction with Isotopic Substitution (NDIS) Experiments

For complex molecular solutes diffraction data is very difficult to interpret in general. A conventional neutron diffraction measurement provides only the neutron-weighted average structure factor $S^N(Q)$,

$$S^N(Q) = \sum_{ij} \frac{c_i c_j b_i b_j}{\langle b \rangle^2} S_{ij}(Q)$$

where c_i and b_i are the concentration and coherent scattering length of element i, $\langle b \rangle$ is the average scattering length over all atoms in the sample, and $S_{ij}(Q)$ is the partial structure factor for the element pair (i,j). Clearly, structural correlations between all pairs of elements enter into the average $S^N(Q)$. As the number n of elements in the sample gets larger, these correlations become increasingly difficult to untangle. The method of neutron diffraction with isotope substitution (NDIS) was developed to help resolve this difficulty, and allows a significant simplification of scattering data[14].

If the isotope substitution $a \rightarrow a'$ can be carried out on a particular element a, in general the scattering length b_a of that element will change. If two measurements are then made with chemically-identical samples but with different isotopes of element a, the appropriately-weighted difference of the two measured structure factors gives a more restricted average,

$$S_a(Q) = \frac{\langle b \rangle^2 S^N(Q) - \langle b' \rangle^2 S'^N(Q)}{\langle b \rangle^2 - \langle b' \rangle^2} = \sum_{ij} \frac{c_j b_j}{\langle b \rangle} S_{aj}(Q)$$

where the primes refer to the measurement with the sample containing the substituted isotope. Since the difference structure factor $S_a(Q)$ is an average over only the n terms involving element a, as opposed to the full number $n(n+1)/2$ of terms involving all element pairs, it is generally much easier to distinguish structural elements arising from the different pairs. Moreover, only pairs involving the element of interest are involved.

For example, in the simple case of KCl in water, there are 10 structure factors, for K-K, Cl-Cl, K-Cl, O-O, H-H, O-H, K-O, K-H, Cl-O, and Cl-H. If however, one did a $^{35}Cl/^{37}Cl$ isotopic substitution for the Cl atom, leaving all of the other atoms as the naturally-occurring isotope, and measured the difference in the scattering between the two solutions, all those terms that did not depend on the Cl would subtract out, leaving only 4 structure factors contributing to the difference intensity. This approach is known as the "first difference method".

The most important atoms in biological molecules are of course C, O, N, and H. Hydrogen atoms are ideal for experiments of this type since they have a large neutron scattering length, and since the coherent scattering length b for H is negative (-3.74 fm) while that for D is positive (6.67 fm). Unfortunately, given the importance of oxygen in biological molecules, ^{16}O, ^{17}O, and ^{18}O all have nearly identical scattering lengths and offer no opportunity for useful isotopic substitution. On the other hand, the difference between ^{12}C and ^{13}C is around 7% (6.65 fm for ^{12}C and 6.19 fm for ^{13}C), which is detectable, and offers a second set of atoms which can be substituted in samples used in scattering experiments.

2.3 MD and NDIS Studies of Sugars

This approach of combining NDIS experiments on specifically-labelled solutes with MD simulations has been used to analyze the structure of several different sugars, including D-glucose and D-xylose. Sugar samples labelled with H/D at single sites, such as that illustrated in Figure 1, were used in a series of experiments that developed a new approach to examining solute structure. The combination of several such experiments contains the information necessary to construct the three dimensional solvent distribution around the molecule, when combined with the appropriate modelling. Furthermore, such data also contains information about the conformation of the solute as well. Such combined studies have been used to determine the position of a hydroxyl group in D-xylose, a previously un-measurable property, and the orientation of the exocyclic hydroxymethyl group in D-glucose.

2.4 Structuring in Solutions of Guanidinium Salts from MD and NDIS

The combination of NDIS and MD studies has also been applied to a series of guanidinium salts in water, including GdmCl, GdmSCN, Gdm_2SO_4, and Gdm_2CO_3. Rich structuring was found in these solutions, as shown in Figure (3) for GdmCl. Direct evidence for hetero-ion pairing and indirect evidence for Gdm^+ ion stacking was found in these studies.

The functions $^nG_N(r)$ and $^nG_{Cl}(r)$ as determined by NDIS and as calculated from the MD simulation are shown in Figure 5. For $^nG_N(r)$ the main difference between these two functions is that the MD produces very sharp peaks for the NH and NC bonds (at 1.0 and 1.4 Å respectively), while the NDIS measurement has considerably broader, resolution-limited features at these positions. The peak at 2.3 Å is due to the intramolecular NN and

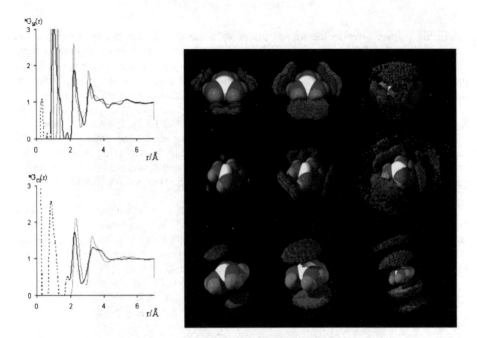

Figure 3   *Right: Densities of water and ions around guanidinium as calculated from MD simulations. The top row shows water oxygen (red) and proton density (green) around the Gdm^+, the middle row shows chloride density, and the bottom row shows Gdm^+ carbon atom density around Gdm^+. Left: A comparison of the total radial distribution functions for all atoms around N and Cl atoms, as calculated from MD simulations and compared to experiment.*

NH correlations, while the peak at 3.1 Å is due, in part, to an intramolecular NH correlation. The general form of the chloride hydration consists of a Cl-H correlation at 2.3 Å, with correlations of the chloride to the remaining OH of the water occurring at about 3.3 Å[15].

The tendency for the guanidinium ions to stack against one another, in a manner similar to hydrophobic association, in spite of their like charges, may explain how this ion functions as a protein denaturant. These planar molecules could stack against extended hydrophobic side-chain groups in the same fashion, preventing them from interacting with the solvent water upon protein unfolding. The ability of the ion to function in this manner is apparently facilitated by the poor hydration of the guanidinium face and by the direction of the hydrogen-bonding capacity of the ion outward in the same plane as the ion.

MD and NDIS studies of GdmSCN, Gdm₂SO₄, and Gdm₂CO₃ found that strong ion pairing was taking place in each of these systems, as in the GdmCl case, but that the topologies of the various anions resulted in very different mesoscopic structures for the overall solutions[16,17]. In the GdmSCN case, the essentially linear guanidinium-thiocyanate ion pairs were uniformly and randomly distributed throughout the system, while in the guanidinium sulphate case, the tetrahedral structure of the sulphate ion allowed it to make several simultaneous "bonds" to guanidinium ions which allowed nanometer-scale clusters to form. Similar nanometer-size clusters were suggested by both the MD and NDIS data for guanidinium carbonate, and the existence of such clusters was subsequently independently confirmed by small angle neutron scattering (SANS) experiments. This difference in solvation may explain the difference in the Hofmeister behaviour of these anions, since thiocyanate is a powerful denaturant while sulphate is only weakly denaturing or even conformation stabilizing.

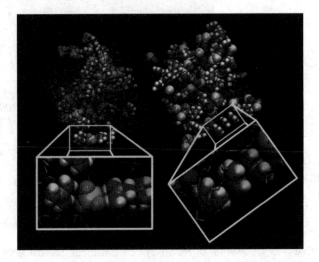

Figure 4 *Right: A snapshot configuration taken from the MD simulation of guanidinium thiocyanate, illustrating the tendency for the solution to remain uniform in the distribution of ions and the stacking of guanidinium faces. Left: A similar snapshot configuration from the guanidinium sulphate simulation, illustrating the tendency for large, nanoscale clusters to form, with the sulphate ions bound to more than one guanidinium ion.*

3 CONCLUSION

Although neutron scattering data from solutions of complex molecular solutes generally cannot be interpreted in terms of microscopic structural detail, when paired with MD simulations of the same system, the results of the modelling can be used to understand the experimental data. In addition, the direct comparison of the MD results with the experimental structure factors can be used to evaluate the adequacy of the computational models and can serve as a guide in revising MM force fields to give more accurate results. Such combined studies have already been used to give the first experimental probe of a sugar hydroxyl conformation in aqueous solution, and have been used to detect the presence of nanometer-sized ion clusters in solutions of guanidinium sulphate and guanidinium carbonate, but not in guanidinium thiocyanate and guanidinium chloride.

Acknowledgements

This project was supported by grant GM63018 from the National Institutes of Health. We gratefully acknowledge grants of beam time at the Institut Laue-Langevin and many useful conversations with C.E. Dempsey.

References

(1) Enderby, J. E.; Neilson, G. W. *Rep. Prog. Phys.* **1981**, *44*, 593-653.

(2) Neilson, G. W.; Enderby, J. E. *Proceedings of the Royal Society of London* **1983**, *A 390*, 353-371.

(3) Neilson, G. W.; Enderby, J. E. *Advances in Inorganic Chemistry* **1989**, *34*, 195-218.

(4) Broadbent, R. D.; Neilson, G. W. *Journal of Chemical Physics* **1994**, *100*, 7543-7547.

(5) Sullivan, D. M.; Neilson, G. W.; Fischer, H. E. *Journal of Chemical Physics* **2001**, *115*, 339-343.

(6) Barnes, A. C.; Neilson, G. W.; Enderby, J. E. *Journal of Molecular Liquids* **1995**, *65/66*, 99-106.

(7) Turner, J. Z.; Soper, A. K.; Finney, J. L. *Journal of Chemical Physics* **1995**, *102*, 5438-5443.

(8) Ueda, K.; Brady, J. W. *Biopolymers* **1996**, *38*, 461-469.

(9) Ueda, K.; Ueda, T.; Sato, T.; Nakayama, H.; Brady, J. W. *Carbohydrate Research* **2004**, *339*, 1953-1960.

(10) Liu, Q.; Brady, J. W. *Journal of the American Chemical Society* **1996**, *118*, 12276-12286.

(11) Mason, P. E.; Neilson, G. W.; Enderby, J. E.; Saboungi, M.-L.; Brady, J. W. *Journal of the American Chemical Society* **2005**, *accepted*.

(12) Naidoo, K. J.; Brady, J. W. *Journal of the American Chemical Society* **1999**, *121*, 2244-2252.

(13) Ueda, K.; Brady, J. W. *Biopolymers* **1997**, *41*, 323-330.

(14) Neilson, G. W.; Enderby, J. E. *Journal of Physical Chemistry* **1996**, *100*, 1317-1322.

(15) Powell, D. H.; Neilson, G. W.; Enderby, J. E. *Journal of Physics: Condensed Matter* **1993**, *5*, 5723-5730.

(16) Mason, P. E.; Dempsey, C. E.; Neilson, G. W.; Brady, J. W. *Journal of Physical Chemistry B* **2005**, *submitted*.

(17) Mason, P. E.; Neilson, G. W.; Dempsey, C. E.; Brady, J. W. *The Journal of Physical Chemistry B* **2005**, *submitted*.

Chemical Reactivity in Biological Surroundings

FROM PRION PROTEIN TO ANTICANCER DRUGS: QM/MM CAR-PARRINELLO SIMULATIONS OF BIOLOGICAL SYSTEMS WITH TRANSITION METAL IONS

M.C. Colombo[1], C. Gossens[1], I. Tavernelli[1] and U. Rothlisberger[1,2]

[1] Laboratory of Computational Chemistry and Biochemistry, Ecole Polytechnique Fédérale de Lausanne, 1015 Lausanne, Switzerland
[2] E-mail: ursula.roethlisberger@epfl.ch

1 INTRODUCTION

Car-Parrinello (CP)[1] molecular dynamics (MD) simulations based on density functional theory (DFT)[2] have become a powerful tool in the investigation of complex chemical and biological systems. This approach allows performing electronic structure calculations at finite temperature and to simulate chemical reactions directly as they occur in gas or in condensed phase environments. Currently, using parallel computers, systems of a few hundred to a thousand atoms can be studied routinely.

By extending this approach to a mixed quantum mechanical-molecular mechanical (QM/MM) hybrid scheme, the system size can be enlarged even further. During the last years, we have developed an interface[3] between the first-principles MD package CPMD,[4] and empirical force field descriptions (GROMOS96[5] or AMBER[6]). Recently, we have extended these hybrid QM/MM Car-Parrinello simulations to adiabatic and nonadiabatic excited state dynamics based on time-dependent density functional theory (TDDFT)[7,8,9,10,11]. We have also been developing several enhanced sampling methods to increase accessible time scales in first-principles-based simulations[12,13,14] and we have proposed a novel approach to capture dispersion forces within standard density functional calculations using purely local approximations to the exchange-correlation functional[15,16,17].

Using hybrid QM/MM Car-Parrinello simulations in combination with classical MD based on empirical force fields and bioinformatics approaches we have been investigating a number of biological problems[18]. In particular, we have characterized the mechanism of action of pharmaceutically important drug targets such as HIV-1 protease[19] and members of the cellular suicide system (caspase-3, 7 and 8)[20,21]. We have also investigated chemically interesting enzymes as models for the rational design of biomimetic green catalysts[22].

Here, we give a short introduction to our QM/MM scheme and review two specific applications of this technique to study the binding of transition metal ions and transition metal based compounds to proteins and DNA.

2 METHOD AND RESULTS

2.1 QM/MM Car-Parrinello Dynamics

In order to describe the interactions of transition metal ions in proteins and nucleic acids we adopt a hierarchical approach based on the partitioning of the system into a chemically active site (computed at the DFT or TDDFT level, QM region) and a classically described environment (MM region). In the hybrid QM/MM MD approach, the total energy of a QM/MM system can be written as [23]

$$E = E_{QM} + E_{MM} + E_{QM/MM} \tag{1}$$

which correspond to the lowest eigenvalue of the Hamiltonian

$$H = H_{QM} + H_{MM} + H_{QM/MM} \tag{2}$$

where $H_{QM/MM}$ describes the interaction between the QM and the MM part. $H_{QM/MM}$, the coupling Hamiltonian, can be divided into a bonded term, if covalent bonds exist between the QM and the MM subsystems, and a non-bonded term

$$H_{nonbonded} = \sum_{i \in MM} q_i \int dr \frac{\rho(\mathbf{r})}{|\mathbf{r} - \mathbf{R}_i|} + \sum_{\substack{i \in MM \\ j \in QM}} v_{vdw}(R_{ij}) \tag{3}$$

where $\mathbf{R}_i$ is the position of MM atom i with the point charge q_i, ρ is the total charge density of the quantum system (electronic and ionic), and $v_{vdw}(R_{ij})$ is the van der Waals interaction between atoms i and j. In the QM/MM scheme employed here, the van der Waals interaction between QM and MM parts is simply described via the classical force field parameters. Alternatively, we have recently developed a method[15,16,17] which is able to account for the effect of dispersion forces within the framework of a standard DFT calculation based on purely local approximations to the exchange-correlation functional. The Coulomb term in the QM/MM interaction Hamiltonian is described at the QM level by replacing the classical point charge of the MM atoms by a suitable (smoother) charge distribution.[3] Computational efficiency is achieved by modeling the long-range electrostatics by a Hamiltonian term that couples the multipole moments of the quantum charge distribution with the classical point charges. For details of the implementation see ref. 3. For further increased computational efficiency, electrostatic interactions for an intermediate layer around the QM region can be taken into account by a scheme based on dynamically restrained electrostatic potential derived (D-RESP) charges[24]. If the QM-MM boundary cuts through a covalent bond, care has to be taken to saturate the valence orbitals of the QM system. In the present implementation, this can be done by "capping" the QM site with a hydrogen atom or an empirically parameterized pseudopotential.[17]

We have recently extended this type of QM/MM ground state dynamics to excited states[7] and nonadiabatic effects[8] by using either a restricted open shell Kohn-Sham (ROKS)[25] or a time-dependent DFT [8,9,10] description for the excited states. For the latter, we use two different schemes based either on linear response formulation (LR-TDDFT)[7] or a time propagation technique (P-TDDFT)[8]. For all TDDFT-MD simulations, the Tamm-Dancoff approximation (TDA)[26] to TDDFT is applied.

2.2 Computational Details

All first-principles based MD simulations are carried out with the CPMD code[4]. We used soft norm-conserving non-local Troullier-Martins pseudopotentials[27] of semicore quality for the transition metals and energy cutoffs for the plane wave expansion of the wave functions of 75-100Ry (75Ry for organoruthenium compounds, 80-Ry for Cu- and Zn-, 100Ry for Mn-loaded prion protein). The inherent periodicity in the plane wave calculations is circumvented solving Poisson's equation for non-periodic boundary conditions[28]. Where not mentioned otherwise, the BLYP[29] functional is employed. First-principles MD runs are performed with a timestep of 0.097 fs at a temperature of 298K maintained via a Nosé-Hoover thermostat. Mixed QM/MM Car-Parrinello MD simulations were performed starting from classically equilibrated structures (for more details see for instance ref. 23).

For the study of the transition metal ion binding to the prion protein, the EPR parameters were calculated with the Amsterdam Density Functional code (ADF2002.01)[30] using the method developed and implemented by Van Lenthe et al..[31] The BP86 [32] functional is used and all electron calculations [33] with the QZ4P basis set are carried out. Relativistic effects are included using the zero order relativistic approximation (ZORA) [34] and the relativistic atomic potentials are calculated with the program DIRAC, supplied with the ADF2002.01 code.

In the case of organoruthenium compounds, DFT *in vacuo* calculations are performed using the Gaussian03[35] and ADF2004[36] packages. For Gaussian03 the B3LYP functional is employed with the SDD ECP for ruthenium and the 6-31+G(d) basis set for the remaining atoms (referred to in the text as *method1*). ADF calculations are carried out using the BP86[32] functional together with the TZP/Zora basis set (*method2*). For more details and comparisons with previously published data see ref. 37. MD runs are carried out using the Amber7[6] package (parm99 force field; time steps of 1-1.5fs; SHAKE bond constraints for bonds to hydrogens; NPT ensemble at 298K and 1bar, Berendsen thermostat, isotropic position scaling, PME). The classical parameterization of two compounds out of the RAPTA series (with η^6-arene: B=benzene, C=p-cymene) and of the RA-en compound (η^6-arene=p-cymene,biphenyl) was performed following the recommended Amber procedure.

2.3 Theoretical Characterization of the Structural and Chemical Properties of Metal Loaded Prion Protein

Prions are infectious agents that play a central role for a group of invariably fatal, neurodegenerative diseases affecting animals such as sheep (scrapie), cattle (BSE), and humans (e.g. variants of the Creutzfeldt-Jacob disease).[38,39] It is now widely established that these diseases are caused by an abnormal isoform PrP[sc] of the normal cellular prion protein PrP[C].[40,41] In spite of the continued efforts made, the physiological function of the prion protein has remained elusive. Nevertheless several contributions have pointed to a key role of metals in PrP metabolism (ref. 42 and refs. therein). In particular, the ability of PrP[c] to bind Cu^{2+} in vitro and in vivo,[43,44] together with the promotion of PrP[c] endocytosis for high copper concentrations[45] suggest a role in copper homeostasis and transport.[45,46] Alternatively, an antioxidant function has been suggested.[47,48,49]

It has generally been assumed that Cu^{2+} ions bind to the histidine-rich unstructured part of the prion protein (ref. 42 and refs. therein) and only a minor fraction of the experiments focused on the C-terminal part (PrP-(123-231)). The Cu-binding sites in the full-length PrP have been analyzed by means of pulse EPR and electron nuclear double resonance (ENDOR) spectroscopy by Van Doorslaer et al..[50,51] Three distinct pH-dependent Cu^{2+} signals were detected for the folded C-terminal domain. For one of them (complex **2**, pH 3-8) it could be ascertained that one histidine is coordinated to the copper ion.

So far attention has mainly focused on copper, however, there is increasing evidence that manganese and zinc ions also bind to PrP, albeit with a lower affinity.[52,53,54,55] However, for these metals, no experimental data is available to elucidate the coordination site and environment. The finding that full length PrP^c refolds in the presence of equimolar concentrations of copper and manganese ions, and binds the same number of both ions[52] suggests that manganese can substitute copper in the holo-form. Moreover, manganese-loaded PrP^c exhibits some superoxide dismutase activity ($\sim$50% of the one of copper-loaded $PrP^{c\ 56}$). It also induces changes in secondary structure (increase in β–sheet content) and proteinase resistance,[52] all features characterizing the PrP^{sc} isoform. In addition, native PrP^{sc} extracted from the brains of infected mice show poor Cu(II) retention but binding of Mn. As the physiological concentration of zinc in the extracellular spaces of the brain is higher than that for copper,[57] the hypothesis that PrP^c is as well involved in zinc homeostasis has also been put forward.

In order to shed some light on the controversial topic of possible metal binding sites in the C-terminal domain, we performed a theoretical characterization of the structural and chemical properties of metal loaded PrP. To this end, we have applied a combination of bioinformatics techniques with classical molecular dynamics (MD) based on empirical force fields and QM/MM first-principles MD simulations.[3] Through the combination of these techniques several candidate binding sites could be identified and the subsequent computation of the corresponding EPR parameters by means of ADF2002.01,[30] enabled a direct comparison with experiments and rendered a narrowing down of the number of putative sites possible.

Computational studies of PrP have so far been limited to classical MD simulations of the entire metal-free C-terminal domain,[58] or portions of it.[59] The binding of open-shell transition-metal ions implies strong polarization and highly directional charge transfer effects that are not easily described within standard force fields. In most cases, an adequate treatment can only be performed with an explicit quantum mechanical electronic-structure calculation. In addition, this problem also necessitates the use of an approach that is able to take all the protein environment into account and to incorporate finite temperature effects that are known to be crucial for biological function.[60] The employed QM/MM Car-Parrinello approach[3,19,20] was able to fulfill all of the three above mentioned requirements.

Specifically, we first performed a statistical analysis of all available high resolution ($\leq$ 2Å) crystal structures of copper proteins, to determine the relative propensity of different amino acids for copper ligation. Using this probability map, we scanned the structure of the mouse prion protein for the most likely Cu^{2+} binding locations. Such an approach is limited to the identification of regions with a high density of amino acids that are likely to bind copper. However, it cannot provide any detailed information about the exact binding partners and the structural changes upon metal ion binding. For this purpose, the resulting candidate structures were subsequently refined in QM/MM Car-Parrinello simulations of several picoseconds around room temperature.

To generate realistic initial models for the high-level calculations, we started from the known NMR structure of the mouse PrP^{61} and equilibrated the system in a box of water using classical MD simulations. After positioning of the metal ion, the binding pockets were allowed to relax via classical MD runs. From this pre-equilibration step, we obtained 4 candidates for likely Cu^{2+} binding sites; one involving His140 (H140_binding site), one involving His177 (H177_binding site) and two in proximity the proximity of H187 (H187_E and H187_D binding sites). All of them where then used as starting structures for the subsequent QM/MM runs. For the latter, we defined sufficiently generous QM regions involving the copper ion and all the possible ligands (amino acids as well as water molecules) for each candidate site. The inclusion or exclusion of amino acid residues as well as water molecules in the QM region was updated manually according to the evolution of the binding site during the dynamics.

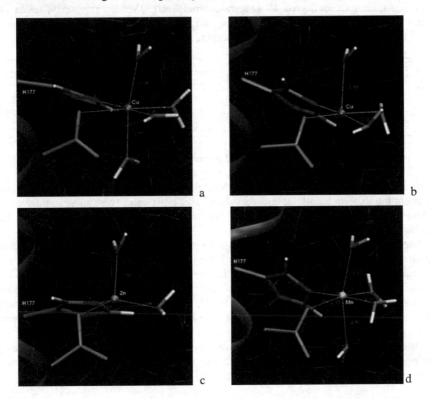

Figure 1   *H177 Cu(II) binding site at the end of the classical MD relaxation (a), at the end of the QM/MM simulation (b); after the substitution with Zn(II) (c) and after the substitution with Mn(II) (d). For panels (b)-(d), atoms drawn in sticks are included in the QM region.*

QM/MM Car-Parrinello simulations were performed at 300K for time periods of 10-15 ps. In spite of the limited time scale, the copper binding sites were able to rearrange significantly (e.g. we were able to observe spontaneous ligand exchanges as well as changes of protonation states) and to ultimately relax to stable coordination geometries. The H140, H177 and H187_E binding sites all assume a square planar geometry, with a fifth weak ligand in axial position, a well known coordination environment for Cu(II). Furthermore the experimental EPR spectra [50,51] are typical for type-II protein-copper complexes, that are known to adopt an axial square planar coordination geometry.[62] The

H187_D binding site instead rearranges into a distorted trigonal bipyramidal configuration. As an illustrative example, the starting and final frames of the QM/MM trajectories of the His177 binding sites are shown in Figure 1 (a and b).

To make direct contact with experiment, we computed the EPR parameters for each binding site using cluster calculations with the ADF code.[30] The choice of the functional, the numerical convergence with respect to the basis set as well as the dimension of the cluster to be included in the calculation, was extensively tested on model systems. For the final calculations, we collected 10 snapshots from each QM/MM trajectory, taken after binding site equilibration at time intervals of 0.35 ps. Some selected EPR parameters are reported in Table 1 in comparison with the experimental data.[50,51]

binding site	$g_{\parallel}$	$g_{\perp}$	$A_{\parallel}^{Cu}$	$A_{\perp}$
H140	2.138(2)	2.038(3)	-351(47)	-35(31)
H177	2.17(1)	2.037(9)	-470(10)	-137(13)
H187_E	2.191(5)	2.051(4)	-465(21)	-139(12)
H187_D	2.149(4)	2.060(9)	-393(8)	-131(20)
exp. [50,51]	2.295 ± 0.005	2.068 ± 0.005	|457|±10	|20|±10

Table 1   *Computed and experimental g and A^{Cu} tensor components.*

The two binding sites H177 and H187_E binding sites exhibit EPR properties in closest agreement the experimentally measured values. These two sites were subsequently probed for Mn(II) and Zn(II) substitution and the resulting structural rearrangements have been characterized in detail.

The simulations were carried out for 6 to 13 ps. Once again the binding sites were able to undergo extensive rearrangements that led away from the initial square planar axial rearrangements to more suitable coordination geometries for the substituted metal ions. Specifically, the presence of Zn(II) induces a reorganization towards tetrahedral coordination (Figure 1.c shows the Zn-loaded H177 binding site as an example), whereas Mn(II) substitution results in an octahedral coordination in the case of the H177 binding site (Figure 1.d) and a tetrahedral coordination for the H187_E binding site.

However, in all cases metal ion substitution can be achieved without any major rearrangements of the overall protein structure supporting the experimentally proposed picture that both Mn(II) and Zn(II) can act as Cu(II) replacements.

2.4 Organoruthenium Anticancer Compounds and their Interactions with DNA

The discovery of cisplatin [Pt(NH$_3$)$_2$Cl$_2$] as an anticancer drug by Rosenberg in 1965[63] resulted in a considerable interest in metallopharmaceuticals.[64] But even 40 years later, problems still remain associated with their use, including low selectivity, general toxicity and intrinsic or acquired drug resistance. In spite of an intense search for alternative compounds, up-to-date, cisplatin still represents the most widely used anticancer drug[65] and is employed in the treatment of a vast majority of all cancer patients.

Although the mechanism by which transition metal compounds exert their medicinal effects still remains a matter of debate, DNA interactions are generally considered to be critical. Several ruthenium complexes have recently attracted attention by showing their potential to overcome some of the above mentioned drawbacks, in particular the general toxicity problem.[66] Besides some inorganic complexes, organometallic ruthenium(II) compounds are currently subject of intense investigations. For the class of ruthenium(II)-

arenes, the most detailed information is available for $[Ru(\eta^6\text{-arene})Cl(en)]^+$ complexes (RA-en) which contain the chelating en=ethylenediamine ligand.[67] These compounds bind preferentially to the N7 of guanine bases and the observed selectivity is suggested to be controlled by the ethylenediamine -NH_2 groups. In fact, crystal structures and 2D-NMR NOE experiments show the formation of H-bonds with the exocyclic oxygens of guanines, but not with the exocyclic amino groups of adenine. Extended aromatic arene ligands were shown to accelerate binding to DNA, and the intercalation of these π-systems into DNA was invoked as possible rationalization for this phenomena.

Especially promising candidates among this class of ruthenium(II)-arene anticancer compounds are complexes that contain the 1,3,5-triaza-7-phosphatricyclo-[3.3.1.1]decane (pta) ligand (RAPTA compounds).[68] Upon DNA binding, the compound $[Ru(\eta^6\text{-}p\text{-cymene})Cl_2(pta)]$ (RAPTA-C) was found to exhibit pH dependent denaturation effects. DNA was damaged at the pH typical of the local environment of hypoxic cells[69], whereas at the pH characteristic of healthy tissue little or no change was observed. This behaviour was ascribed to the property of the pta ligand which contains three tertiary amine nitrogens as potential proton acceptors at low pH, and the protonated form was proposed as the active agent. It has also been demonstrated *in vitro* that RAPTA-C exhibits highly selective anticancer activity, targeting cancer cells without harming healthy cells. In contrast, the model compound for the protonated derivative, $[Ru(\eta^6\text{-}p\text{-cymene})Cl_2(pta\text{-Me})]^+$, has an indistinguishable toxicity against both cancer and healthy cells. However, information concerning the mode of binding to DNA at the molecular level and the structural changes induced upon drug binding, are still missing.

We investigated RAPTA and RA-en complexes and their interactions with DNA using a combination of classical MD,[6] quantum mechanical electronic structure calculations (QM),[35,36] and the hybrid quantum/classical (QM/MM) approach that has been recently developed in our group.[3] Focusing on double-stranded DNA (dsDNA), we try to rationalize the binding mode of RAPTA (**1**) and RA-en (**2**) compounds (Figure 2) with the final goal being the rational design of compounds with increased binding affinity and selectivity.

It is known experimentally, that both cisplatin and RA-en bind preferentially to the N7 atom(s) of guanine in dsDNA. Therefore, starting from a crystal structure of a cisplatin 1,2-intra-strand adduct to duplex DNA,[70] we docked the monofunctional $[Ru(\eta^6\text{-arene})Cl(en)]^+$ and the bifunctional RAPTA $[Ru(\eta^6\text{-arene})Cl_2(pta)]$ series of compounds in analogy to cisplatin to the major groove of the DNA sequence d(CCTCTG*G*TCTCC)/ d(GGAGACCAGAGG). G* are guanosine bases that coordinate to the ruthenium compounds through their N7 atom by replacing the halogen ions therein. As reference, the same DNA sequence was also simulated in a setup where cisplatin was completely removed in order to study the relaxation of the DNA towards its unperturbed B-DNA form.

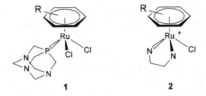

Figure 2 *Investigated Ruthenium(II)arene complex.*

2.4.1   The protonation state of the pta ligand.   It is known that cancer tissue is more acidic than healthy tissue.[69] The protonation state of the pta ligand in these different environments is therefore of particular interest.[68] The pta ligand might get protonated in the more acidic environment of cancer cells exhibiting different properties compared to the unprotonated state. Using *method1* we estimated *in vacuo* proton affinities as enthalpy differences H(B)-H(BH$^+$). As reference, the values for the isolated pta ligand and two of its derivatives were also calculated as their experimental pK$_a$ are known. We found good correlation of exp. pK$_a$ values and calc. proton affinities for the free pta ligand and its derivatives (Table 2). In agreement with experiments, the pta ligand turned out to be more basic than hexamethylenetetramine (nta) and the oxidized pta derivative (pta=O).[71] The nitrogen atoms were found to represent the most basic site in all investigated pta-like molecules. Our results suggest for the RAPTA series pK$_a$ values similar to those of the isolated pta ligand. This prediction is in good agreement with experiments carried out by Laurenczy *et al.*[72] who observed notable protonation of the pta ligand below pH 6.5 for compounds of the type of Ru(II)(H$_2$O)$_x$(pta)$_y$. However, it contradicts the ^{31}P-NMR measurements in aqueous solution by Scolaro *et al.* who report pK$_a$ values in the range of 2.99-3.31 for the chloride species of the RAPTA compounds. But the same authors speculate that the pK$_a$ of the RAPTA compounds, upon binding to DNA, could increase to a value that might make the pta ligand susceptible for protonation at moderate pH. We calculated the protonation energy for the di-aqua species of RAPTA-B to be 104 kcal/mol and the one of the di-N7-guanine species to be 129 kcal/mol. As the protonation of the di-N7-guanine species is clearly favoured over the di-aqua species we expect the pK$_a$ of the di-guanine species to be much higher than that of the aqua species. The two latter systems carry a charge of +2 which decreases the calc. *in vacuo* proton affinity in contrast to the situation in aqueous solution where the charge is stabilised by hydration.

	pta/H$^+$	nta/H$^+$	pta=O/H$^+$	RAPTA-B/H$^+$	RATPA-C/H$^+$
Exp. pK$_a$	5.6-6.1	4.9	2.5	3.2	3.1
H$^+$ affinity	225.7	224.6	216.7	226.6	225.6

Table 2   *Experimental pK$_a$ and calculated proton affinities at 298K (kcal/mol).*

2.4.2   Arene versus pta loss. Mass spectrometry measurements on DNA binding to the single-stranded (ss) oligomer (5'-ATACATGGT ACATA-3') reveal a different behaviour of the RAPTA compounds compared to the data reported for the related RA-en complexes. Whereas for the RAPTA series a loss of arene has been observed, this phenomena did not occur for the RA-en compounds. Using our computational scheme, we were able to provide a simple rationalization of the different behaviour of these two classes of drugs in terms of the ligand binding energies. The binding energy (BE) of arene was found to be at least 20 kcal/mol higher for [Ru(η^6-p-cymene)Cl(en)]$^+$ than for any of the investigated RAPTA compounds. Moreover, the experimentally observed relative tendency to arene loss among the RAPTA series of compounds correlates well with the differences in arene-ruthenium binding energies.[37]

On the other hand, calculations on the chloride model compounds showed a lower BE for ruthenium-pta compared to the arene, suggesting preferred loss of pta (which is indeed observed in an MS/MS experiment for the chloride species in the absence of ssDNA[73]). Since the experiments with ssDNA were carried out in aqueous solution for the hydrolysed RAPTA species, we calculated (*method2*) the ruthenium-pta/benzene interaction energies for all species that might be involved in the reaction with guanine (Table 3). Since we neglect solvent effects in this simple model, a reasonable comparison of the gas phase results with respect to the aqueous phase data is only possible for systems carrying the same charge. These simple calculations yield the remarkable conclusion that only in the case of the di-guanine derivative does the ruthenium-benzene BE favour arene loss (17.7 kcal/mol) over pta loss (34.7 kcal/mol). The di-aqua species shows significant higher BEs for both the benzene ($\Delta E=60.8$ kcal/mol) and the pta ($\Delta E=36.7$ kcal/mol) when compared to the di-guanine species. The same effect can be seen if one compares the mixed aqua-chloride species with the guanine-chloride species for which the pta and the benzene BEs are 13.0 and 26.5 kcal/mol, respectively, larger than for the aqua species. Consistently with our previous calculations,[37] we obtain higher BEs for the corresponding osmium derivatives. An increase of the BE by 8.4 kcal/mol for the di-guanine species provides an explanation why in the case of osmium no experimental arene loss is observed.

We specially focussed our attention to the protonated pta species. In contrast to the chloride containing compounds, for which we observe little effect on the ruthenium-phosphine BE, large variations in the BEs are seen for the di-aqua and di-guanine species. While we calculate a smaller Ru-pta BE (63.5 vs. 71.4 kcal/mol) for the protonated di-aqua species, we calculate a strong increase in the case of the protonated di-guanine species (73.1 vs. 34.7 kcal/mol). For the ruthenium-benzene BE the effect is inversed, 13.4 kcal/mol for the Ru-pta BE of the protonated di-guanine species (unprotonated 17.7 kcal/mol) vs. 101.6 kcal/mol for the di-aqua species (unprotonated 78.5 kcal/mol). Like the latter, the protonated chloride species show an increase of the ruthenium-arene BE compared to the unprotonated form. We therefore conclude that the strength of the ruthenium-arene interaction is not only a strong function of the net charge of the system, but also of the nature of the coordinated co-ligands.

	Charge	σ-pta/pta-H$^+$	η^6-benzene
[Ru(benzene)Cl$_2$(pta)]	0	29.2	33.1
[Ru(benzene)Cl$_2$(pta-H$^+$)]$^+$	+1	30.5	43.5
[Ru(benzene)Cl(H$_2$O)(pta)]$^+$	+1	43.9	62.1
[Ru(benzene)Cl(GuaN7)(pta)]$^+$	+1	30.9	35.6
[Ru(benzene)(H$_2$O)(GuaN7)(pta)]$^{2+}$	+2	51.6	61.7
[Ru(benzene)(H$_2$O)$_2$(pta)]$^{2+}$	+2	71.4	78.5
[Ru(benzene)(H$_2$O)$_2$(pta-H$^+$)]$^{3+}$	+3	63.5	101.6
[Ru(benzene)(GuaN7)$_2$(pta)]$^{2+}$	+2	34.7	17.7
[Ru(benzene)(GuaN7)$_2$(pta-H$^+$)]$^{3+}$	+3	73.1	13.4
[Os(benzene)(GuaN7)$_2$(pta)]$^{2+}$	+2	36.8	26.1

Table 3 *pta and benzene binding energies (kcal/mol) in RATPA-B derivatives.*

By using H$_2$O as a probe ligand which approaches the ruthenium center in RAPTA-B in between the two chloride ions, we identified different intermediates which are likely to be involved in the mechanism of arene loss. Constraining the water oxygen to bind to ruthenium we observed a decrease in the arene hapticity from η^6 to η^4 to η^2 coordination, while bonding to the chloride and pta ligands remains unaffected. We computed the η^4-intermediate to be 13.3 kcal/mol (*method2*) and the η^2- intermediate to be 14.1 and 14.9 kcal/mol (two conformations observed) less stable than the η^6-coordinated species. The binding energy of benzene in the η^2-intermediate was calculated to be 15.5 kcal/mol. As reported in ref. 37, the binding energy of e.g. guanine bound via its N7 to the [RuCl$_2$pta] fragment is of the order of 22.3 kcal/mol. From these results, we can envision a model in which, assuming an associative mechanism, an initial energy barrier has to be overcome in order to displace the arene but in a second step, the sum of three new σ bonds may compensate for the loss of the η^6- coordinated arene.

Finally, we estimated the overall thermodynamic stability of the ruthenium-benzene interaction in aqueous environment. For this purpose, we substituted the benzene with three water molecules in the di-aqua and di-guanine species. Whereas the benzene BE in $[Ru(benzene)(H_2O)_2(pta)]^{2+}$ was calculated to be 78.5 kcal/mol we obtained for the sum of the three water molecules in $[Ru(H_2O)_3(H_2O)_2(pta)]^{2+}$ a similar but slightly smaller value, (77.6 kcal/mol). However, this order does change upon protonation. The BE of benzene in $[Ru(benzene)(H_2O)_2(pta-H^+)]^{3+}$ was determined to be only 101.6 kcal/mol while the sum of the three H_2O in $[Ru(H_2O)_3(H_2O)_2(pta-H^+)]^{3+}$ amounts to 108.1 kcal/mol. The stability changes even more significantly in the case of the di-guanine species for which we calculate a benzene BE of 17.7 kcal/mol in $[Ru(benzene)(GuaN7)_2(pta)]^{2+}$ which is significantly smaller than the sum of three water molecule contributions in $[Ru(H_2O)_3(GuaN7)_2(pta)]^{2+}$ (45.5 kcal/mol). These results show that the influence of two coordinating guanine ligands changes significantly the BEs of the arene ligand. From a thermodynamic point of view, arene loss is possible. However, $[Ru(arene)(H_2O)_2(pta)]^{2+}$ species are known to be stable in aqueous solution.[68] Therefore, the experimentally frequently observed inertness of the arene seems to be of kinetic origin.

The question remains which ligand(s) may saturate the uncoordinated ocatahedral positions following arene loss upon reaction with ssDNA. Our DFT calculations show that it is very unlikely that the arene is replaced by a π-bound nucleobase.[37] However, our results suggest the formation of an adduct, in which the vacant arene-ruthenium coordination sites are occupied by nucleobases that are σ-bonded via their nitrogen atoms. In this model, the highly flexible ssDNA could wrap around the ruthenium center to occupy five coordination sites in the octahedral adduct leaving only the pta ligand unchanged. Classical MD studies show that the phosphate backbone of the DNA has no tendency to interact with the ruthenium(II) center. These ongoing studies suggest the formation of either chelates (e.g. in the case of guanine involving N7 and O6) and/or σ-coordination of a nearby base. Taking into account the stiffness of ssDNA and our results on the aqua-species (see above), a possible participation of water molecules in order to saturate the Ru-centered octahedral complex is also conceivable.

2.4.3 Binding to dsDNA. Even though arene loss is a possible phenomena for drug binding to ssDNA, such an effect appears less likely in the case of double-stranded (ds)DNA. The two DNA strands do not have the flexibility to wrap around a ruthenium coordination centre without breaking an extensive amount of Watson-Crick hydrogen bonds. Moreover, in NMR and MS experiments on mixtures of RAPTA compounds with nucleosides only a ligand exchange between chloride, water and nucleobases has been observed.[73] Since inside the cell, DNA is present most of the time in its double-strand form, dsDNA has to be considered as the most important target for ruthenium(II)-arene compounds with the only readily accessible nitrogen donor atoms being the N7 of adenine and guanine[74] in the major groove. As the latter is known to be the most nucleophilic site and constitutes the main target for anti-cancer drugs like cis-platin[74] and **2**[67], we further focused our analysis on ruthenium-guanine-N7 interactions. All investigated chloride species are known to undergo hydrolysis at the low chloride concentration of ~4mM which is typical for the intracellular milieu. We therefore calculated (*method2*) the *in vacuo* binding energies for the aqua and guanine(N7) adducts which are both charged +2. Starting with RAPTA-B we calculated the BE for guanine coordinated via its N7 atom in

$[Ru(benzene)Cl(GuaN7)(pta)]^+$ to be 25.8 kcal/mol whereas coordinated water in $[Ru(benzene)Cl(H_2O)(pta)]^+$ is with 11.1 kcal/mol significantly less strong bound. This difference in BE of 14.7 kcal/mol increases further to 53.8kcal/mol in the mixed complex $[Ru(benzene)(H_2O)(GuaN7)(pta)]^{2+}$ in which we determined the BE of guanine to be 66.6 kcal/mol and the BE of water to be only 12.8 kcal/mol. Finally, we compared the complexes containing two aqua or two guanine ligands. The coordination of the two water molecules in $[Ru(benzene)(H_2O)_2(pta)]^{2+}$ represents 42.6 kcal/mol compared to 111.1 kcal/mol for the two guanines in $[Ru(benzene)(GuaN7)_2(pta)]^{2+}$ (or even 119.2 kcal/mol in the case of the osmium compound $[Os(benzene)(GuaN7)_2(pta)]^{2+}$). We also compared the protonated di-aqua and di-guanine species $[Ru(benzene)(GuaN7)_2(pta-H^+)]^{3+}$ and $[Ru(benzene)(H_2O)_2 (pta-H^+)]^{3+}$. Both guanines resulted in a BE of 158.1 kcal/mol compared to 61.7 kcal/mol for the two aqua ligands. Here the difference between the two types of ligands is significantly higher (96.4 kcal/mol) than for the unprotonated species (68.5 kcal/mol). In summary, we can conclude that the difference in BE between a coordinated aqua or guanine ligand for RAPTA-B increases with an increase of the positive net-charge on the system and with the number of guanine bonds present. In the benzene derivative of RA-en the difference of a coordinated water in $[Ru(benzene)en(H_2O)]^{2+}$ (26.8 kcal/mol) vs. guanine in $[Ru(benzene)en(GuaN7)]^{2+}$ (76.1 kcal/mol) amounts to 49.3 kcal/mol. This difference is comparable to the one observed for the mixed RAPTA-B complex $[Ru(benzene)(H_2O)(GuaN7)(pta)]^{2+}$. Overall, the BE for all investigated compounds show that the Ru-guanine interaction is much stronger than the one with water. The higher the net charge and the more N-donors are coordinated to the ruthenium centre, the less favoured is an aqua-ruthenium complex compared to guanine-ruthenium interactions. These results quantify the experimentally observed preference of ruthenium(II)-arene compounds to coordinate to nucleobases upon hydrolysis.

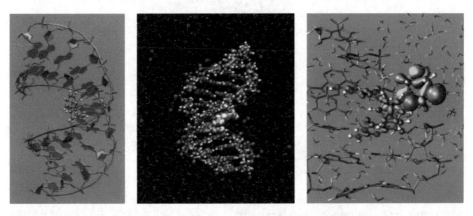

Figure 3   *RAPTA-B placed into the major-groove of dsDNA started from the corresponding crystal structure of cisplatin (left), MD-setup with explicit solvent and counter ions (middle), HOMO of RAPTA-B in the QM/MM-setup (right).*

Based on a molecular graphics investigation of $[Cp_2M]^{2+}$ coordinating to a rigid model oligonucleotide duplex, Marks *et al.* argued against cisplatin-like d(pGpG) complexation motifs[75]. This led to the widespread perception that transition metal arene compounds with two bulky ligands cannot bind to the major groove of DNA. However, using QM/MM molecular dynamics and allowing for a full structural relaxation, we could show that both investigated ruthenium compounds **1** and **2** can bind to the major groove and form stable adducts. We docked the ruthenium compounds in a cisplatin-like fashion and relaxed the DNA and the surrounding explicit water molecules and counter ions. The DNA turns out to be highly flexible, adapts very fast and widens to accommodate the ruthenium complex. After this initial equilibration, we let the full system evolve freely in time, describing at the quantum level the ruthenium complex as well as the one/two guanine bases that it binds to, while describing the rest of the DNA and the surrounding solvent in a classical scheme (Figure 3).

The local and global structural changes of DNA observed for the RAPTA series are similar to those reported for cisplatin.[76] Using a parameterization of the ruthenium complexes, derived from DFT *in vacuo* and verified by QM/MM explicit solvent simulations, we used classical MD to extend the time scale of our investigation. Whereas we did not observe any dramatic conformational changes in the case of the RAPTA compounds, we observed a severe perturbation of the Watson-Crick base pairing adjacent to the binding site of $[Ru(\eta^6\text{-}p\text{-cymene})G^*(en)]^{2+}$ (Figure 4). This defect proved to be stable in further QM/MM simulations as well as for very extended MM simulations[77] and has recently been suggested based on experimental observations.[78]

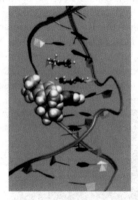

Figure 4   *p-cymeneRu(II)en having caused a Watson-Crick base-pair break in the major-groove of ds-DNA. The QM part of the system is shown in VDW, the broken base pair in CPK representation, water molecules and counter ions are omitted for clarity.*

In summary, loss of arene, although possible to occur with ssDNA assisted by nucleobases binding through σ bonds is unlikely to occur with dsDNA. Binding of RAPTA compounds to the major groove of dsDNA via two adjacent guanine N7 was shown to be possible and to lead to stable adducts.

With the help of the atomistic picture that we have obtained, we hope to get a deeper understanding of the molecular mechanisms of the anti-cancer activity that is exhibited by these intriguing class of compounds.

3 CONCLUSION

We have presented two examples of the characterization of transition metal ion interactions in biological systems. Due to the highly directional and often fluxional nature of bonding, the description of transition metal ions in proteins and nucleic acid systems is a particularly difficult task that often necessitates the use of a quantum mechanical electronic structure approach. At the same time, the protein environment and solvent have to be taken into account, as well as dynamical finite temperature effects. We show here that by a suitable combination of bioinformatics approaches, classical empirical force field based MD and first-principles based QM/MM hybrid simulations it is possible to gain a comprehensive molecular picture of transition metal ion interactions in natural systems.

References

1   R. Car, M. Parrinello, *Phys.Rev.Lett.* 1985, **55**, 2471.
2   R. Parr and W. Yang, *Density-Functional Theory of Atoms and Molecules*, Oxford University Press, New York, 1989.
3   A. Laio, J. VandeVondele, U. Rothlisberger, *J. Chem. Phys.*, 2002, **116**, 6941; A. Laio, J. VandeVondele, U. Rothlisberger, *J. Phys. Chem. B*, 2002, **106**, 7300.
4   CPMD, IBM Corp. (1990-2001). Copyright MPI fuer Festkoerperforschung Stuttgard (1997-2001). Available online at: http://www.cpmd.org.
5   W.F. van Gunsteren, S. R. Billeter, A.A. Eising, P.H. Hünenberger, P. Krueger, A. E. Mark, W. R. P. Scott and I. G. Tironi, *Biomolecular Simulation: The GROMOS96 Manual and User Guide*, Hochschulverlag AG an der ETH Zuerich, Zuerich, Switzerland, 1996.
6   D.A. Case, D. A. Pearlman, J.W. Caldwell, T.E. Cheatham III, J. Wang, W.S. Ross, C.L. Simmerling, T.A. Darden, K.M. Merz, R.V. Stanton, A.L. Cheng, J.J. Vincent, M. Crowley, V. Tsui, H. Gohlke, R. J. Radmer, Y. Duan, J. Pitera, I. Massova, G.L. Seibel, U.C. Singh, P.K. Weiner, P.A. Kollman, *AMBER 7*, University of California, San Francisco, 2002.
7   J. Hutter, *J. Chem. Phys.* 2003, **118**, 3928.
8   I. Tavernelli, U.F. Röhrig, U. Rothlisberger, *Mol. Phys.*, 2005, **103**, 963.
9   U.F. Röhrig, I. Frank, J. Hutter, A. Laio, J. VandeVondele, U. Rothlisberger, *ChemPhysChem.*, 2003, **4**, 1177.
10   M. Sulpizi, P. Carloni, J. Hutter, U. Rothlisberger, *Physical Chemistry Chemical Physics*, 2003, **5**, 4798.
11   M. Sulpizi, U.F. Röhrig, J. Hutter, U. Rothlisberger, *Intl. J. Quant. Chem.*, 2005, **101**, 671.
12   L. Guidoni, U. Rothlisberger, (submitted).
13   J. VandeVondele, U. Rothlisberger, *J. Am. Chem. Soc.*, 2002, **124**, 8163.
14   J. VandeVondele, U. Rothlisberger, *J. Phys. Chem. B*, 2002, **106**, 203.
15   A.O. von Lilienfeld, I. Tavernelli, D. Sebastiani, U. Rothlisberger, *Phys. Rev. Lett.*, 2004, **93**, 153004.
16   O.A. von Lilienfeld, I. Tavernelli, U. Rothlisberger, D. Sebastiani, *Phys. Rev.B.*, 2005, **71**, 195119.
17   A.O. von Lilienfeld, I. Tavernelli, D. Sebastiani, U. Rothlisberger, *J. Chem. Phys.*, 2005, **122**, 14113.
18   P. Carloni, U. Rothlisberger, M. Parrinello, *Acc. Chem. Res.*, 2002, **35**, 455.

19   S. Piana, D. Bucher, P. Carloni, U. Rothlisberger, *J. Phys. Chem. B*, 2004, **108**, 11139.
20   M. Sulpizi, A. Laio, J. VandeVondele, U. Rothlisberger, A. Cattaneo, P. Carloni, *Proteins- Structure, Function and Genetics*, 2003, **52**, 212.
21   S. Piana, U. Rothlisberger, 2005, *Proteins: Structure, Function and Genetics* **59**, 765.
22   L. Guidoni, K. Spiegel, M. Zumstein, U. Rothlisberger, *Angew. Chem.*, 2004, **116**, 3348.
23   For recent reviews see: M.C. Colombo, L. Guidoni, A. Laio, A. Magistrato, P. Maurer, S. Piana, U.F. Röhrig, K. Spiegel, M. Sulpizi, J. VandeVondele, M. Zumstein and U. Rothlisberger, *CHIMIA*, 2002, **56**, 11-17; P.Sherwood, in 'Modern Methods and Algorithms of Quantum Chemistry', Ed. J. Grotendorst, John von Neumann Institute for Computing, Jülich, NIC Series, **1**, 2000, p.257.
24   A. Laio, J. VandeVondele, U. Rothlisberger, *J. Phys. Chem. B*, 2002, **106**, 7300.
25   I. Frank, J. Hutter, D. Marx, M. Parrinello, 1998, **108**, 4060.
26   S. Hirata, M. Head-Gordon, *Chem. Phys. Lett.*, 1999, **314**, 291.
27   N. Trouiller, and J.L. Martins, *Phys. Rev. B* ,1991, **43**, 1993.
28   R.W. Hockney, *Methods Comput. Phys.*, 1970, **9**, 136. G. J. Martyna, M. E. Tuckerman, *J. Chem. Phys.*, 1999, **110**, 2810.
29   A.D. Becke, *Phys. Rev. A*, 1988, **38**, 3098; C. Lee, W. Yang, R. G. Parr, *Phys. Rev. B*, 1988, *37*, 785.
30   *ADF2002.01*, SCM, Vrije Universieit, Amsterdam, The Netherlands, 2002, http://www.scm.com.
31   E. van Lenthe, P.E.S. Wormer, A. van der Avoird, *J. Chem. Phys.*, 1997, **107**, 2488; E. van Lenthe, A. van der Avoird, P.E.S. Wormer, *J. Chem. Phys.*, 1998, **108**, 4783.
32   A.D. Becke, *Phys. Rev. A.*, 1988, **38**, 3098; J.P. Perdew, *Phys. Rev. B*, 1986, **33**, 8822.
33   P. Belanzoni, E.J. Baerends, M. Gribnau, J. Phys. Chem. A,1999, **103**, 3732.
34   E. van Lenthe, E.J. Baerends, J.G. Snijders, *J. Chem. Phys.*, 1993, **99**, 4597; E. van Lenthe, E.J. Baerends, J.G. Snijders, *J. Chem. Phys.*, 1994, **101**, 9783; E. van Lenthe, J.G. Snijders, E. J. Baerends, *J. Chem. Phys.*, 1996, **105**, 6505; E. van Lenthe, R. van Leeuwen, E.J. Baerends, J.G. Snijders, *Int. J. Qua. Chem.*, 1996, **57**, 281; E. van Lenthe, A. Ehlers, E.J. Baerends, *J. Chem. Phys.*, 1999, **110**, 8943.
35   J. A. Pople *et al.*, *Gaussian 03*, Revision B.03; Gaussian, Inc.: Pittsburgh PA, 2003.
36   *ADF2004.01*, SCM: Vrije Universiteit, Amsterdam, The Netherlands, 2004, http://www.scm.com.
37   A. Dorcier, P.J. Dyson, C. Gossens, U. Rothlisberger, R. Scopelliti, I. Tavernelli *Organometallic*, 2005, **24**, 2114.
38   S.B. Prusiner, *Science*, 1991, **252**, 1515.
39   S.B. Prusiner, *Science*, 1997, **258**, 245.
40   C. Weissmann, *FEBS Lett.*, 1996, **389**, 3.
41   A.L. Horwich, C. Weissmann, *Cell*, 1997, **89**, 499.
42   D.R. Brown, *Biochem. Soc. Symp.*, 2004, **71**, 193.
43   M.L. Kramer, H.D. Kratzin, B. Schmidt, A. Romer, O. Windl, S. Liemann, S Hornemann, H. Kretzschmar H, *J.Biol.Chem.*, 2001, **276**, 16711.
44   K.F. Qin, Y.Yang, P. Mastrangelo, D. Westaway, J. Biol. Chem., 2002, **277**, 1981.
45   D.R., Brown, F. Hafiz, L.L. Glasssmith, B.S. Wong, I.M. Jones, C. Clive, S.J. Haswell, *EMBO J.*, 2000, **19**, 1180.
46   R.M. Whittal, H.L. Ball, F.E. Cohen, A.L. Burlingame, S.B. Prusiner, M.A. Baldwin, *Protein Science*, 2000, **9**, 332.
47   D.R. Brown, B.S. Wong, F. Hafiz, C. Clive, S.J. Haswell, I.M. Jones, *Biochem. J.*, 1999, **344**, 1.

48   D.R. Brown, C. Clive, S.J. Haswell, *J. Neurochem.*, 2001, **76**, 69.
49   D.R. Brown, R.S.J. Nicholas, L. Canevari, *J. Neurosci. Res.*, 2001, **67**, 211.
50   G.M. Cereghetti, A. Schweiger, R. Glockshuber, S. Van Doorslaer, *Biophys.J.*, 2001, **81**, 516.
51   S. Van Doorslaer, G.M. Cereghetti, R. Glockshuber, A.Schweiger, *J. Phys. Chem. B*, 2001, **105**, 1631.
52   D.R., Brown, F. Hafiz, L.L. Glasssmith, B.S. Wong, I.M. Jones, C. Clive, S.J. Haswell, *EMBO J.*, 2000, **19**, 1180.
53   R.M. Whittal, H.L. Ball, F.E. Cohen, A.L. Burlingame, S.B. Prusiner, M.A. Baldwin, *Protein Science*, 2000, **9**, 332.
54   G.S. Jackson, I. Murray, L.L.P. Hosszu, N. Gibbs, J. P., Waltho, A.R. Clarke, J. Collinge, 2001, *Proc. Natl. Acad. Sci. U.S.A*, **98**, 8531.
55   K.F. Qin, Y. Yang, P. Mastrangelo, D. Westaway, 2002, *J. Biol. Chem.*, **277**, 1981.
56   D.R. Brown, B.S. Wong, F. Hafiz, C. Clive, S.J. Haswell, I.M. Jones, *Biochem. J.*, 1999, **344**, 1.
57   C.S. Burns, E. Aronoff-Spencer, C.M. Dunham, P. Lario, N.I. Avdievich, W.E. Antholine, M.M. Olmstead, A. Vrielink, G.J. Gerfen, J. Peisach, W.G. Scott, Millauser G.L., *Biochemistry*, 2002, **41**, 3991.
58   D.O.V Alonso, S.J. DeArmond, F.E. Cohen, V. Daggett, *Proc. Natl. Acad. Sci. U.S.A*, 2001, **98**, 2985; J. Zuegg, J.E. Greedy, *Biochemistry*, 1999, **38**, 13862.
59   H.F. Ji, H.Y. Zhang, L.A. Shen, J. Biomol. Struct. Dyn., 2005, **22**, 563; R.I. Dima, D. Thirumalai, *Proc. Natl. Acad. Sci. U.S.A*, 2004, **101**, 15335; M. Pappalardo, D. Milardi, C. La Rosa, C. Zannoni, E. Rizzarelli, D. Grasso, *Chem. Phys. Lett.*, 2004, **390**, 511.
60   M. Karplus, G.A. Petsko, *Nature*, **347**, 1990.
61   R. Riek, S. Hornemann, G. Wider, M. Billeter, R. Glockshuber, K. Wuthrich, *Nature*, 1996, **382**, 180.
62   J. Peisach, W. E. Blumberg, *Arch. Biochem. Biophys.*, 1974, **165**, 691.
63   B. Rosenberg, L. Vancamp, T. Krigas, *Nature*, 1965, **205**, 698.
64   M.J. Clarke, F. Zhu, D. R. Frasca, *Chem. Rev.*, 1999, **99**, 2511; E. Wong, C. M. Giandomenico, *Chem. Rev.*, 1999, **99**, 2451.
65   M.A. Fuertes, C. Alonso, J. M. Perez, *Chem. Rev.*, 2003, **103**, 645.
66   G. Sava, I. Capozzi, A. Bergamo, R. Gagliardi, M. Cocchietto, L. Masiero, M. Onisto, E. Alessio, G. Mestroni, S. Garbisa, *Int. J. Cancer*, 1996, **68**, 60; M. Galanski, V.B. Arion, M.A. Jakupec, B. K. Keppler, *Pharm. Des.*, 2003, **9**, 2078.
67   R.E. Aird, J. Cummings, A. Ritchie, M. Muir, R. Morris, H. Chen, P.J. Sadler, D.I. Jodrell, *Br. J. Cancer,* 2002, **86**, 1652; H. Chen, J. A. Parkinson, R.E. Morris, P.J. Sadler, *J. Am. Chem. Soc.,* 2003, **125**, 173; H. Chen, J.A. Parkinson, S. Parsons, R.A. Coxall, R. O. Gould, P.J. Sadler, *J. Am. Chem. Soc.,* 2002, **124**, 3064; R. Morris, A. Habtemariam, Z. Guo, S. Parsons, P.J. Sadler *Inorg. Chim. Acta,* 2002, **339**, 551; R.E. Morris, R.E. Aird, P.d.S. Murdoch, H. Chen, J. Cummings, N.D. Hughes, S. Parsons, A. Parkin, G. Boyd, D.I. Jodrell, P.J. Sadler, *J. Med. Chem.,* 2001, **44**, 3616; F. Wang, H. Chen, J.A. Parkinson, P.d.S. Murdoch, P.J. Sadler, *Inorg. Chem.,* 2002, **41**, 4509; F. Wang, H. Chen, S. Parsons, I.D.H. Oswald, J.E. Davidson, P.J. Sadler, *Chem. Eur. J.,* 2003, **9**, 5810; R. Fernandez, M. Melchart, A. Habtemariam, S. Parsons, P.J. Sadler, *Chem. Eur. J.,* 2004, **10**, 5173; F. Wang, J. Bella, J.A. Parkinson, P.J. Sadler, *J. Biol. Inorg. Chem.,* 2005, **10**,147; H. Chen, J.A. Parkinson, O. Novakova, J. Bella, F. Wang, A. Dawson, R. Gould, S. Parsons, V. Brabec, P.J. Sadler, *Proc. Natl. Acad. Sci., USA* 2003, **100**, 14623.
68   C.S. Allardyce, P.J. Dyson, D.J. Ellis, S.L. Heath, *Chem. Commun.*, 2001, **15**, 1396; C.S. Allardyce, P.J. Dyson, D.J. Ellis, P.A. Salter, R. Scopelliti, *J. Organomet. Chem.*

2003, **668**, 35; C. Scolaro, A. Bergamo, L. Brescacin, R. Delfino, M. Cocchietto, G. Laurenczy, T. J. Geldbach, G. Sava, P. J. Dyson, *J. Med. Chem.*, 2005, **48**, 4161.

69   R.A. Gatenby, R.J. Gillies, *Nature Rev. Cancer*, 2004, **4**, 891.

70   P.M. Takahara, C.A. Frederick, S.J. Lippard, *J. Am. Chem. Soc.*, 1996, **118**, 12309.

71   D.J. Darensbourg, J.B. Robertson, D.L. Larkins, J. H. Reibenspies, *Inorg. Chem.* 1999, **38**, 2473.

72   J. Kovacs, F. Joo, A. C. Benyei, G. Laurenczy, *Dalton Trans.* 2004, **15**, 2336.

73   P.J. Dyson personal communication.

74   J. Reedijk, *Chem. Commun.* 1996, **7**, 801.

75   L.Y. Kuo, M.G. Kanatzidis, M. Sabat, A.L. Tipton, T. J. Marks, *J. Am. Chem. Soc.* 1991, **113**, 9027.

76   K. Spiegel, U. Rothlisberger, P. Carloni, *J. Phys. Chem. B* 2004,**108**, 2699.

77   C. Gossens, I. Tavernelli, U. Rothlisberger, manuscript in preparation.

78   O. Novakova, H. Chen, O. Vrana, A. Rodger, P.J. Sadler, V. Brabec, *Biochemistry* 2003, **42**, 11544.

SIMULATIONS OF ENZYME REACTION MECHANISMS IN ACTIVE SITES:
accounting for an environment which is much more than a solvent perturbation

Jill E. Gready[1], Ivan Rostov[1,2] and Peter L. Cummins[1]

[1]Computational Proteomics Group, John Curtin School of Medical Research, Australian National University, Canberra ACT 0200, Australia
[2]Supercomputer Facility, Australian National University, Canberra ACT 0200, Australia

1 INTRODUCTION

The goal of this chapter is to outline the special challenges faced in using computational methods to define the reaction mechanisms of enzymes. Our approach is to set the problem in the context of the special chemical and structural features which differentiate enzymes from other types of catalysts, and the limitations of experiment in probing these features and defining their contributions to the reaction facilitation. We then briefly discuss the types of computational methods available and how they can be applied to an enzyme problem. Two major features – the roles of active-site protons and water molecules, and enzyme conformational flexibility – are illustrated by results from our simulation work on dihydrofolate reductase. After briefly reviewing published work, we focus on two novel simulation protocols and methods we have recently developed, combining the use of ONIOM *ab initio* QM/MM calculations and MD simulations with semiempirical QM/MM potentials, to address issues of transition state structure and the reaction free energy surface for hydride-ion transfer.

1.1 Enzymes as Unique Protein Catalysts

Elucidating the mechanisms of enzyme reactions is a very difficult *chemical* problem. At physiological temperature and pH, enzymes catalyze, with ease, quite fantastic chemistry. Uniquely for a catalyst, the structure, properties and dynamics of an enzyme have been constrained and selected by biological evolution to perform particular reactions. While their catalytic powers have been "fashioned" by this process, how and why enzymes work is still largely a mystery, despite much study and debate.[1-3] Several theories have been proposed which impute special properties to explain the power of enzymes, such as transition-state stabilization, electrostatic preorganization, low-energy H bonds, pK_a matching, entropy-enthalpy compensation, desolvation, steric strain, and near-attack complexes.[4-6] However, the results of increasing numbers of detailed studies for a range of enzyme-reaction types do not indicate a need for special chemical concepts. Rather the issues are of scale and degree, and derive from the molecular complexity of proteins coupled with evolutionary refinement, which enable unusual solvation, electrostatic and dynamical properties to be imposed on the active site and the chemistry which takes place within it. As an enzyme is an evolutionary solution, i.e. it is not "designed', it cannot be assumed that it is the *best possible* solution for effecting a particular reaction. These points suggest that there is no universal explanation of enzymic catalytic power, but rather that different enzyme-protein properties may be selected to different extents in the evolution of a given enzyme.

1.2 Challenges for defining enzyme mechanisms

Investigation of a particular enzyme aims to understand how its structure, energetics and dynamics combine to determine the chemical mechanism of the *particular* reaction – how the enzyme environment binds substrates, products and transition states (TSs), and facilitates the reaction.

Starting from the concept that the efficiency of enzymes derives from the way evolution has shaped the way the active site "works", we can pose a number of questions based on the corollary that there is an "orchestration" of contributions during the overall process of substrate binding, reaction and product release:

What are the specific roles of active-site groups?

How do they work in concert or sequentially during reaction?

What are the origins and fates of water molecules and protons?

What is the influence of the wider enzyme protein, particularly electrostatic interactions, and dynamical motions and conformational flexibility?

In considering facilitation of the reaction, we are usually implicitly comparing it with reaction in solution. But here we need to consider not only factors contributing to lowering of the effective barrier height but also whether the shape of the reaction surface, and, hence the effective mechanism or transit pathway, may be different in the enzyme. An obvious case is the sequence of steps, for example protonation and hydride-ion transfer as relevant to dihydrofolate reductase, our enzyme example in this chapter. However, in considering the reaction-cycle energetics, we also need to keep in mind the possibility that the effective enzyme mechanism may be flexible due to the complexity of the reaction surface. A manifold of protein conformational states contributes to the reaction free energy surface, and to energetically accessible pathways.

1.3 Experiment? Simulation?

Proposal of so many explanatory theories (1.1) has largely arisen from the lack of experimental tools to address the issues raised in 1.2 directly. The static picture of an enzyme from x-ray crystallography, the major experimental "window" of the last decades, has dominated thinking on the molecular catalytic mechanisms. While it provides a detailed view of the "visible players", the amino acid residues and their interactions with substrates in the active site, it misses key "invisible" players, protons and water, and dynamical and electrostatic properties of the wider protein (and solvent), and provides only limited information on conformational flexibility. As noted, the question is how to decipher and quantitate the possible roles for these features pertinent to the active-site environment for the reaction by a particular enzyme. While structural definition is variably approachable by experiment, *energetic* definition of the roles of all these potential players is not. Both experimentalists and theoreticians agree (!) that elucidation of the question requires energetic and thermodynamic analysis by computer simulations.[4,7] Many experimental studies have provided evidence for significant changes in enzyme conformation during the course of reaction, and suggested that such changes affect the efficiency of catalysis. Experimental data include x-ray structures of enzyme complexes which mimic the stages of the reaction,[8] and NMR relaxation data indicating altered motions, in both near and far regions of the enzyme, from binding of the ligands at the active site.[9] Conversely, major effects on enzyme activity from mutations quite far from the active site have been reported, and it is suggested these perturb coupled motions which promote reaction.[10]

1.4 Computational issues for enzyme simulations

Following on from these arguments, we can identify a number of factors which are likely to be variably important in aiding enzyme catalysis, and, thus, would need to be represented reliably, and in a balanced way, in a simulation of reaction:

(1) Electrostatic effects, both polarization of reactive species (substrates, TS, products) at the active site and back polarization of the enzyme, as well as actual charge transfer (CT) between these species and the enzyme;

(2) Effects from constraining the relative orientation of reactive species with each other, and with anchoring active-site residues, on geometries, especially the TS geometry;

(3) Statistical averaging of energies for individual enzyme complex configurations resulting from conformational flexibility of the enzyme;

(4) Desolvation effects — or rather resolvation effects as the active-site becomes the effective new solvent — which may contribute substantially to, particularly, the entropic component of relative solution and enzyme free energies.

As a potential qualifier of (3), we note that some authors (e.g. ref. 10) take a stronger view of the effects of protein flexibility by imputing a direct dynamical role in reaction promotion from transient co-ordinated enzyme motions "pushing" reactants at the active site across barriers towards reaction (e.g. by compressing bonds on one side and/or lengthening them on the other). The issue is whether reaction is promoted by a sequence of dynamical events or statistically by key residues simultaneously being in effective positions with favourable geometric parameters. With the limitations of current computational methodologies, it is difficult to address this distinction in an unbiased way.

A defining feature of (1) is that as the active-site environment is non-isotropic, electrostatics will be particularly important. Enzyme proteins may be highly polarized directionally, with these effects manifest by net electric fields across the active-site region. Also, these effects may be very sensitive to enzyme conformation, including during different stages of the overall reaction. From the computational viewpoint, there is a challenge in representing enzyme electrostatics as the usual assumptions of them being only short range do not apply; they do not cancel in 3-D, as in the isotropic case. This has major consequences in use of computational methods and models involving molecular truncation (e.g. active-site fragment complexes) and partition (e.g. QM/MM), as discussed below. As may be readily appreciated, the electrostatic effect is compounded by the enzyme flexibility, requiring them to be considered jointly.

1.5 Goals for enzyme simulations

Specifically, within a computational simulation our goal is to define a complete mechanistic profile with sufficient accuracy to discriminate the states, roles and energetic contributions of molecular features not definable, or readily definable, by experiment. Foremost among these, and pertinent to the majority of enzyme problems, are the "invisibles" — protons and water molecules. We wish to define: the states of ionizable active-site residue sidechains and protonation states for reaction species, during the course of the reaction; the identities of the enzyme proton donors and abstracters for the various steps; the networks for channelling protons into or away from the active site; and the origin of water molecules consumed or produced. Energetic definition at the required level is very challenging computationally as typical stable H-bond strengths are 5-10 kcal/mol, and relative H-bond strengths may be even less for two realistic alternatives. Although the power to discriminate close alternatives is very limited using current methods this may be irrelevant as relatives energies will be modulated by conformational flexibility and both alternatives may contribute to the energetic manifold of the reaction.

1.6 Simulation methods for enzyme reactions

While, in principle, the physicochemical factors and molecular roles contributing to reaction facilitation just outlined are quantifiable using simulations, accurate representation of all of them simultaneously is still beyond the feasibility of any one current computational method.

   The common approach adopted in the literature is to divide the problem into computational "windows" accessible by methods capable of treating aspects of the overall problem separately. In general these fall into three main classes which depend on the quality of the electronic description of the active site, the representation of its wider enzyme environment, and the degree of conformational sampling. The size of the molecular systems which can be treated scales accordingly. Typical size-dependent limitations for *ab initio* QM calculations (e.g. HF, DFT) on isolated active-site fragment complexes or analogous QM regions embedded within an MM representation for the rest of the enzyme and solvent system, such as in an ONIOM QM/MM calculation,[11,12] are up to about 200 atoms or 100 atoms, respectively, with full geometry optimisation. Whole-enzyme complexes up to 20,000 atoms or more can be treated with linear-scaling semiempirical QM (SE-QM) calculations, such as for the localized MO implementation of Stewart[13] in MOPAC/MOZYME[14] or LOCALSCF,[15,16] with limited or full geometry optimisation, respectively. While characteristics of MD simulations vary significantly depending on the application, their scope has increased dramatically recently due to increased power from cluster computing and a shift to multi-trajectory simulations to improve sampling. Systems of 30,000 atoms or more (enzyme + solvent) for multiple (10-100) trajectory MD (MMD) simulations of several nanoseconds (ns) each would now be routine. In our work, covered also by the examples in this chapter, we treat enzyme plus solvent systems of about 20,000 atoms for MMD or REMD[17] (replica exchange MD) simulations of about 1 ns each with a SE-QM/MM potential in which the QM region is up to about 140 atoms. Although Rothlisberger and colleagues have now reported several QM/MM Car-Parrinello MD (using plane-wave DFT) simulations of enzyme reactions (e.g. ref. 18), this methodology is still being developed and its range of applicability is still unclear.

1.7 QM/MM methods in enzyme simulations

For reasons outlined in 1.6 and despite several inherent problems, hybrid QM/MM methods have become popular in the last decade in tackling the issue of representing both particular chemical events in the active site, as well as longer-range perturbations.[19,20] These methods provide a computational compromise between accuracy and computational feasibility in enzyme simulations by dividing the complete molecular system into two regions treated by two levels of theory. Thus, they allow representation of the reaction chemistry by a QM method, and the perturbation and physical restraint from the remainder of the enzyme and solvent system by a force-field (MM) description. There are two main types of QM/MM methods — linking and extrapolation —, which differ depending on how the QM and MM interactions are counted.

In the more common linkage scheme,[19,20] the QM region is linked to the MM system by explicit calculation of the interactions between atoms of the QM and MM parts. The properties of the whole system, such as total energy, are constructed as a sum of QM- and MM-treated parts and a special QM/MM interaction term:

$$E = E_{QM} + E_{MM} + E_{QM/MM} \tag{1}$$

The simplest 2-layer extrapolation scheme involves applying QM to the primary fragment, applying MM to both the primary fragment and the whole system, and then combining all three calculations to calculate the energy as:

$$E = E_{MM}(\text{whole}) \ E_{MM}(\text{fragment}) + E_{QM}(\text{fragment}) \tag{2}$$

Applications of the extrapolation method to biological problems, including enzyme simulations, are few. There are only two methods, ONIOM[11] and SIMOMM,[21] both implemented in major *ab initio* QM programs (GAUSSIAN-03[12] and GAMESS-US, respectively). In contrast to the more numerous linkage QM/MM methods which mostly employ semiempirical QM, their main practical uses are for treating the primary subsystem by *ab initio* or DFT methods.

1.8 Approaches for enzyme simulations

As in any scientific investigation, no one method or protocol fits all and, as noted in 1.6, for simulation of enzyme reactions there is no method capable of a total unbiased description. An objective analysis of requirements suggests that the computational protocol needs to be able to provide: (1) an adequate description of electronic changes in the ligands and immediate active site during the course of the reaction, including perturbations from whole-enzyme electrostatics; (2) calculation of free energies, not enthalpies, to allow comparison with experiment; and (3) incorporation of the effects of conformational flexibility, implicit in free energies. Accounting for (1) requires an *ab initio* QM model supplemented by an MM or other lower-level description, while (2) and (3) require MD simulation. Currently the level of QM within MD simulations with QM/MM potentials is effectively restricted to semiempirical QM. However, SE-QM has well known deficiencies,[22,23] notably in representing H bonds, which are almost universally important in enzyme active sites, and π-delocalisation effects, which are critically important for our application to DHFR.

In previous work on DHFR we attempted to address the requirements with a multi-method approach using *ab initio* QM active-site fragment complex calculations complemented by SE-QM/MM MD simulations.[24-26] However, there are problems in combining results from the two types of method; e.g. we used a rather *ad hoc* procedure in ref. 25 to correct semiempirical QM energies with *ab initio* estimates. As illustrated by the example results in this chapter, we are now using a more integrated approach in which output from one type of study is used as input to another, specifically large sets of ONIOM *ab initio* QM/MM calculations using multiple coordinate sets sampled from SE QM/MM MD simulations. This approach is now computationally tractable as it is possible to do thousands of ONIOM calculations using large cluster computers, managed by scripts and analysed by spreadsheets. We also use the MOZYME method within MOPAC[14] and

LOCALSCF[15] to obtain independent checks on several aspects of the performance of our SE-QM/MM model[27,28] such as electrostatic interactions,[29] to check that the QM region is well chosen by calculating the extent of charge transfer between the QM and MM parts (Bliznyuk, Cummins and Gready, unpublished results), and to estimate effects due to lack of MM polarization.[30]

2 APPLICATION TO DIHYDROFOLATE REDUCTASE

2.1 Mechanistic challenges of DHFR

DHFR catalyses the reduction of folate to dihydrofolate (DHF) and tetrahydrofolate (THF) *via* a two-step reaction of protonation, and hydride-ion transfer from the cofactor NADPH:[31,32]

$$Folate + H^+ + NADPH \rightarrow DHF + NADP^+$$
$$DHF + H^+ + NADPH \rightarrow THF + NADP^+$$

Many structures of DHFR complexes have been solved crystallographically: a schematic for the overall enzyme structure with bound ligands is shown in Figure 1, while the structures of the ligands and some interacting active-site groups, especially the ordered H-bonded network around the pyrimidine moiety of the pterin ring of substrate, is shown in Figure 2. The protein is a small monomer of ~160-190 residues (159 (*E. coli*) or 187 (human) residues; ~3000 protein atoms).

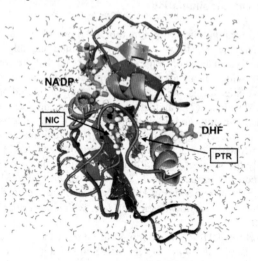

Figure 1 *Cartoon of the structure of the E. coli DHFR complex with DHF and NADP$^+$ (pdb ID: 1rx2[8]), showing the opposed buried reactive rings (pterin (PTR) and nicotinamide (NIC)) of substrate and cofactor with their long anionic tails (Fig. 2) snaking out to more solvent-exposed regions near the protein surface.*

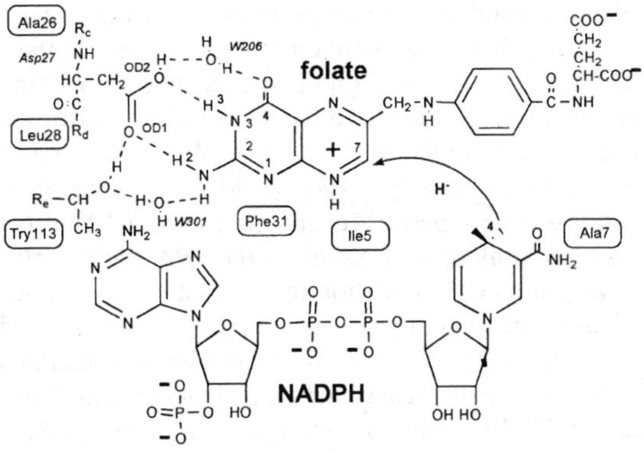

Figure 2   *The structures of the protonated folate substrate and cofactor NADPH with key interacting enzyme groups in the DHFR active site. The hydride ion to be transferred from C4 of the nicotinamide ring to C7 of the pterin ring is shown.*

Despite intensive study, key aspects of the DHFR mechanism remain elusive. These are whether the carboxylate group of the conserved active-site Asp (bacterial DHFR) or Glu (eukaryotic DHFR) residue is protonated or ionized during the reaction, the source of the protons, and the role of active-site water molecules. In the DHFR active site, several water molecules, especially that positioned between the Asp OD2 and O4 groups of the pterin ring of substrates (labelled W206 in *E. coli* x-ray structures[8]), are located so precisely and consistently that they are effectively a part of the genetic blueprint. Thus, it is likely they have an essential role, either structural or mechanistic.

In addressing the DHFR mechanism computationally several other challenging issues need to be addressed in addition to the Asp protonation state and the roles of water molecules. The reactive pterin and nicotinamide rings are unusual π-delocalized heterocycles, especially the pterin ring which shows complex preferred protonation patterns which vary depending on the oxidation state (folate, DHF, THF).[33,34] Also, both substrate and cofactor are large molecules with highly charged (anionic) tails. Recent experimental and simulation studies have also shown complex conformational movements during ligand binding and resulting from mutations, suggesting that conformational channelling or selection of preferred conformations which proceed to reaction with lower barriers may be important in the DHFR mechanism.[9,10]

2.2 Previous background results

As the two studies for which we present results arose from earlier work, we briefly summarize the scope of the more recent earlier work and the conclusions drawn. These studies addressed the problem of protonation activation for *both* substrates (folate and DHF) to form the reactive ternary complex, including the roles of water molecules, and also the effect of alternative protonation-state patterns on the reaction energetics. Using *ab initio* QM calculations on active-site fragment complexes, we found for both the folate and DHF active ternary complexes that the active-site Asp carboxylate group needs to be protonated first, and that a second proton then directly protonates N8 (folate) or N5 (DHF).[24] The presence of water W206 H-bonded to the protonated Asp and O4 of the pterin ring is critically important for correct protonation of the enzyme-bound substrates. This result provided, for the first time, a coherent explanation for the conserved presence of the water molecule and *basic* Asp27 group (i.e. it acts as a base, not as an acid, to prepare the active site for reaction). In more recent work we confirmed the primary result at the free-energy level, i.e. that initial protonation of Asp is favoured over protonation of N5 of DHF, using semiempirical QM/MM REMD free energy simulations. These simulations incorporated our new perturbed quantum atom (PQA) method which allows, effectively, creation and annihilation of atoms in the QM Hamiltonian within the closed shell RHF formalism, and allowed calculation of the relative protonation free energy of the Asp27 and N5 DHF proton acceptor sites using thermodynamic integration methods.[26]

In following up the results of ref. 22 for the preferred protonation state of the reactive complexes, we used *ab initio* QM calculations on active-site fragment complexes together with semiempirical QM/MM MD simulations and the LRA (linear response approximation) method to investigate the effect of the protonation state of Asp27 on the reaction energetics.[25] However, results for both reactions (folate and DHF) did not show an unambiguous preference for the neutral state of Asp27 over the ionised form, and the necessity to apply constraints in the fragment-complex calculations gave poor TS structures with overestimated activation energies.

3 RESULTS

3.1 ONIOM QM/MM for transition-state structures

3.1.1 Motivation and capabilities. Geometry optimizations of enzyme active-site fragment-complex models often encounter problems due to the floppy nature of such complexes. Fixing edge atoms or applying other restraints to mimic the natural constraints of the enzyme environment is often necessary to allow optimization to complete, as we found for the DHFR fragment-complex optimizations from ref. 25, just noted. However, this introduces artefacts, particularly for TS structures which often show relatively small but important contractions compared with reactant and product complexes. Consequently we were interested in assessing the usefulness of the ONIOM QM/MM implementation in GAUSSIAN[11] for performing the DHFR fragment-complex optimizations within, initially, a realistic active-site physical "mould". To gauge the effects of protein conformational flexibility on the enzyme-complex geometries and activation and reaction energies, input coordinates were taken from 20 snapshots of trajectories of the DHFR reactant complexes from separate (semiempirical) QM/MM MD simulations.

For calculation of meaningful energy differences, i.e. activation and reaction enthalpies, optimizations of three stationary points (reaction (RC), TS and product (PC) complexes) within the same local MM conformational well are required. To meet this stringent requirement we used the Development Version of GAUSSIAN containing the quadratic coupled optimization algorithm (QuadMacro)[35] not yet available in the public release version of G03.[12] As implemented, this allows optimizations of minima (RC, PC) of large QM/MM systems using both the Mechanical (ME) and Electronic (EE) Embedding schemes, which provide physical restraint only or additionally electrostatic polarization of the QM region by the MM region, respectively. However, the current implementation allows TS optimizations only at the ME level.

3.1.2 Computational details. The solvated protein model was generated as follows. Input coordinates were taken as 20 snapshots from QM/MM (PM3/Amber) MD trajectories for the *E. coli* DHFR.DHF$^+$.NADPH reactant complex generated by the MOPS program[36] at T = 300K (initial coordinates from the *E. coli* DHFR.DHF.NADP⁺ complex 1rx2[8]). The QM region consisted of the complete protonated DHF and NADPH ligands and the Asp27 side chain (-CH₂-COOH), as for the simulations reported next for the free energy surface (3.2). The system consisted of enzyme (2617 atoms), DHF (52 atoms) and NADPH (74 atoms) solvated in a 33 Å sphere of TIP3P water molecules (3742), giving 13,843 atoms in total. For the ONIOM calculations, the QM region, as shown in Figure 3, consisted of the 6-CH2-dihydropterin (PTR 160) and *p*-aminobenzoylamide (FOL 161) residues of DHF, the nicotinamide ring (NIC 164) of NADPH, the Asp27 and Thr113 residues of the protein chain, and 2 water molecules making H-bond links between the substrate and protein (total 81 atoms plus 5 hydrogen-type link). To reduce the computational expense of the interactions between MM atoms, the system was trimmed to a 30 Å sphere which removed waters only (total system ~8,500 atoms). Boundary water molecules beyond 25 Å from the complex centre were fixed, leaving ~5,500 atoms, including the complete enzyme and ligands, marked for optimisation. Atomic MM charges for atoms of substrate and cofactor were taken from Merz-Kollman analysis of fragment QM calculations (HF/6-31G(d,p)) of the whole separate substrate and the whole separate cofactor.

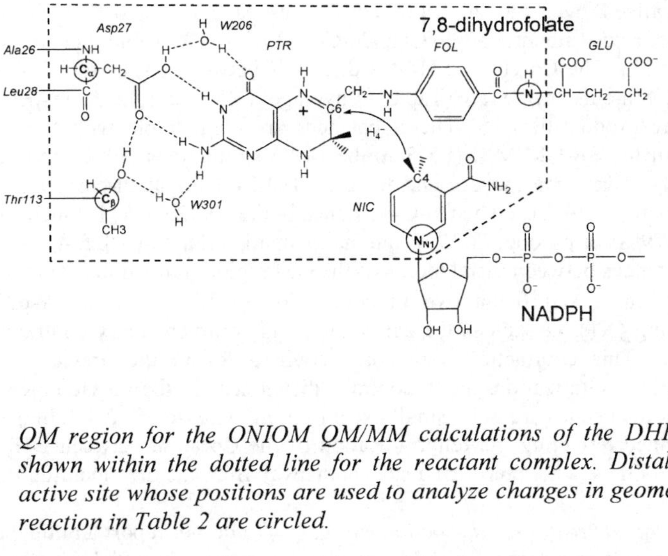

Figure 3 *QM region for the ONIOM QM/MM calculations of the DHFR reaction is shown within the dotted line for the reactant complex. Distal atoms of the active site whose positions are used to analyze changes in geometry during the reaction in Table 2 are circled.*

As input geometries were representative of the reactant state, and quite far from the TS in the coordinate space, the TS optimization was set up as a multi-step process. In the first stage, ONIOM-ME (HF/3-21G:Amber) with the default microiteration procedure[37] (no explicit coupling of the QM and MM coordinates) was used with constraints on the C4-H4 and H4-C6 distances to bring the complex closer to the expected TS geometry. C6-H4 and C4-H4 distances were fixed and gradually varied towards their expected TS values, while the rest of the non-fixed coordinates in the QM and MM regions were optimised. Three to four such constrained optimizations were required to shift initial values of hydride-donor R(C4-H4) and hydride-acceptor R(C6-H4) distances to their expected TS values of ~1.43 Å and ~1.22 Å, respectively. In total, this stage of the geometry optimization took 60-90 QM steps to converge. In the second stage, the constraints on the bond distances R(C4-H4) and R(C4-H6) were lifted, and the TS optimization and QuadMacro[35] options were turned on. Initial force constants for coordinates of the QM atoms were calculated using the CalcFC option. On average, the second stage of TS optimization took 7-11 steps. To speed up calculations, first stage optimizations were performed with loose convergence criteria, and then for the second stage TS optimizations the convergence criteria were switched to normal.

We then used the quadratic coupled algorithm[35] to calculate reactant and product minima for the given TS. The TS coordinates were used as input with a slight shift in position of H4 of 0.1 Å towards either reactant or product minima. As previously, starting with the loose convergence criteria and switching to the normal criteria allowed reductions in the total number of steps in the optimisations: reactant and product minima were reached in 50-60 steps, on average.

3.1.3 Results: TS geometry and active-site contraction. From the RC, TS and PC optimised geometries of the 20 input coordinate sets, we can analyze the details of the configurations at the reaction centre (Table 1) during the reaction, as well as changes over a wider region of the active site (Figure 4 and Table 2). As can be seen from the arrangement of atoms within the QM region in Figure 4, all three stationary points are quite well localized over the 20 input configurations considered and the orientation between the two rings (dihedral angle C8a-C6-C4-N1 of ~153°; Table1) changes very little during the reaction. The C6-H4 and H4-C4 distances between the hydride H4, donor C4, and acceptor C6 atoms in the TS (Table 1) average 1.25 and 1.42 Å, respectively, and indicate a more product-like TS. These distances are comparable with those we found previously from the SE-QM/MM (PM3/Amber) MD simulations (~1.25 and 1.40 Å) but are significantly lower than those found for the HF/3-21G fragment-complex constrained optimization (both ~1.49 Å).[25] The C6-C4 distance in the TS (2.65 Å) is much smaller than for reactant (3.79 Å) or product (3.35 Å) and has a small variation (0.025 Å). The averaged interatomic distances between distal atoms of the QM region (circled in Figure 3) as shown in Table 2, reveal a significant overall contraction (~0.2 Å) in the Asp27 (Cα) to nicotinamide ring (N1), i.e. across the active site, and complementary contractions in the other distances. This contraction was not accounted for in the previous constrained fragment-complex optimizations.[25] By contrast, differences in the measurements between reactant and product states are quite small except for an increase of ~0.4 Å in the N(FOL)–N_{N1}(NIC) distance reflecting the rehybridisation of the C6 atom between DHF and THF which effectively moves the pterin-ring sidechain away from the nicotinamide-ring plane.

3.1.4 Results: activation and reaction energies. We do not report an in depth analysis of the energy results from the ONIOM calculations here: this will be given elsewhere (Rostov, Cummins, Kobayashi, Vrevren, Frisch and Gready, to be published). However, for comparison with results for the free energy surface calculations, we note the following activation and reaction energies (averaged over the 20 configurations): 37.3±4.4 and 19.5±4.1 kcal/mol for ONIOM-ME (HF-3-21G:Amber) and 28.4±4.3 and 9.5±4.1 kcal/mol for single-point ONIOM-EE (HF-3-21G:Amber) calculations using the ONIOM-ME (HF-3-21G:Amber) geometries. These values were calculated from the QM part of the total ONIOM energy: the latter energies contain additional MM terms (eqn (2)) and are ~3 and 5 kcal/mol higher for ONIOM-ME and ONIOM-EE, respectively. The difference, i.e. lowering of the activation and reaction energies with ONIOM-EE, reflects the effects of polarization of the QM region by the enzyme.

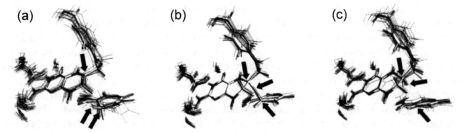

Figure 4 *3-D-representation of 20 superimposed geometries of the ONIOM QM region (Fig. 3) at the (a) reactant, (b) TS and (c) product stationary points of the DHFR hydride transfer to DHF obtained from the ONIOM-ME (HF/3-21G:Amber) geometry optimizations. One structure is shown in stick representation for easy viewing; the rest in wireframe. Black arrows point to the C6 atom of substrate DHF, the H4 atom of the hydride ion being transferred and the C4 atom of the cofactor NADPH.*

Table 1 *The average arrangement, over 20 configurations, of the C6, H4 and C4 atoms and inter-ring dihedral angle in reactant, TS, and product states of the DHFR hydride transfer to DHF obtained from the ONIOM-ME (HF/3-21G:Amber) geometry optimizations.*

	Reactant	TS	Product
R(C6-H4), Å	3.071±0.307	1.245±0.018	1.088±0.005
R(C4-H4), Å	1.080±0.003	1.419±0.026	2.473±0.137
R(C6-C4), Å	3.788±0.195	2.648±0.025	3.346±0.117
α(C6-H4-C4), deg.	126.3±14.7	168.7±5.0	136.9±6.0
γ(C8a-C6-C4-N1), deg.	153.8±11.4	152.6±8.3	153.3±8.5

Table 2 *The average distances in Å, over 20 configurations, between terminal atoms[a] of the QM region in reactant, TS, and product states of the DHFR hydride transfer to DHF obtained from the ONIOM-ME (HF/3-21G:Amber) geometry optimizations.*

	Reactant	TS[b]	Product[c]
C_α(Asp27)–C_β(Thr113)	6.01±0.14	5.96±0.16 (-0.05)	6.00±0.15 (-0.01)
C_α(Asp27)–N(FOL)	9.47±0.24	9.37±0.22 (-0.10)	9.46±0.25 (-0.01)
C_α(Asp27)–N_{N1}(NIC)	13.79±0.38	13.60±0.22 (-0.19)	13.82±0.29 (0.03)
C_β(Thr113)–N(FOL)	13.08±0.22	13.03±0.23 (-0.05)	13.20±0.28 (0.12)
C_β(Thr113)–N_{N1}(NIC)	12.31±033	12.19±0.23 (-0.12)	12.28±0.25 (-0.03)
N(FOL)–N_{N1}(NIC)	11.50±0.51	11.43±0.23 (-0.07)	11.87±0.37 (0.37)

[a] The terminal atoms of the QM region used for measurements are circled in Figure 3.
[b,c] In brackets, the difference between TS and reactant, and product and reactant values, respectively.

3.1.5   Assessment of ONIOM calculations.   We found that the ONIOM QM/MM method implemented in GAUSSIAN allowed efficient geometry optimisation of the QM active-site models, without the need to impose the artificial restraints commonly used for optimization of isolated fragment-complex models. This allowed TS-complex structures to contract. The method provides an attractive option for modelling enzyme reactions at a level of theory which can account for electronic (i.e. an *ab initio* model), enzyme-perturbation (i.e. with electronic embedding (EE), currently only with mechanical embedding (ME) optimized geometries), and enzyme-conformational effects (i.e. with configurational sampling from separate SE-QM/MM simulations) simultaneously. We have extended the core calculations, as reported here, to examine the reliability of the results with respect to variation of several parameters of the protocol such as higher electronic basis sets, different composition of the QM region, and different *ab initio* model (MP2, DFT) (Rostov, Cummins, Kobayashi, Vrevren, Frisch and Gready, to be published).

3.2 Generation of reaction free energy surface with QM/MM+MD and ONIOM QM/MM calculations

3.2.1 Motivation.   As critiqued in 1.8, MD simulations with SE-QM/MM potentials allow calculation of activation and reaction free energies with conformational sampling, but the semiempirical method provides an inadequate electronic representation of the QM region, particularly for π-delocalization and H bonding. To address these issues we have devised a method which combines SE-QM/MM MD simulations with single-point ONIOM-EE QM/MM (HF/6-31G* or B3LYP/6-31G*) energy calculations, using configurations sampled from points on the reaction hypersurface. Thus, we compute the statistical ensemble using computationally efficient SE-QM/MM MD simulations, and then compute the energy derivatives required for the free energy over a smaller subspace of R using more computationally expensive *ab initio* QM methods.

3.2.2 Method.   First we use SE-QM/MM+MD (MOPS[36]) to generate a free energy map, $G(r)$, as a function of two interatomic distances, r_1 and r_2, which define the reaction coordinate. For the DHFR reduction of DHF, r_1 and r_2 correspond to the C6-H4 and H4-C4 distances, respectively. The free energy changes, ΔG $(r, r \pm \Delta r)$, can be calculated as an ensemble average over the remaining R degrees, using either perturbation or integration formalisms, with determination of the free energies of activation and reaction requiring location of the stationary points of $G(r)$, i.e. $\nabla G(r) = 0$.

As shown in Figure 5, several simulations along curved paths, x, are undertaken to generate $G(x,\alpha)$ at discrete values of x and α: in this case we have used 21 values of α (i.e. $\Delta\lambda = 0.05$) for the numerical integration along pathways with x values 1.3, 1.5, 2.0, 2.5, 3.0 and 4.0. In earlier work we employed a similar scheme to scan the free energy surface.[27b] These points are then projected onto a grid in r_1 and r_2 to locate the stationary points of $G(r)$.

$$r_1 (x,\lambda) = r_1 (0) + [r_1 (1) - r_1 (0)][1 - (1 - \lambda)^x] \qquad r_1 (0) = 1.0 \text{ Å}, r_1 (1) = 3.4 \text{ Å} \qquad (3)$$

$$r_2 (x,\lambda) = r_2 (0) + [r_2 (1) - r_2 (0)]^{l^x} \qquad\qquad r_2 (1) = 3.4 \text{ Å}, r_2 (1) = 1.0 \text{ Å} \qquad (4)$$

We calculate the ensemble average of the λ derivative of E over the remaining (R) degrees of freedom and integrate to get the free energy in the $\{x, \alpha\}$ coordinate system:

$$dG(x,\alpha) = \qquad <\partial_\lambda E(R, r_1(x,\lambda), r_2(x,\lambda))/\partial\lambda > d\lambda \qquad (5)$$

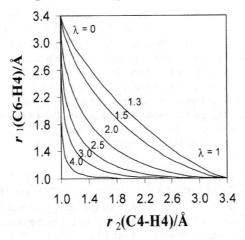

Figure 5 *Plot showing strategy for scanning the free energy surface $G(x,\alpha)$ for the DHFR hydride transfer to DHF in terms of two reaction coordinates, r_1 and r_2, and incremental calculation along curved paths x between $\lambda=0$ and $\lambda=1$.*

Next, we used the ONIOM-EE QM/MM (HF/6-31G* or B3LYP/6-31G*) method to improve the reliability of the SE-QM calculation. Single-point energies and finite-difference energy gradients at the $G(x,\alpha)$ points were calculated by ONIOM using sets of configurations sampled for each of the simulation points:

$$\frac{\delta E(r(\mathrm{x}, \lambda))}{\delta\lambda} = \frac{E(r(\mathrm{x}, \lambda + \delta)) - E(r(\mathrm{x}, \lambda - \delta))}{2\delta} \tag{6}$$

If E_1 and E_2 correspond to the SE-QM/MM and ONIOM energies, respectively, then, by integration, ΔG may be better estimated by:

$$dG = \left[\frac{1}{N}\sum_{i=1}^{N} \frac{\partial E_1(R_i)}{\partial\lambda} + \frac{1}{n}\sum_{j=1}^{n} \frac{\partial E_2(R_j)}{\partial\lambda} - \frac{1}{n}\sum_{j=1}^{n} \frac{\partial E_1(R_j)}{\partial\lambda} \right] d\lambda \tag{7}$$

where $N >> n$ & $n \in N$, if $\Delta(E_1 - E_2)$ is relatively small, or otherwise simply by:

$$dG = \left[\frac{1}{n}\sum_{j=1}^{n} \frac{\partial E_2(R_j)}{\partial\lambda} \right] d\lambda \tag{8}$$

3.2.3 Computational details. Our recent capability to perform multi-trajectory simulations routinely represents a major enhancement as it allows us to address the problem of configuration space sampling more effectively than in our previously reported studies.[25] In the present calculations we used 12 trajectories, sampling each at 1 ps intervals for 12 ps after at least 100 ps of equilibration, and simply used eqn (8) to evaluate the ensemble averages with the $n = 12 \times 12 = 144$ configurations. Independent multi-trajectory simulations were carried out for the 21 values of λ, giving rise to $21 \times 144 = 3024$ sample points for each of the six pathways, x (see Fig. 5), used to span the free energy surface. The details for the protein model are as for the ONIOM calculations described in 3.1.2. However, the solvation layer was extended from 33 to 36 Å (5115 water molecules and 17962 atoms in total). All atoms within 32 Å were included in the MD, leaving a 4 Å outer boundary layer of frozen water molecules. The QM region consisted of the complete NADPH and DHF$^+$ ligands, and the Asp27 side chain (-CH$_2$-COOH) with a hydrogen link atom. No cutoff was applied to the interaction between the QM and MM atoms. The calculations were performed with MOPS.[36]

3.2.4 Results. The plots in Figure 6 show ΔG for 3 values of x (1.5, 2.5 and 4.0) with E computed using PM3 (MOPS), HF/6-31G* (ONIOM-EE) and DFT-B3LYP/6-31G* (ONIOM-EE) for 21 values of λ. The average of the energy derivative (eqn (6)) at each λvalue was computed over 144 configurations, i.e. each x curve contains data from 6048 ($21 \times 2 \times 144$) single-point energy evaluations at PM3, HF or B3LYP level. The results highlight the importance of correlation (DFT *vs* HF) in lowering the activation barrier. PM3 performs much better than HF in reproducing the general shape of the DFT curve, providing some justification for using PM3 to generate the configurations from MD.

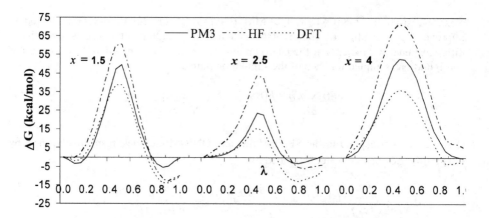

Figure 6   *Plots of ΔG for three curved slices (x=1.5, 2.5 and 4.0) of the free energy surface of the DHFR hydride transfer reaction to DHF. λ=0 and λ=1 represent the reactant and product wells, respectively; see Figure 5.*

The free energy surface, shown in Figure 7, contains all the information necessary to characterize the reactant, TS and product states. The raw path (x) data used to generate the free energy surface were least-squares fitted to a low order (quadratic or cubic) polynomial for each value of λ between 0.10 and 0.85. It was found inappropriate to fit the free energy profiles at λ values outside this range due to the fact that paths close to the end points ($\lambda = 0$ and $\lambda = 1$) rapidly converge (see Figs 5 and 7) forcing the actual small free energy differences to be comparable with statistical sampling errors. In these cases we simply averaged the free energy over x. Thus, by using these averages and the minimum of the polynomials together with a linear interpolation between λ we can follow the minimum free energy path and determine quite precisely values for x and λ that are appropriate for reactants, TS and products. The distances in Table 3 were obtained from eqns (3) and (4) using such values of x and λ. Further analysis of the ensemble average structures to extract values for other geometrical parameters, such as in Table 2, will be reported elsewhere (Cummins, Rostov and Gready, to be published).

As shown in Table 3, the agreement between the R(C6-H4) and R(C4-H4) distances obtained by MD and ONIOM geometry optimization is mostly quite good, although there is a significant deviation for R(C6-H4) at the TS. Further analysis is required to define the origin of such differences. Important issues to be addressed include differences in the underlying QM treatment (semiempirical *vs ab initio* QM), polarization effects, the choice of QM region, and the treatment of configurational sampling.

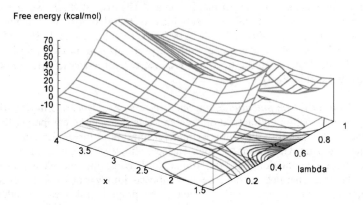

Figure 7   *3-D representation of the PM3 free energy surface. Free energy (kcal/mol) plotted as function of reaction pathway (x) and λ coupling parameter (lambda) for the hydride transfer to DHF in DHFR. The iso-energy contours (2-D projections in the x, lambda plane) show the reactant (λ ~ 0.1) and product ((λ ~ 0.8) potential wells.*

Table 3   *Distances and free energies (relative to reactants) obtained from the PM3 QM/MM free energy surface.*

	Reactant	TS	Product
R(C6-H4), Å[a]	3.05 (-0.02)	1.50 (0.25)	1.09 (0.00)
R(C4-H4), Å[a]	1.08 (0.00)	1.41 (-0.01)	2.51 (0.04)
Relative free energy (kcal/mol)[b]	0	26 (12, 2)	-5 (25, 15)

[a] Distances in parentheses are deviations from the *ab initio* ONIOM-ME optimized geometry results given in Table 1. [b] Relative energies in parentheses are for ONIOM-ME and ONIOM-EE HF/3-21G calculations from 3.1.3.

The same considerations apply to the comparisons of the free energies with the energy differences obtained from the ONIOM geometry optimizations. The results in Table 3 (and section 3.1.3) show an ONIOM-EE (HF/3-21G) activation energy of ~28 kcal/mol, which is not much different from the PM3 free energy, but a reaction energy of ~10 kcal/mol compared with the PM3 free energy of ~-5 kcal/mol. However, we can also obtain initial estimates of reaction and activation free energies, respectively, for the ONIOM-EE HF/6-31G* and ONIOM-EE B3LYP/6-31G* surface calculations from the x=1.5 and 2.5 curves in Figure 6. If the former are compared with the ONIOM-EE (HF/3-21G) geometry optimization results then a different pattern is evident: ~45 and ~28 kcal/mol, and ~-10 and 10 kcal/mol for the activation and reaction energies, respectively. The energies are similar for the HF and DFT reaction energies (about –10 to –15 kcal/mol) but very different for the HF (~45 kcal/mol) and DFT (~15 kcal/mol) activation energies, indicating a strong contribution from correlation. For reference, activation and reaction free energies of 13.4 and -4.4 kcal mol, respectively, for the DHFR-catalyzed reduction of DHF can be estimated from experimental data.[25] Clearly although we now have the capacity to perform such extensive calculations as reported for the two new example protocols combining *ab initio* ONIOM QM/MM calculations and SE-QM/MM MD simulations calculations, convergence with respect enzyme model (particularly QM region), QM model and sampling treatment requires extensive analysis (Cummins, Rostov and Gready, to be published).

3.2.5 Perspectives on potential of enzymic free energy surfaces. Enzymic free energy surfaces provide reaction and activation free energies that can be compared directly with experiment. Consequently, they are extremely important for understanding enzymic reaction mechanisms. Generation of such surfaces, however, requires many simulations and a very large computational effort. To gain further insights into what drives enzymic reactions, the free energies need to be decomposed into various components, and the approximations used to generate them need to be properly assessed. To address these issues, we propose initially to perform additional free energy simulations along a single pathway, the minimum free energy pathway obtained at the PM3-QM/MM level, with decomposition into QM and QM/MM (residue) interaction contributions. These simulations can be carried out using different choices for the QM region and energy derivatives at higher levels of QM theory in order to gauge the importance of charge transfer and polarization contributions to the free energy.

3 CONCLUSION

Our aim in this chapter has been to analyze the challenge of simulating enzymic reactions realistically, by reviewing the complexity of the molecular problem in the context of the current computational tools and computing power available. Our goals are not merely to be able to reproduce experimental data, although, of course, this is the ultimate test of the reliability of a simulation, but to go beyond experiment in providing a more detailed description of "how enzymes work", which is not accessible by experiment (1.3). Key issues identified are the nature of the "solvent" interaction of the substrates with the active-site and wider enzyme, particularly polarization and charge transfer (i.e. electrostatics), and the implications of its anisotropy, as well as the contributions of protein conformation flexibility and dynamics. To explore these issues computational methods are required which allow computation of free energies, and decomposition of these energies into component contributions (1.2 and 1.4).

To this end we have developed a number of approaches and tools and trialled them on several enzyme problems, most notably DHFR. In this chapter we introduce two new methods and protocols, in application to the DHFR mechanism, which integrate ONIOM *ab initio* QM/MM calculations with semiempirical QM/MM MD simulations and exploit the recent power of cluster computing to perform multi-trajectory simulations, geometry optimization of large QM complexes at *ab initio* level, and thousands of single-point *ab initio* QM calculations. For both methods, particularly the free energy surface method, the initial results we have presented here allow insights not previously available. Both have potential for further development and extension to study convergence against several parameters (3.1.5 and 3.2.4), and for analysis of the magnitudes of possible contributions (both molecular and physical) to reaction facilitation from energy decomposition.

References

1   A. Warshel, *J. Biol. Chem.* 1998, **273**, 27035.
2   W.W. Cleland, P.A. Frey PA and J.A. Gerlt, *J. Biol. Chem.* 1998, **273**, 25529.
3   W.R. Cannon and S.J. Benkovic, *J. Biol. Chem.* 1998, **273**, 26257.
4   A. Warshel, *Annu. Rev. Biophys. Biomol. Struct.* 2003, **32**, 425.
5   P.A. Kollman, B. Kuhn, O. Donini, M. Perakyla, R. Stanton and D. Bakowies, *Acc. Chem. Res.* 2001, **34**, 72.
6   K. Sharp, *Prot. Sci.* 2001, **10**, 661.
7   D. Blow, *Struct. Fold Des.* 2000, **8**, R77.
8   M.R. Sawaya and J. Kraut, *Biochemistry* 1997, **36**, 586.
9   J.R. Schnell, H.J. Dyson and P.E. Wright, *Annu. Rev. Biophys. Biomol. Struct.* 2004 **33**, 119.
10 S.J. Benkovic and S. Hammes-Schiffer, *Science* 2003, **301**, 1196.
11 S. Dapprich, I. Komaromi, K.S. Byun, K. Morokuma and M.J. Frisch, *THEOCHEM - J. Mol. Struct.* 1999. **462**, 1.
12 M.J. Frisch, et al., *Gaussian 03*, 2003, Gaussian, Inc., Pittsburgh PA.
13 J.J.P Stewart, *Int. J. Quant. Chem.* 1996, **58**, 133.
14 J.J.P. Stewart, *MOPAC 2000*, 1999, Fujitsu Limited, Tokyo.
15 V.L. Bugaenko, V.V. Bobrikov, A.M. Andreyev, N.A. Anikin, V.M. Anisimov, *LocalSCF*, 2003 version 1.0, Fujitsu Limited, Tokyo.
16 N.A. Anikin, V.M. Anisimov, V.L. Bugaenko, V.V. Bobrikov, A.M. Andreyev, *J. Chem. Phys.* 2004, **121**, 1266.
17 Y. Sugita and Y. Okamolto, *Chem. Phys. Lett.* 1999, **314**, 141.
18 S. Piana, D. Bucher, P. Carloni and U. Rothlisberger, *J. Phys. Chem. B* 2004, **108**, 11139.
19 J.Gao and M.A. Thompson, eds., *Combined Quantum Mechanical and Molecular Mechanical Methods*, ACS Symposium Series, 1998, Vol. 712.
20 G. Monard, X. Prat-Resina, A. Gonzales-Lafont and K.M. Lluch, *Int. J. Quant. Chem.* 2003, **93**, 229.
21 J.R. Shoemaker, L.W. Burggraf and M.S. Gordon, *J. Phys. Chem.* 1999, **103**, 3245.
22 T. Clark, *THEOCHEM-J. Mol. Struct.* 2000, **530**, 1.
23 W. Weber and W. Thiel, *Theor. Chem. Acc.* 2000, **103**, 495.
24 P.L. Cummins and J.E. Gready, *J. Am. Chem. Soc.* 2001, **123**, 3418.
25 P.L. Cummins, S.P. Greatbanks, A.P. Rendell and J.E. Gready, *J. Phys. Chem. B* 2002, **106**, 9934.
26 P.L. Cummins and J.E. Gready, *J. Comput. Chem.* 2005, **26**, 561.

27 P.L. Cummins and J.E. Gready, *J. Comput. Chem.* a) 1997, **18**, 1496; b) 1998, **19**, 977; c) 1999, **20**, 1028.

28 F.J. Luque, N. Reuter, A. Cartier and M.F. Ruiz-Lopez. *J. Phys. Chem. A* 2000, **104**, 10923.

29 S.J. Titmuss, P.L. Cummins, A.P. Rendell, A.A. Bliznyuk and J.E. Gready, a) *Chem. Phys. Lett.* 2000, **320**, 169; b) *J. Comput. Chem.* 2002, **23**, 1314.

30 J. Zuegg, A.A. Bliznyuk and J.E. Gready, *Mol. Phys.* 2003, **101**, 2437.

31 K.A. Brown and J. Kraut, *Faraday Discuss.* 1992, **93**, 217.

32 R.L. Blakely, *Adv. Enzymol.* 1995, **70**, 23.

33 J.E. Gready, *J. Comput. Chem.* 1985, **6**, 377.

34 J.E. Gready, *Biochemistry* 1985, **24**, 4761.

35 T. Vreven, M.J. Frisch, K.N. Kudin, H.B. Schlegel and K. Morokuma. *Mol. Phys.* submitted.

36 P.L. Cummins, *Molecular Orbital Programs for Simulations (MOPS)*, 1996, ANU, Canberra.

37 T. Vreven, K. Morokuma, O. Farkas, H.B. Schlegel and M.J. Frisch, *J. Comput. Chem.* 2003, **24**, 760.

THEORETICAL STUDIES OF PHOTODYNAMIC DRUGS AND PHOTOTOXIC REACTIONS

Rita C. Guedes,[1,2] Xiao Yi Li,[1] Daniel dos Santos[1,3] and Leif A. Eriksson[1,4]

[1]Department of Natural Sciences and Örebro Life Science Center, Örebro University, 701 82 Örebro, Sweden
[2]currently at: CECF, Faculdade de Farmácia da Universidade de Lisboa, Av. Prof. Gama Pinto, 1649-003 Lisboa, Portugal
[3]currently at: International University Bremen, School of Engineering and Science, 28725 Bremen, Germany.
[4]corresponding author: leif.eriksson@nat.oru.se

1 INTRODUCTION

Photodynamic therapy (PDT) has over the past two decades evolved into an important tool in the treatment of a large variety of diseases, including various skin disorders such as psoriasis, vitiligo and mycosis fungoides; tumors in skin, head and neck; or as antibacterial and antiviral agent.[1-3] Photochemical treatment is based on topical application of a photosensitive drug – generally a polycyclic heteroatomic aromatic compound – followed by irradiation in the UV-A and/or visible regions of the spectrum (320-400 and 400-760 nm, respectively). The photosensitizer will interact with the radiation and serve as an "antenna" leading excitation energy into the tissue in question and inducing a variety of reactions with the target molecules, often involving the generation of reactive oxygen species (ROS).

The initial step in the photochemical reaction sequence involves the excitation of the sensitizer to the first (or second) singlet excited state. The subsequent oxygen dependent or oxygen independent reactions may be divided as follows, and act in competition depending on the local surrounding of the photosensitizer:

i. Oxygen independent photoinduced binding to the target (*e.g.* DNA or lipids) directly through a low-lying excited singlet state. This occurs if the photosensitizer and the target system are in close contact, and primarily involves concerted C_4-cyclisation reactions between C=C double bonds on the sensitizer and the target. Direct photobinding has been observed to DNA, lipid membranes, and membrane receptors.[3-7]

ii. Oxygen dependent type I reactions. These are initiated by intersystem crossing from the S_1 state to the long-lived first excited triplet state T_1, followed by reduction of the system in the T_1 state to form the corresponding radical anion. This may subsequently transfer its excess electron to molecular oxygen, thereby generating the superoxide radical anion. This reaction occurs if there is a good electron donor nearby, and the oxygen concentration is relatively low.

iii.   Oxygen dependent type II reactions. In this case the excitation energy of the T_1 state of the photosensitizer is transferred directly to molecular oxygen, thereby generating reactive singlet oxygen ($^1\Delta_g$ state). This reaction will dominate if oxygen concentration is high, and no reducing agents are readily available.

A number of different photosensitizers have been proposed, the most common currently in use being based on the psoralen family (furocoumarins), or various porphyrin derivatives such as photophrin. Also other large compounds including anthrapyrazoles, isoquinoline alkaloids, phylloerythrins or perylenequinones have been suggested.[8] Among the latter, the compound hypericin, consisting of two fused anthracene moieties, has been extensively studied.[9] In addition, molecules can serve as unwanted photosensitizers. For example, a number of common non-steroid anti-inflammatory drugs (NSAIDs) such as ibuprofen, ketoprofen and diflunisal display UV-induced phototoxic reactions by initiating the generation of various ROS. These in turn give rise to cutaneous symptoms such as rashes, blisters, erythemia, and urticaria, as well as more serious problems such as mutations and melanoma.

Despite extensive research in the field, the exact reactions of each compound are still largely unknown, giving room for theory to assist in the elucidation of their properties and the possible mechanisms by which the different species act. In addition, knowing the mechanisms involved, computational chemistry can be employed in order to fine tune the photosensitizer's properties and to explore the chemistries of possible new compounds and their derivatives. We herein report upon our work involving excitation properties and possible oxygen dependent reactions of hypericin and forucoumarin derivatives, and upon the distribution of the latter in membranes and their direct photobinding to lipids.

2   METHODOLOGY

With the exception of the membrane distribution studies (section 4.1), all systems were explored using density functional theory (DFT), employing the B3LYP exchange-correlation functional.[10-12] Geometries were optimized at the B3LYP/6-31G(d,p) or B3LYP/6-31+G(d,p) levels, followed by time-dependent (TD-)DFT calculations of excited states at the B3LYP/6-31+G(d,p) or B3LYP/6-311+G(d,p) levels.[13-14] Zero-point vibrational energy calculations were performed at the level of optimization. All quantum chemical studies were performed using the Gaussian 03 program.[15]

For the MD studies we used the Gromacs program[16,17] to simulate the psoralen distribution in two different lipid bilayer models: a saturated membrane consisting of 64 dipalmitoyl-phosphatidylcholine (DPPC) lipids solvated by 1474 water molecules (potentials and initial bilayer patch by Söderhäll and Laaksonen[18]) and an unsaturated membrane containing 128 1-palmitoyl-2-linoleoyl-*sn*-glycero-3-phosphatidylcholines (PLPC) solvated by 2453 water molecules (potentials and initial bilayer patch by A. Rouk and co-workers[19]).

The Gromacs force field was used throughout. In the cases where oxygen interaction parameters in the psoralen heterocycles were lacking, potentials between chemically similar atoms of nitrogen containing heterocycles were employed (after initial testing). Partial charges on the atoms of these molecules were obtained through B3LYP/6-311+G(2df,p) calculations using the Gaussian 03 program.[15] The simulation parameters (NPT simulations at T=300 K, Nose-Hoover T-coupling, semi-isotropic Parinello-Rahman P-coupling and particle-mesh Ewald summation for the electrostatics) are similar to those employed in previous work.[19] The different systems were constructed by inserting a psoralen molecule in the bilayer middle (the lowest density zone) and equilibrated for 2 ns. For each system, a 20 ns production run followed in which the system trajectories were collected every 0.2 ps.

3 PSORALEN COMPOUNDS and REACTIVE OXYGEN SPECIES

Psoralens, or furocoumarins, are a class of compounds found in many different plants throughout the world. They are, *e.g.*, utilized in photochemotherapy of skin disorders, in pathogen eradication technology and in extracorporeal treatment of leukemia. In a recent study,[20] we explored the photochemical properties of a large set of psoralen derivatives, some of which are displayed in Fig. 1.

The computed singlet and triplet excitation data was found to agree to within ca 0.2 eV with experimental data, and revealed two bands of low-lying singlet excitations with weak to intermediate transitions at around 3.6-4 eV and 4.1-4.3 eV, and a band of stronger, degrading, transitions around 5 eV (250 nm). The first excited triplet lies ca 2.8 eV above the ground state for all systems. The data is summarized in Tab.1. The increased conjugation resulting from the additional furan ring on the psoralens as compared with coumarin itself, lowers the energy of the singlet excitations, whereas the triplets are highly similar for all systems.

The substituent patterns on the different furocoumarins have slight effects on the singlet transitions, and may be applied as a means to fine tune the excitation energy by up to ca 0.3 eV. Comparing the data of the psoralens to the excitation energy of molecular oxygen (ca 0.98 eV[21]), as a means to generate $^1\Delta_g(O_2)$, we note that energetically all of the systems explored will have the capacity to transfer excitation energy and thus to generate singlet oxygen.

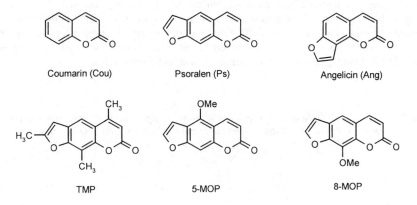

| Coumarin (Cou) | Psoralen (Ps) | Angelicin (Ang) |

| TMP | 5-MOP | 8-MOP |

Figure 1. *Set of furocoumarin (psoralen) derivatives described in the current work.*

Table 1.   *Low-lying vertical singlet and triplet excitation energies (eV and nm), and corresponding oscillator strengths f, for Cou, Ang, Ps, 8-MOP and TMP of Figure 1. The data is given as E (λ) [f], with the first row displaying the theoretical [20],and the second row (in italics) the experimental data [2].*

	Singlet excitations			T_1
Cou	4.13 (300) [0.12]	4.59 (270) [0.17]	5.39 (230) [0.02]	2.78 (446)
	3.96 (313) [0.06]			*2.70 (in water)*
Ang	3.86 (321) [0.04]	4.30 (289) [0.16]	5.09 (243) [0.43]	2.80 (443)
		4.13 (300) [0.17]	*5.04 (246) [-]*	
Ps	3.76 (330) [0.08]	4.34 (286) [0.22]	5.14 (241) [0.40]	2.79 (444)
	3.76 (330) [-]		*5.08 (244) [-]*	*2.72 (in water)*
8-MOP	3.57 (348) [0.02]	4.16 (298) [0.23]	4.98 (249) [0.25]	2.75 (450)
	3.54 (350) [0.04]	*4.09 (303) [0.22]*	*5.04 (246) [0.38]*	*2.72 (in ethanol)*
TMP	3.66 (339) [0.09]	4.28 (290) [0.20]	4.96 (250) [0.18]	2.84 (437)
	3.70 (335) [-]	*4.16 (298) [0.14]*	*4.96 (250) [0.27]*	*2.78 (in water)*

In addition to the generation of singlet oxygen, one of the key phototoxic reactions is that of superoxide generation. In Tab. 2 we report the vertical electron affinities of the subset of systems, computed in aqueous solution. The electron affinities of the ground states lie in the 2.0-2.4 eV range for the systems listed, and are hence moderately active oxidizing agents. However, when raised to the first excited triplet, the oxidizing power is considerably increased; to ca 5 eV. The excited psoralens will hence rather easily be able to abstract an electron from surrounding molecules. Having formed the furocoumarin radical anion, its reduction potential is only in the 2-2.4 eV range. Comparing this to the electron affinity of oxygen in aqueous solution, computed to be as high as 3.9 eV,[20] it is obvious that the reduced psoralens can contribute to superoxide radical formation. Interestingly, however, the oxidizing power of the T_1 states of the psoralens are such that these may become reduced by superoxide (unless direct reactions take place instead), thus giving a chain reaction of oxidations and reductions between excited furocoumarins and molecular oxygen. This has the net effect that phototoxiticity of the psoralens is quenched, which allows the psoralens to a larger extent to reach and interact with the target systems such as DNA, which promotes a predominance of direct photobinding.

Table 2.   *Vertical electron affinities (eV) of the ground and first excited triplet states of the systems in Tab. 1. All data computed at the IEFPCM-B3LYP/6-31+G(d,p) level.*

	VEA_{aq}	$VEA(T_1)_{aq}$
Cou	2.34	5.12
Ang	2.23	5.03
Ps	2.30	5.09
8-MOP	2.32	5.07
TMP	2.15	4.99

4   PSORALENS AND LIPID MEMBRANES

4.1   Distribution in membranes

For the drug to enter into the cell and reach DNA, it first needs to penetrate the lipid bilayer membrane. In order to explore some of the key features of furocoumarin interaction with lipid membranes, and the role of the substituent pattern of the compounds, MD simulation studies were undertaken of Ps, Ang, 5-MOP, 8-MOP and TMP in DPPC and PLPC membrane models.[22] The furocoumarins were initially all placed in the center of the lipid bilayer; however, within the 2 ns equilibration run, all systems moved from the middle of the bilayer to one of the water/phospholipid interfaces. This is unlike the behaviour of recently studied small molecules inside lipid bilayers,[23-25] and displays the amphiphilic character of the psoralens.

Although the molecules moved towards the interface, the probability to find one of these molecules near the heads of the phospholipids is very low, since this is the most dense region. The distributions of the molecules in the lipid bilayer models are shown in Figure 2. The density distributions have Gaussian shapes with maximum probabilities just inside the polar head group regions, close to the maximum distance of penetration for water molecules in the bilayer (the molecules move between regions 2 and 3 in the classification of Marrink and Berendsen[23]).

The large polarities of the furocoumarins imply that they will not diffuse near the apolar bilayer middle; instead, they are attracted by the polar medium near the interface. On the other hand, the large sizes of the molecules makes them avoid the most dense regions; the resulting distribution is a balance between these two contributions. Within this general trend we can nevertheless observe some differences between the distributions.

For all molecules studied, the maximum probability is located at roughly the same distance from the bilayer middle. The TMP molecule, with its larger amount of aliphatic substituents, is able to move more deeply inside both bilayers. This is also seen for the psoralen molecule in the DPPC bilayer. Within the DPPC bilayer the distribution of the angelicin molecule is shifted away from the center, indicating that the molecule prefers the more polar environment near the interface. Since the molecules move over a fairly large region (ca 17.5 Å) from the water/bilayer interface towards the center of the bilayer in a low frequency movement (it takes about 1-1.5 ns for the molecule to span this distance), long simulations are needed to correctly sample the distributions.

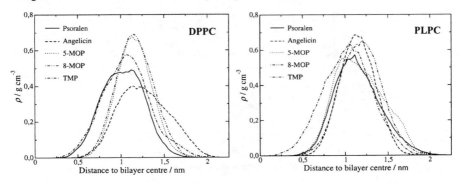

Figure 2.   *Density distribution for the psoralen molecules inside the DPPC and PLPC bilayers.*

Table 3.   *Self diffusion coefficients of the psoralen molecules inside the DPPC and PLPC lipid bilayers, in units of 10^{-5} cm^2 s^{-1}.*

	DPPC	PLPC
Ps	0.066 ± 0.001	0.060 ± 0.013
Ang	0.070 ± 0.007	0.055 ± 0.009
5-MOP	0.053 ± 0.003	0.045 ± 0.001
8-MOP	0.046 ± 0.001	0.052 ± 0.006
TMP	0.060 ± 0.014	0.076 ± 0.006

From Fig. 2 we note that in both bilayers the TMP molecule has the largest mean square displacement (MSD) whereas 5-MOP and 8-MOP have the lowest. Using the Einstein relation,[26] and fitting the MSD values between 50 and 100 ps to a straight line, we obtained the diffusion coefficients presented in Table 3. The diffusion coefficients follow the same general trend as the MSD data: the lowest values are found for the 5-MOP and 8-MOP molecules in both bilayers, and the TMP molecule has a large value in both.

The free energy profiles as a function of the distance to the bilayer centre was calculated using a potential of mean force formalism[27] and is displayed in Fig. 3. The free energy increases when the molecules move from the water/bilayer interface towards the middle of the bilayer. The largest barriers are found for the MOP molecules while TMP presents the lowest value. The methyl substituents in the TMP molecule gives a more lipid-like character giving more favourable interactions with the lipid hydrocarbon chains. Another feature of the free energy profiles is the fact that the free energy increases when the molecules move away from the water/bilayer interface towards the water layer. This behaviour is again consistent with a surfactant character and is in striking difference with the behaviour found for other molecules inserted in lipid bilayers where the free energy monotonically increases as the molecule moves from the water layer towards the centre of the bilayer, or increases at the interface and then decreases as the molecule moves towards the bilayer centre.[23-25]

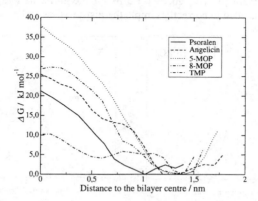

Figure 3.   *Free energy profiles for the psoralen molecules inside the DPPC bilayer.*

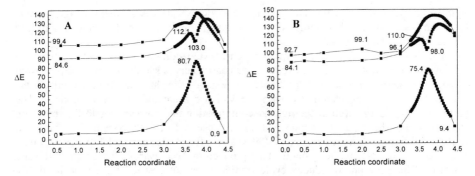

Figure 4. *Energy curves (kcal/mol) for thermal and photochemical cycloadditions: (A) tmp-fu-oame; (B) tmp-py-oame.*

The MD simulation studies show that the psoralens easily penetrate into lipid membranes, and that the more hydrophobic substituents give rise to very low barriers (ca 10 kcal/mol) for the molecules to traverse the lipid bilayer middle. This information opens up for the study of novel psoralen derivatives (currently under way in our laboratory), where membrane permeability is further enhanced, enabling the compounds to easily diffuse into the cells and interact with target DNA.

4.2 Direct photoinduced reactions

Once inside the membranes, it is clear that the compounds will be in close contact with the hydrocarbon tails of the fatty acids. Hence, should the drugs be exposed to UV-radiation while being inside the membranes, direct photoaddition to double bonds in the lipids, or excitation energy transfer to generate singlet oxygen, are likely events. The singlet oxygens can, in turn, react with unsaturated bonds of the fatty acids and give rise to lipid peroxidation reactions.[28] Photoinduced binding between psoralen compounds and lipid membrane constituents have been observed experimentally. In order to elucidate these reactions in more detail, experimental studies of photoreaction between trimethyl psoralen (TMP) and model system oleic acid methyl ester (OAME) have been reported, generating four isomeric products.[29] Two of these, formed in higher yield, were derived from the C_4-cycloaddition with OAME and show a local *cis, cis*-configuration at the site of addition, while the other two derive from eadilic acid methyl ester (EAME) formed by initial isomerization of OAME to EAME, and display a local *trans, cis*-configuration.

A theoretical explanation of the observed product yields was presented in our recent study, in which the excited state reaction surfaces for model systems consisting of TMP and *cis*- and *trans*-hex-3-ene were explored.[30] In these studies, the ground state addition reactions were fully characterized, followed by single point TD-DFT calculations along the ground state reaction coordinate in order to map the excited state surfaces. The systems will be referred to as tmp–fu/py–oame/eame, where fu/py indicates bonding to the furan or pyrone side of TMP.

The reaction energies (ΔE) for the concerted cycloadditions of the different systems showed that the reactions are essentially thermoneutral (furan adducts) or slightly endothermic (pyrone adducts). The ground state energy surfaces all display high barriers (between 73.5 and 82 kcal/mol), which means thermally induced cycloadditions are inaccessible for these systems.

The S_0, S_1 and S_2 energy curves for the photocycloaddition of two of the systems are displayed in Fig. 4. Although the excited state energy curves should not be strictly viewed as potential energy curves, since the excited state geometries have not been optimized, the overall features are in accordance with the van der Lugt and Oosteroff model.[31] The presence of an energy well at the nuclear configuration that corresponds to the transition structure along the thermal reaction pathway is observed in each system. This potential energy well traps the excited system and serves as a funnel through which the system may either go 'forwards' to the ground state product or 'backwards' to the ground state reactants. Thus, the photo-induced cycloaddition does not proceed via an electronically excited product, but rather through de-excitation from the decay channel to the ground state TS structure. Strictly speaking, the photochemical decay channel corresponds to a conical intersection where the ground- and excited-state potential energy surfaces are degenerate.[32,33]

The intrinsic photoreactivity of the psoralens is determined by the electronic structure of the lowest excited states and primarily involves the 3,4 (3',4') double bond of the pyrone (furan) moiety. The $S_0 \rightarrow S_1$ excitation energies at the reactant complexes are all ca 84 kcal/mol, with the excitations of the tmp-fu series being slightly higher than those of the tmp-py series (albeit only by ca 0.5 kcal/mol). On the S_1 surfaces the systems must all overcome a barrier of 25-31 kcal/mol to reach the decay channel, which is probably too large.

Inspection of the computed data reveals that also the S_2 surfaces can be involved in the photoaddition processes. We furthermore note that these are distinctively different for the furan and pyrone systems. For the pyrone adducts, the S_2 surfaces are lower-lying, and reach a degenerate intersection with S_1 by overcoming a barrier of only 6-12 kcal/mol. The subsequent energy barriers separating the intersection points and the energy wells on the S_1 surfaces are 14-15 kcal/mol. For the furan systems, no degenerate intersections are observed between the S_1 and S_2 surfaces. A large barrier of 27-31 kcal/mol has to be surmounted on the S_1 surface to reach the decay channel to the ground state, whereas the S_2 surface lies significantly higher in energy throughout (and possibly also involves interaction with the S_3 surface).

Of all systems studied, the smallest barrier to reach the degenerate intersection on S_2 (6.5 kcal/mol), the smallest barrier to then reach the decay channel to the ground state on S_1 (13.9 kcal/mol), and the lowest lying decay channel, are all observed for tmp-py-oame. This explains the experimental result[29] that this adduct is formed in the highest yield. In the actual membrane, it is reasonable to assume that the systems will be closer packed than what is given by the free reactants in vacuum. The computed surfaces show that this will not influence adduct formation to the furan moieties. For the pyrone adducts, however, if the compounds are closer already from the start, they may be inside the well of the S_1/S_2 intersection point upon excitation, which thus increases the product yield even further.

The UV absorption spectra of the tmp-py-oame product derived from computed[30] and experimental[34] data are shown in Fig. 5. Subsequent photobinding from the first adduct, to form cross-linked membrane structures, will involve excitation at ca 270 nm for initial pyrone adducts (315 nm for initial furan adducts). Hence, the formation of a di-adduct from the main product of the first photoaddition requires energy in the UV-B regime, albeit still relatively low-lying.

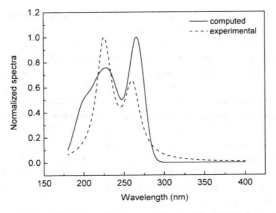

Figure 5. *Computed (solid line) vs experimental (dashed line) UV absorption spectra of tmp-py-oame.*

5   HYPERICIN DERIVATIVES – A NEW FAMILY OF PHOTOSENSITIZERS

5.1 Hypericin properties

Present limitations of porphyrin-based photodynamic photosensitizers have pressed the search for more appropriate compounds with low dark toxicity, selective accumulation in malignant tissues, proper retention time, absorption in the phototherapeutic window, and high triplet yield.[9,35,36] Hypericin (Fig. 6) and its derivatives hold very promising photochemical properties, which render them possible candidates as new phototherapeutic medicines.[37,38] Recent motivation for the study of hypericins is a result of their enormous pharmaceutical potentials as antiviral, antiretroviral, antibacterial, antipsoriatic, antidepressant and antitumor agents.

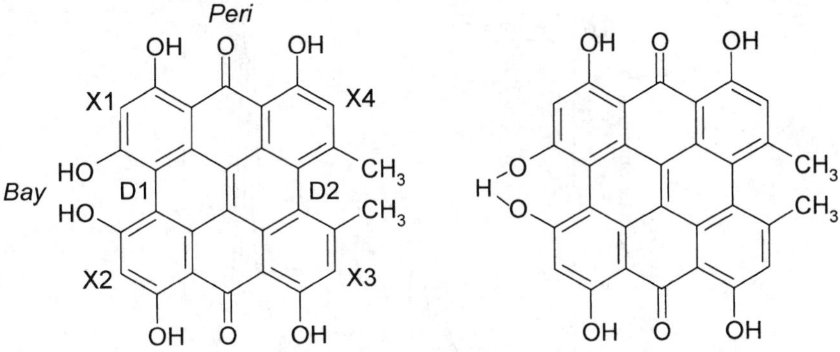

Figure 6. *The hypericin molecule (left) and its phenolate ion obtained from bay region deprotonation (right).*

The hypericin molecule is a widespread natural photoactive pigment that has long been used in traditional medicine, in the treatment of depression and wound healing,[9] and is a known antiviral agent against a large number of viruses such as hepatitis B,[39] herpes,[40,41] papilloma[42] and human immunodeficiency virus – HIV.[43] It is also effective in the case of T-lymphocyte-mediated diseases, such as multiple sclerosis, myasthenia gravis, rheumatoid arthritis, scleroderma, polymyositis, pemphigus, psoriasis, and transplant rejection.[44-46] The anticancer activity of hypericin has been demonstrated both in vitro and in vivo,[47,48] and it is currently being used in clinical trials for HIV.

Several mechanisms have been proposed for the photodynamic activity following irradiation of hypericin, including oxygen-dependent type I and type II reactions, and photoinduced excited state proton or hydrogen atom transfer (see Ref. 37 and references therein), but its mode of action still remains unclear. For example, the proximity of the enol and keto groups provides an environment that suggests excited-state intramolecular proton transfer or hydrogen atom transfer.[49-51]

We have studied the electronic properties and phototoxic reactions of the hypericin molecule (including determination of the most stable conformers and solvent effects) and the hypericin phenolate ion, and the effects of halogenation (with Cl, F, and Br) on their properties.[37,38] The most stable conformer is that with a hydrogen bond in the bay region (*cf.* Fig. 6), ca 3.7 kcal/mol lower in energy than the non hydrogen-bonded hypericin.[37] The carbon skeleton is highly rigid, with a slight twist introduced at the bay regions induced by the repulsion of the substituents (dihedrals D1 and D2 in Fig. 6, are of the order of 30 degrees). Deprotonation is possible at the phenolic hydroxyl groups in the peri or bay regions. However, these groups have very different acidities due to the formation of vinylogous carboxylic acid in the case of bay region deprotonation. The resulting phenolate ion (Fig. 6) is stabilized by a very strong hydrogen bond in contrast to the formation of a peri-phenolate ion which is destabilized by the interaction of the lone pairs of the carbonyl oxygen atoms. The standard Gibbs energy of reduction of hypericin in aqueous solution is associated with an energy gain of 42.7 kcal/mol, whereas the deprotonation reaction (giving $H^+(aq)$) is exoergic by close to 5 kcal/mol.[37]

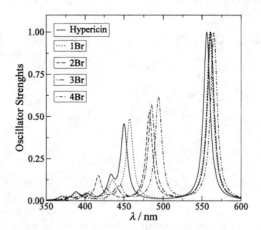

Figure 7.   *Calculated absorption spectra of bromo-substituted hypericins.*

The absorption spectrum of hypericin shows that the first singlet absorption band is in the phototherapeutic window 500-600 nm and is assigned to the first $\pi \rightarrow \pi^*$ transition (polarized along the short molecular C_2 axis of the hypericin molecular skeleton).[37,52] The first two transitions have the largest probabilities. Bay deprotonation has a small effect on the absorption band, which is shifted to slightly longer wavelengths.[37] This is concomitant with a reduction in twist angle D1 to 16 degrees. Investigation of a nearly planar hypericin derivative without the methyl group substituents in the D2 region revealed that the increased planarity did not affect the absorption properties to any significant degree.[53]

Our studies furthermore suggest that, besides the abovementioned direct reduction or deprotonation in aqueous solution, triplet hypericin is readily reduced (with an energy gain of ca 5 eV, as for the psoralens), followed by electron transfer from the ionized drug to molecular oxygen as a source of superoxide radical anions. In addition, triplet excitation energy transfer from hypericin to molecular oxygen generating singlet molecular oxygen is also a very viable reaction for these compounds.

5.2 Effects of halogenation

It is believed that the photosensitive properties of hypericin can be improved by either increasing the quantum efficiency of singlet oxygen/superoxide radical formation or by shifting the main absorption band of hypericin to higher wavelengths, in order to not be obscured by absorption of heme proteins. One suggestion to modify hypericin properties involves halogen substitution of the aromatic protons in positions X1-X4 (Fig. 6).

In our studies of F, Cl and Br substituted hypericin, it was concluded that the excitation energies generally decrease with increasing halogen substitution, as displayed in Fig. 7.[38] In the case of bromo substitution the shift in the absorption spectra is very similar to available experimental data.[54] In particular, the lowest lying (main) excitation is only blue shifted a few nm, whereas the second excitation is more strongly influenced through substitution, and gives an increasingly large red-shift with increasing halogenation. In addition, halogen substitution influences the electronic properties of the compounds, in that both the electron affinities and ionization potentials increase by 0.1-0.2 eV. Hence, the halogenated hypericins are more easily reduced than the parent compound. In addition, the triplet reduction potentials increase with increasing number of substituents (for example, 1Br < 2Br < 3Br < 4Br). Our results suggest that halogen substitution will influence neither the main excitation properties nor the triplet energy level, and thus not increase the quantum yield of $^1O_2$. However, given the increased reduction potential of the compound, halogen substitution should be able to provide an increase in superoxide levels through electron transfer reactions.[38]

Acknowledgements

The authors gratefully acknowledge funding from the Swedish science research council (VR) and the Portuguese foundation for science and technology (FCT). We also gratefully acknowledge generous grants of computing time at the national supercomputing facilities in Linköping (NSC) and Stockholm (PDC).

References

1   *The Science of Photomedicine*; J.D. Regan and J.A. Parrish, eds; Plenum, N.Y., 1982.
2   *CRC Handbook of Photochemistry and Photobiology*, CRC Press Inc.; Boca Raton, 1995.
3   S. Caffieri, D. Vedaldi, A. Daga and F. Dell'Acqua, in *Frontiers in Photobiology*, A. Shima, M. Ichahasci, Y. Fujiwara and H. Takebe, eds.; Excerpta Medica, Amsterdam, 1995.
4   F.P. Gasparro, in Ref 2.
5   G. Rodhigiero, F. Dall'Acqua and D. Averbeck, in *New Psoralen DNA Photobiology, Vol. I*, F.P. Gasparro, ed.; CRC Press, Boca Raton, 1988.
6   J.D. Laskin et al, *Proc. Natl. Acad. Sci. USA* 1985, **82**, 6158.
7   J.D. Laskin and E. Lee, *Biochem. Pharmacol.* 1991, **41**, 125.
8   R. Ebermann, G. Alth, M. Kreitner and A. Kubin, *J. Photochem. Photobiol. B: Biology* 1996, **36**, 95.
9   H. Falk, *Angew. Chem. Int. Ed.*, 1999, **38**, 3116, and references therein.
10   A.D. Becke, *J. Chem. Phys.* 1993, **98**, 5648.
11   C. Lee, W. Yang and R.G. Parr, *Phys. Rev. B* 1988, **37**, 785.
12   P.J. Stephens, F.J. Devlin, C.F. Chabalowski and M.J. Frisch, *J. Phys. Chem.* 1994, **98**, 11623.
13   M.E. Casida, in *Recent Advances in Density Functional Methods, Part I*, D.P. Chong, ed., World Scientific, Singapore 1995, 155.
14   R.E. Stratmann, G.E. Scuseria, and M.J. Frisch, *J. Chem. Phys.*, 1998, **109**, 8218.
15   M.J. Frisch, et al. (2003) *GAUSSIAN 03 Revision B.02*, Gaussian Inc., Pittsburgh, PA.
16   H.J.C. Berendsen, D. van der Spoel, and R. van Drunen, *Comp. Phys. Comm.* 1995, **91**, 43.
17   E. Lindahl, B. Hess and D. van der Spoel, *J. Mol. Mod.* 2001, **7**, 306.
18   J.A. Söderhäll and A. Laaksonen, *J. Phys. Chem.B* 2001, **105**, 9308.
19   M. Bachar, P. Brunelle, D.P. Tieleman and A. Rauk, *J. Phys. Chem.B* 2004, **108**, 7170.
20   J. Llano, J. Raber and L.A. Eriksson, *J. Photochem. Photobiol. A: Chemistry* 2003, **154**, 235.
21   K.P. Huber, G. Herzberg, *Molecular spectra and molecular structure. Part IV. Constants of diatomic molecules*. Van Nostrand, N.Y., 1979.
22   D. dos Santos and L.A. Eriksson, in preparation (2005).
23   S.-J. Marrink and H.J.C. Berendsen, *J. Phys. Chem.* 1994, **98**, 4155.
24   S.-J. Marrink and H.J.C. Berendsen, *J. Phys. Chem.* 1996, **100**, 16729.
25   D. Bemporad, J.W. Essex and C. Luttmann, *J. Phys. Chem.* 2004, **108**, 4875.
26   M.P. Allen and D.J. Tildesley *Computer Simulation of Liquids*, Oxford University Press: Oxford, 1990.
27   E. Paci, G. Ciccotti, M. Ferrario, R. Kapral, *Chem. Phys. Lett.* 1991, **176**, 581.
28   I. Tejero, A. González-Lafont, J.M. Lluch and L.A. Eriksson, *Chem. Phys. Lett.* 2004, **398**, 336.
29   K.G. Specht, W.R. Midden and M.R. Chedekel, *J. Org. Chem.* 1989, **54**, 4125.
30   X.Y. Li and L. A. Eriksson *Photochem. Photobiol.* 2005, submitted.
31   W.T.A.M. van der Lugt and L.J. Oosterhoff *J. Am. Chem. Soc.* 1969, **91**, 6042.
32   M.A. Robb, F. Bernardi and M. Olivucci, *Pure Appl. Chem.* 1995, **67**, 783.
33   M.A. Robb, and M. Olivucci, *J. Photochem. Photobiol. A: Chem.* 2001, **144**, 237.
34   S. Caffieri, D. Vedaldi, A. Daga and F. Dall'Acqua, in *Psoralens in 1988, Past, Present and Future*; T.B. Fitzpatrick, P. Forlot, M.A. Pathak and F. Urbach, eds; John Libbey Eurotext, Montrouge, France, 1988, 137.
35   Z. Diwu and J.W. Lown, *Pharmacol. Ther.*, 1994, **63**, 1.
36   Z. Diwu, R.P. Haugland, J. Liu, J.W. Lown, G.G. Miller, R.B. Moore, K. Brown, J. Tulip and M.S. Mcphee, *Free Radical Biology & Medicine*, 1996, **20**, 589.
37   R.C. Guedes and L.A. Eriksson, *J. Photochem. Photobiol. A: Chemistry*, 2005, in press.
38   R.C. Guedes and L.A. Eriksson, *J. Photochem. Photobiol. A: Chemistry*, 2005, submitted.
39   G. Moraleda, T.T. Wu, A.R. Jilbert, C.E. Aldrich, L.D. Condreay, S.H. Larsen, J.C. Tang, J.M. Colacino and W.S. Mason, *Antiviral Res.*, 1993, **20**, 235.
40   I. Lopez-Bazzocchi, J.B. Hudson and G.H.N. Towers, *Photochem. Photobiol.*, 1991, **54**, 95.

41 P.A. Cohen, J.B. Hudson and G.H.N. Towers, *Experientia*, 1996, **52**, 180.
42 D. Meruelo and G. Lavie, US-A 5506271 A, 1996 [*Chem. Abstr.* 1996, **124**, 307566g].
43 D. Meruelo, S. Degar, A. Nuria, Y. Mazur, D. Lavie, B. Levin and G. Lavie, in *Natural products as Antiviral Agents*; C.K. Chu and H.G. Cutler, eds; Plenum: New York, 99, and references therein.
44 D. Meruelo and G. Lavie, WO-A 9203049, 1992, [Chem. Abstr. 1992, **117**, 20506g].
45 D. Meruelo and G. Lavie, WO-A 9308797, 1993, [Chem. Abstr. 1993, **119**, 86035a].
46 D. Meruelo and G. Lavie, US-A 5514714 A, 1996, [Chem. Abstr. 1996, **125**, 26266r].
47 C.J. Thomas , L. Pardini, R.S. Pardini, in *Proceedings of the Third Biennial Meeting of the International Photodynamic Association,* Buffalo, NY, 1990.
48 C.D. Liu, D. Kwan, R.E. Saxton and D.W. Mcfadden, *J. Sur. Res.* 2000, **93**, 137.
49 D.S. English, K. Das, K.D. Ashby, J. Park, J.W. Petrich and E.W. Castner, Jr., *J. Am. Chem. Soc.* 1997, **119**, 11585.
50 D.S. English, K. Das, J.M. Zenner, W. Zhang, G.A. Kraus, R.C. Larock and J.W. Petrich, *J. Phys. Chem. A* 1997, **101**, 3235.
51 N.J. Wills, J. Park, J. Wen, S. Kesavan, G.A. Kraus, J.W. Petrich and S. Carpenter, *Photochem. Photobiol.* 2001, **74**, 216.
52 J.L. Wynn and T.M. Cotton, *J. Phys. Chem.*, 1995, **99**, 4317.
53 L.A. Eriksson, unpublished data.
54 H. Falk and W. Schmitzberger, *Monatsh. Chem.,* 1993, **124**, 77.

ACID/BASE-PROPERTIES OF RADICALS INVOLVED IN ENZYME-MEDIATED 1,2-MIGRATION REACTIONS

K. Nakata[1,2] and H. Zipse[1]

[1]Department of Chemistry and Biochemistry, LMU München, Butenandtstrasse 5-13, D-82131 München, Germany
[2]Current address: Science Research Center, Hosei University, Tokyo, Japan

1 INTRODUCTION

A growing number of enzymes have in recent times been shown to use radical intermediates in substrate turnover. In a number of the proposed substrate radicals the radical center is located adjacent to an electron withdrawing substituent such as a thioester or carboxylate group as depicted in structure **A** (Scheme 1).[1-3] We have recently shown that, due to the combined influence of the radical center and the thioester moiety, the acidity of the C-H bond located in β-position to the carbonyl group is significantly enhanced as compared to the α-C-H bond in closed shell thioesters, resulting in a pK_a for the β-C-H bond of 14 for R = CH$_3$S at 25 °C.[4]

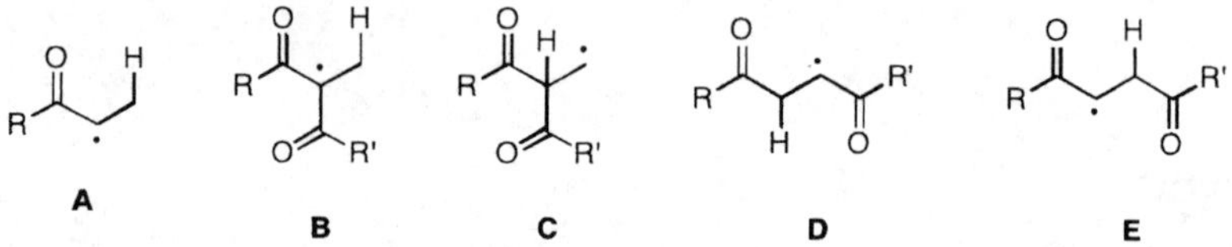

Scheme 1

This raises the question of even larger C-H bond acidities in radicals containing a second activating group. From a general point of view the combination of two electron withdrawing substituents with an ethyl radical fragment can give rise to four different combinations (**B** - **E** in Scheme 1). Radicals of structure **D** have been proposed as intermediates in the benzylsuccinate synthase reaction (R, R' = OH),[5] while radicals of structure **C** and **D** have been proposed as intermediates in the methylmalonyl-CoA mutase reaction (R = SCoA, R' = OH).[6] For the latter reaction a number of model systems have also been studied in basic alcohol solution (R = EtS or EtO, R' = EtO) in order to explore mechanistic aspects of this carbon skeleton rearrangement and the involvement of organocobalt intermediates.[7] Retey and coworkers have proposed that the enhanced C-H bond mobility in radicals of type **D** (R = SCoA, R' = OH) is responsible for the washing-in of deuterium at the C-2 and C-3 positions in the methylmalonyl-CoA reactant and the succinyl-CoA product during long-term incubation with methylmalonyl-CoA mutase in

D_2O.[8] The proposed 1,2-H migration leads to radicals of type **D** with interchanged R and R' groups (R = OH, R' = SCoA). The experiments have, in part, been performed with the methylmalonyl-carba-(dethia)-CoA substrates, in which the thioester sulfur atom has been replaced by a methylene group. This suggests that radicals of type **D** (R = CH_2CoA, R' = OH) have similar acidity properties as compared to the parent thioester radicals. This observation parallels the similarity of C-H bond acidities of closed shell ketones and thioesters of related structure.[9] In order to provide a quantitative basis for mechanistic models involving the deprotonation of radicals of general structure **B** - **E** we have now studied the acidity of these species in aqueous solution using a combination of theoretical methods.

2 THEORETICAL METHODS

The difference of the acidity of two species BH and AH in aqueous solution can conveniently be calculated from the free energy of reaction $\Delta G_{rxn,sol}$ for the thermodynamic cycle shown in Figure 1 using equation 1. It has recently been shown in a series of theoretical studies on the acidity of carboxylic acids and alcohols that the free energy difference in solution can conveniently be calculated from the gas phase reaction free energy $\Delta G_{rxn,gas}$ and the solvation free energies ΔG_{solv} for all reactants and products (equation 2).[10-14]

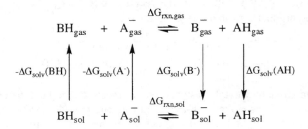

Figure 1 *Thermodynamic cycle for the calculation of relative pK_a values of acids BH and AH*

$$\Delta pK_a = pK_a(BH) - pK_a(AH) = \Delta G_{rxn,sol}/2.303RT \tag{1}$$

$$\Delta G_{rxn,sol} = -\Delta G_{solv}(BH) - \Delta G_{solv}(A^-) + \Delta G_{rxn,gas} + \Delta G_{solv}(B^-) + \Delta G_{solv}(AH) \tag{2}$$

$$\Delta G_{rxn,gas} = G_{gas}(B^-) + G_{gas}(AH) - G_{gas}(BH) - G_{gas}(A^-) \tag{3}$$

The gas phase reaction free energy can, in principle, be calculated using any quantum mechanical procedure. Very good results at acceptable computational effort have in the past been obtained using model chemistries of the G2, G3, or CBS type.[10,12,14] For carboxylic acids very convincing results have also been obtained using DFT methods.[13] For the open shell systems studied here particular care must be taken to minimize the effects of spin contamination. We are therefore using a variant of the G3(MP2)B3 method[15,16] whose performance has been described in detail by the original authors. The particular variant used here termed "G3(MP2)(+)-RAD(p)" is a minor modification of the method defined by Radom et al. for open shell systems.[17] Results obtained with this method for closed shell systems should be very similar to those obtained with the G3MP2B3 method. Our variant[4] includes B3LYP/aug-cc-pVDZ optimized geometries and unscaled frequencies instead of those at B3LYP/6-31+G(d,p) level, but uses the original single point calculations at the ROMP2/6-31+G(d) and ROMP2/G3MP2large level together with the UCCSD(T)/ROHF/6-31+G(d) single point energy. The G3(MP2)(+)-RAD(p) Gibbs free energy is thus calculated as defined in equation 4.

$$
\begin{aligned}
G_{gas}(G3(MP2)(+)\text{-}RAD(p)) = {} & E(UCCSD(T)/ROHF/6\text{-}31+G(d)) \\
& + (E(ROMP2/G3MP2large) - E(ROMP2/6\text{-}31+G(d))) \\
& + (G(B3LYP/aug\text{-}cc\text{-}pVDZ) - E(B3LYP/aug\text{-}cc\text{-}pVDZ))
\end{aligned} \tag{4}
$$

The higher level correction (HLC) as well as the spin orbit coupling correction (SO) also considered in the original formulation of G3 theory are of no relevance for the thermochemical cycle shown in Figure 1 and are thus not included in the free energy values reported here. Solvation free energies have been calculated using the CPCM continuum model in combination with UAHF radii.[18,19] In order to allow for structural relaxation of all systems in the presence of a polar medium such as water, all structures have first been optimized at the CPCM/UAHF/B3LYP/aug-cc-pVDZ level. Solvation energies have then been calculated from single point calculations at the CPCM/UAHF/BHandHLYP/6-31G(d) level (using the solution-optimized geometries) and gas phase BHandHLYP/6-31G(d)//B3LYP/aug-cc-pVDZ energies. The expression in equation 2 is fully equivalent to calculating the solution free energy $G_{sol}(X)$ of each species X as the sum of its gas phase free energy $G_{gas}(X)$ and its free energy of solvation $\Delta G_{solv}(X)$ (equation 5), and then expressing the reaction free energy $\Delta G_{rxn,sol}$ as a sum over the reactant and product contributions (equation 6).

$$
G_{sol}(X) = G_{gas}(X) + \Delta G_{solv}(X) \tag{5}
$$

$$
\Delta G_{rxn,sol} = -G_{sol}(BH) - G_{sol}(A^-) + G_{sol}(B^-) + G_{sol}(AH) \tag{6}
$$

For conformationally flexible systems such as those studied here the use of equation 6 has certain advantages as Boltzmann-averaged free energies $<G_{sol}(X)>$ for all species can be calculated in a straight forward manner from the solution free energies of all single conformers (equations 7 + 8):

$$
<G_{sol}> = \sum_{i=1}^{n} w_i G_{sol,i} \tag{7}
$$

$$
w_i = \frac{\exp(-G_{sol,i}/RT)}{\sum_{i=1}^{n} \exp(-G_{sol,i}/RT)} \tag{8}
$$

If not noted otherwise all resulted cited in the text are based on Boltzmann-averaged free energies at T = 298.15 K. The UCCSD(T)/ROHF/6-31+G(d)//B3LYP/aug-cc-pVDZ single point calculations have been performed with MOLPRO[20] and all other calculations have been performed with Gaussian 03.[21]

3 RESULTS AND DISCUSION

The pK$_a$ calculations reported here have been performed with dimethyl malonate **1** as the reference system (Scheme 2). For this system we can assume that the literature value for the acidity of diethyl malonate with pK$_a$ = 12.9 is practically identical to that of the (structurally simpler) dimethyl ester.[22]

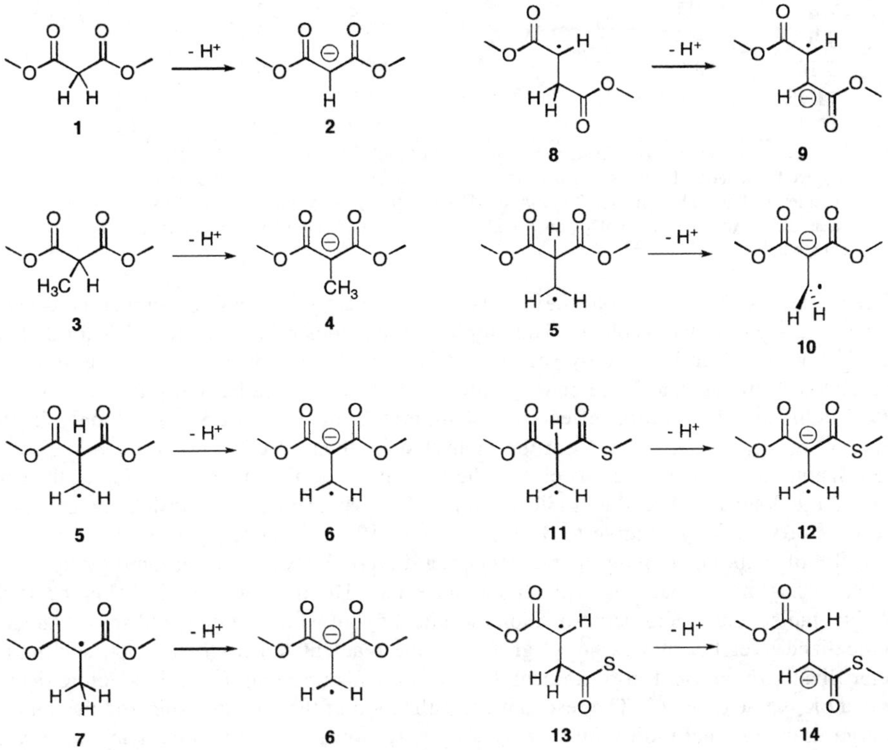

Scheme 2

As the first step in approaching the acidity of **D**-type substrate radicals we have calculated the α-C-H bond acidity in 2-methyl dimethyl malonate **3**. Somewhat surprisingly the C-H bond acidity for this seemingly simple system has not yet been studied experimentally. Using a method based on empirical increments Mayr and Bug estimate the pK$_a$ values of diethyl malonate and diethyl 2-methylmalonate to be rather small (12.9 vs. 13.1 in aqueous solution).[23] A somewhat larger difference between these two systems of 2.3 pK$_a$ units results from acidity measurements in DMSO (16.4 vs. 18.7).[24] Arnett et al. have studied the corresponding dimethylesters in DMSO and found a pK$_a$ difference of 2.2 between malonate and 2-methyl malonate (15.88 vs. 18.04).[25]

Table 1   *C-H Bond Acidities of the Carbonyl Compounds Shown in Scheme 2 as Calculated from the Thermodynamic Cycle in Figure 1 and Using Dimethyl Malonate **1** as the Reference System.*

System	$\Delta G_{rxn,gas}{}^{a}$	$\Delta pK_{a,gas}{}^{b}$	$\Delta G_{rxn,sol}{}^{c}$	$\Delta pK_{a}{}^{d}$	$pK_{a}{}^{e}$
1/2	0.0	0.0	0.0	0.0	+12.9
3/4	+7.1	+1.3	+13.7	+2.4	+15.3
5/6	-54.5	-9.5	-45.1	-7.9	+5.0
7/6	-15.5	-2.7	-14.8	-2.6	+10.3
8/9	-62.5	-10.9	-43.1	-7.6	+5.4
5/10	+7.1	+1.2	+12.0	+2.1	+15.0
11/12	-86.9	-15.2	-65.9	-11.6	+1.4
13/14	-95.5	-16.7	-62.0	-10.9	+2.1

[a]Reaction free energies as defined by equation 3 in kJ/mol. [b]Differences in pK_a values as defined by equation 1, but gas phase reaction free energies only. [c]Reaction free energies as defined by equation 2 in kJ/mol. [d]Differences in pK_a values as defined by equation 1. [e]Aqueous phase pK_a values at 298.15 K calculated using **1** as the reference compound.

These latter values are close to the results obtained here with a pK_a difference between **1** and **3** of 2.4 pK_a units (Table 1). This implies an aqueous pK_a value of +15.3 for **3**. This value is not much influenced by solvation effects as the pK_a difference between **1** and **3** calculated from gas phase free energies alone (+1.3 units) is rather close to that obtained after inclusion of solvation effects (+2.4 units). Radical **5** differs from **3** only in the removal of a hydrogen atom from the methyl substituent at C2. This primary radical is significantly more acidic than its closed shell parent **3** mainly due to formation of the spin and charge delocalized product radical anion **6**. This latter type of delocalization enhances the C-H bond acidity relative to **3** by more than 10 units, yielding a predicted pK_a for radical **5** of +5.0. Positioning the radical center directly between the carbonyl groups as in radical **7** has, in comparison, a much smaller effect. Deprotonation of **7** yields the same product radical anion **6** as deprotonation of radical **5**, but while the unpaired spin is already delocalized over the two carbonyl groups in the reactant radical **7**, no such stabilizing delocalization is present in **5**. Radical **7** therefore ends up being much less acidic than **5** with a pK_a value of +10.3. The resonant delocalization of the unpaired spin and the anionic charge in the product radical anion may also be responsible for the remarkable acidity of dimethyl succcinyl radical **8** with a pK_a value of +5.4. While spin delocalization is only possible within the adjacent carbonyl group in reactant radical **8**, a much larger resonance space exists in the product radical anion **9**. Whether resonant spin delocalization is indeed responsible for the acidity enhancements predicted for radicals **5**, **7**, and **8** can be assessed using geometry-constrained product radical anions such as **10**. The latter species is a true transition state for rotation of the methylene group on the B3LYP/aug-cc-pVDZ potential energy surface and is characterized through exclusive localization of the unpaired spin at the methylene carbon atom. Constrained radical **10** is 57 kJ/mol less stable than radical **6**. If deprotonation of radical **5** were to lead to **10** (instead of **6**), the α-C-H bond acidity would be practically identical to that of the closed shell parent system **3** (Table 1).

Closed shell thioesters are known to be more acidic than the corresponding oxygen-esters by ca. 4 pK_a units in aqueous solution.[9] Not surprisingly this trend is also visible in the systems studied here. Radical **11** differs from radical **5** only in the substitution of one of the ester groups through a thioester moiety. This substitution leads to a pK_a enhancement of 3.6 units and a predicted aqueous phase pK_a value for **11** of +1.4. The introduction of a thioester moiety in succinyl radical **8** lowers the pK_a by 3.3 units, yielding a predicted pK_a of +2.1 for radical **13**. The pK_a difference between **1** and **11** or **1** and **13** calculated for the gas phase is even larger than that calculated for aqueous solution (Table 1). This implies that the larger acidity of the thioester radicals is intriniscally a gas phase effect, whose consequences are moderated through the better solvation of a more concentrated charge distribution (as in **2**) as compared to that of a delocalized charge distribution (as in **12** or **14**). A similar statement can also be made for the corresponding oxygen-containing ester radicals **7** and **8** (Table 1).

The structures of the energetically most favorable conformers for the species **11 - 14** as optimized at the CPCM/UAHF/B3LYP/aug-cc-pVDZ level of theory in water are shown in Figure 2.[26] The most pronounced structural changes observed upon deprotonation of **11** at C2 are the elongation of the carbonyl C-O bonds (e.g. 1.214 to 1.246) as well as the shortening of the C-C bonds connecting the carbonyl groups and the center of deprotonation (e.g. 1.549 to 1.446). This can readily be understood as a consequence of the delocalization of anionic charge created at C2 to the two ester carbonyl groups. A similar degree of shortening can also be observed for the C-C bond connecting the formal radical center with the center of deprotonation (1.491 to 1.424). This may also be interpreted as a consequence of charge delocalization into the radical center as well as spin delocalization into the malonyl π-system. However, this latter effect appears not to be very dominant as the spin density located at the terminal methylene group is only marginally smaller in radical anion **12** (0.93) as compared to neutral radical **11** (1.16).

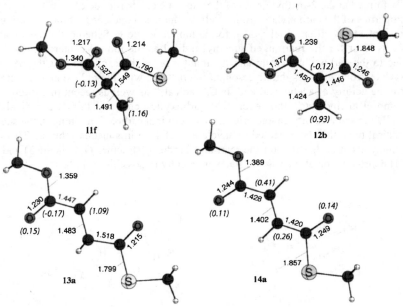

Figure 2 *Structures of the energetically most favorable conformers of radicals 11 - 14 as optimized at the CPCM/UAHF/B3LYP/aug-cc-pVDZ level of theory in water. Bond distances are given in Ångstroms. Atomic spin densities are given (in brackets) for the most relevant centers*

Similar structural changes accompany the deprotonation of succinyl radical **13**. The pronounced elongation of the formal carbonyl C-O double bonds as well as contraction of the connecting carbon framework can again be understood as a consequence of the delocalization of anionic charge towards the carbonyl oxygen atoms. The delocalization of negative charge is now accompanied by an equally pronounced delocalization of unpaired spin density, which can mainly be found at the central two carbon atoms in radical anion **14**. The spin density distributions shown in Figure 2 have been derived from the Kohn-Sham orbitals calculated within the framework of the CPCM/UAHF/B3LYP/aug-cc-pVDZ solvation model. The spin densities calculated at the corresponding gas phase level are, however, largely similar and deviate usually less than 10% from the solution phase values. From the analysis of the gas and solution phase pK_a differences (Table 1), the structural characteristics of the neutral radicals and their radical anions, and the spin density distributions (Figure 2) we may thus conclude that the reasons for the radical-induced enhancements of C-H bond acidity in malonic and succinic ester derivatives can mainly be found in the possibility of charge delocalization into (in **5** and **11**) and through (in **8** and **13**) the radical center.

How relevant are these findings for the substrate reaction of methylmalonyl-CoA mutase? All available structural data of the active site of methylmalonyl-CoA indicates that the substrate carboxylate group forms a salt bridge to an adjacent arginine side chain (Arg207).[27-30] While the mechanism of migration of the thioester moiety may be largely unaffected by the state of protonation of the substrate carboxylate group,[31] the presence of a negatively charged carboxylate group as present in a salt bridge must be expected to be largely detrimental to the C-H bond acidity at the C2 position. Since the radical center initially formed at the 2-methyl group of the substrate can hardly be expected to change the pK_a properties of the carboxylate group itself, we have to conclude that the salt-bridge will likely be intact also after initial radical formation (structure **F**, Scheme 3). The situation is much less clear for the rearranged succinyl radical as the radical center is now directly attached to the carboxylate group and it cannot *a priori* be excluded that the pK_a of the carboxylate group changes due to resonant interactions with the radical center. Literature data for the acidity of acetic acid and its C2 radical,[32] as well as recent measurements by Newcomb et al. indicate,[33] however, that carboxylic acids and their C2 radicals have practically the same acidity characteristics. The terminal carboxylate group in the succinyl-CoA radical must thus be expected to have a pK_a of 4 - 5 in aqueous solution. This would again imply a salt bridge also for the rearranged radical (structure **G**, Scheme 3) and make the C-H deprotonation at the anionic stage practically impossible.

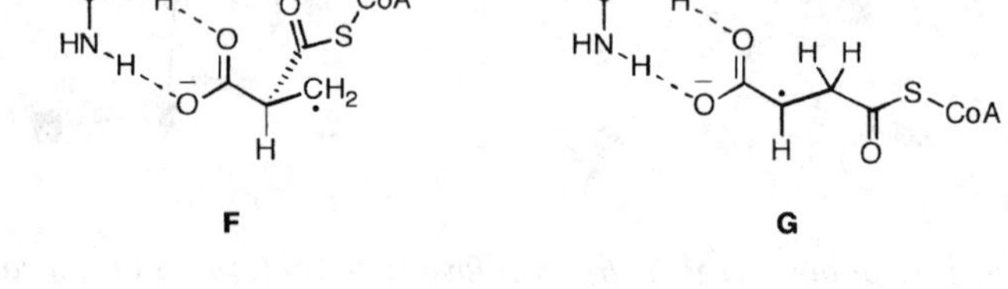

Scheme 3

4 CONCLUSION

The C-H bond acidity of radicals **B** - **E** generated from methylmalonyl or succinyl esters far exceeds that of the closed shell parent compounds, with particularly low pK_a values being predicted for the hemithioesters. The acidities of these species also exceed those of the related monothioester radicals of type **A** by more than 10 pK_a units. The acidity enhancement derives mainly from the added possibility for delocalization of the negative charge and to a minor extent also to the possibility of delocalization of the unpaired spin.

References

1   W. Buckel, B. T. Golding, *FEMS Microbiol. Rev.* **1999**, *22*, 523 - 541.
2   P. A. Frey, *Annu. Rev. Biochem.* **2001**, *70*, 121 - 148.
3   R. Banerjee, *Biochemistry* **2001**, *40*, 6191 - 6198.
4   D. M. Smith, W. Buckel, H. Zipse, *Angew. Chem.* **2003**, *115*, 1911 - 1945; *Angew. Chem. Int. Ed.* **2003**, *42*, 1867 - 1870.
5   B. Leuther, C. Leutwein, H. Schulz, P. Hörth, W. Haehnel, E. Schlitz, H. Schräger, J. Heider, *Mol. Microbiol.* **1998**, *28*, 615 - 628.
6   R. Banerjee, *Chem. Rev.* **2003**, *103*, 2083 - 2094.
7   (a) G. Choi, S. Choi, A. Galan, B. Wilk, P. Dowd, *Proc. Natl. Acad. Sci. USA* **1990**, *87*, 3174 - 3176. (b) P. Dowd, B. Wilk, B. K. Wik, *J. Am. Chem. Soc.* **1992**, *114*, 7949 - 7951. (c) M. He, P. Dowd, *J. Am. Chem. Soc.* **1996**, *118*, 711. (d) M. He, P. Dowd, *J. Am. Chem. Soc.* **1998**, *120*, 1133 - 1137.
8   W. E. Hull, M. Michenfelder, J. Retey, *Eur. J. Biochem.* **1988**, *173*, 191 - 201.
9   T. L. Amyes, J. P. Richard, *J. Am. Chem. Soc.* **1992**, *114*, 10297 - 10302.
10 I. A. Topol, G. J. Tawa, S. K. Burt, A. A. Rashin, *J. Phys. Chem. A* **1997**, *101*, 10075 - 10081.
11 G. Schüürmann, M. Cossi, V. Barone, J. Tomasi, *J. Phys. Chem. A* **1998**, *102*, 6706 - 6712.
12 (a) M. D. Liptak, G. C. Shields, *Int. J. Quantum Chem.* **2001**, *85*, 727 - 741. (b) A. M. Toth, M. D. Liptak, D. L. Phillips, G. C. Shields, *J. Chem. Phys.* **2001**, *114*, 4595 - 4606. (c) M. D. Liptak, G. C. Shields, *J. Am. Chem. Soc.* **2001**, *123*, 7314 - 7319. (d) M. D. Liptak, K. C. Gross, P. G. Seybold, S. Feldgus, G. C. Shields, *J. Am. Chem. Soc.* **2002**, *124*, 6421 - 6427. See also the comments on this approach in (e) J. R. Pliego, Jr., J. M. Riveros, *J. Phys. Chem. A* **2002**, *106*, 7434 - 7439. (f) J. R. Pliego, Jr., *Chem. Phys. Lett.* **2003**, *367*, 145 - 149.
13 G. A. A. Saracino, R. Improta, V. Barone, *Chem. Phys. Lett.* **2003**, *373*, 411 - 415.
14 A. M. Magill, K. J. Cavell, B. F. Yates, *J. Am. Chem. Soc.* **2004**, *126*, 8717 - 8724.
15 L. A. Curtiss, K. Raghavachari, P. C. Redfern, V. Rassolov, J. A. Pople, *J. Chem. Phys.* **1998**, *109*, 7764 - 7776.
16 A. G. Baboul, L. A. Curtiss, P. C. Redfern, K. Raghavachari, *J. Chem. Phys.* **1999**, *110*, 7650 - 7657.
17 (a) S. D. Wetmore, D. M. Smith, B. T. Golding, L. Radom, *J. Am. Chem. Soc.* **2001**, *123*, 7963 - 7972. (b) S. D. Wetmore, D. M. Smith, J. T. Bennett, L. Radom, *J. Am. Chem. Soc.* **2002**, *124*, 14054 - 14065.
18 V. Barone, M. Cossi, *J. Phys. Chem. A* **1998**, *102*, 1995 - 2001.
19 V. Barone, M. Cossi, J. Tomasi, *J. Chem. Phys.* **1997**, *107*, 3210 - 3221.
20 MOLPRO, a package of *ab initio* programs designed by H.-J.Werner and P. J. Knowles, version 2002.6, R. D. Amos, A. Bernhardsson, A. Berning, P. Celani, D. L. Cooper, M. J. O. Deegan, A. J. Dobbyn, F. Eckert, C. Hampel, G. Hetzer, P. J. Knowles, T. Korona, R. Lindh, A.W. Lloyd, S. J. McNicholas, F. R. Manby, W. Meyer, M. E. Mura, A. Nicklass, P. Palmieri, R. Pitzer, G. Rauhut, M. Schütz, U. Schumann, H. Stoll, A. J. Stone, R. Tarroni, T. Thorsteinsson, and H.-J. Werner.

21 Gaussian 03, Revision B.03, M. J. Frisch, G. W. Trucks, H. B. Schlegel, G. E. Scuseria, M. A. Robb, J. R. Cheeseman, J. A. Montgomery, Jr., T. Vreven, K. N. Kudin, J. C. Burant, J. M. Millam, S. S. Iyengar, J. Tomasi, V. Barone, B. Mennucci, M. Cossi, G. Scalmani, N. Rega, G. A. Petersson, H. Nakatsuji, M. Hada, M. Ehara, K. Toyota, R. Fukuda, J. Hasegawa, M. Ishida, T. Nakajima, Y. Honda, O. Kitao, H. Nakai, M. Klene, X. Li, J. E. Knox, H. P. Hratchian, J. B. Cross, V. Bakken, C. Adamo, J. Jaramillo, R. Gomperts, R. E. Stratmann, O. Yazyev, A. J. Austin, R. Cammi, C. Pomelli, J. W. Ochterski, P. Y. Ayala, K. Morokuma, G. A. Voth, P. Salvador, J. J. Dannenberg, V. G. Zakrzewski, S. Dapprich, A. D. Daniels, M. C. Strain, O. Farkas, D. K. Malick, A. D. Rabuck, K. Raghavachari, J. B. Foresman, J. V. Ortiz, Q. Cui, A. G. Baboul, S. Clifford, J. Cioslowski, B. B. Stefanov, G. Liu, A. Liashenko, P. Piskorz, I. Komaromi, R. L. Martin, D. J. Fox, T. Keith, M. A. Al-Laham, C. Y. Peng, A. Nanayakkara, M. Challacombe, P. M. W. Gill, B. Johnson, W. Chen, M. W. Wong, C. Gonzalez, and J. A. Pople, Gaussian, Inc., Wallingford CT, 2004.

22 A. Albert, E. P. Serjeant, *The Determination of Ionization Constants: A Laboratory Manual*, 3rd ed.; Chapman and Hall: London, **1984**; pp 137 - 160.

23 T. Bug, H. Mayr, *J. Am. Chem. Soc.* **2003**, *125*, 12980 - 12986.

24 F. G. Bordwell, J. A. Herrelson, A. V. Satish, *J. Org. Chem.* **1989**, *54*, 3101 - 3105.

25 E. M. Arnett, S. G. Maroldo, S. L. Schilling, J. A. Harrelson, *J. Am. Chem. Soc.* **1984**, *106*, 6759 - 6766.

26 A larger number of structures has been optimized for each of the four systems. The suffix added to the number of each structure (e.g. "f" for structure **11**) relates to the numbering system for all conformers. The structures of all conformers are available from the authors upon request.

27 N. H. Thomä, P. R. Evans, P. F. Leadlay, *Biochemistry* **2000**, *39*, 9213 - 9221.

28 M. D. Vlasie, R. Banerjee, *Biochemistry* **2004**, *43*, 8410 - 8417.

29 N. H. Thomä, T. W. Meier, P. R. Evans, P. F. Leadlay, *Biochemistry* **1998**, *37*, 14386 - 14393.

30 F. Mancia, N. H. Keep, A. Nakagawa, P. F. Leadlay, S. McSweeney, B. Rasmussen, P. Bösecke, O. Diät, P. R. Evans, *Structure* **1996**, *4*, 339 - 350.

31 S. D. Wetmore, D. M. Smith, L. Radmore, *ChemBioChem* **2001**, *2*, 919 - 922.

32 E. Hayon, M. Simic, *Acc. Chem. Res.* **1974**, *7*, 114 - 121.

33 N. Miranda, P. Daublain, J. H. Horner, M. Newcomb, *J. Am. Chem. Soc.* **2003**, *125*, 5260 - 5261.

DEVELOPMENT OF A HETEROGENEOUS DIELECTRIC GENERALIZED BORN MODEL FOR THE IMPLICIT MODELING OF MEMBRANE ENVIRONMENTS

M. Feig[1,2] and S. Tanizaki[1]

[1]Department of Biochemistry and Molecular Biology, Michigan State University, East Lansing, MI, 48824, USA
[2]Department of Chemistry, Michigan State University, East Lansing, MI, 48824, USA
feig@msu.edu

1 INTRODUCTION

The structure and dynamics of biological macromolecules is intimately linked to their interactions with the surrounding environment. In computer simulations of biomolecules in atomic detail the environment can be represented either explicitly or implicitly. Explicit environments are easily realized by placing a given molecule in a box of explicit water and/or other solvent molecules. Techniques such as periodic boundaries[1] and Ewald summation[2-5] have been developed to simulate such systems accurately and efficiently within the limits of the classical approximations.[6] The main drawback of explicit solvent simulations is the substantial cost for calculating solvent-solvent interactions in addition to solute-solute and solute-solvent interactions which limits the application of such methods with respect to feasible system sizes and simulation times.

As an alternative, implicit representations of the environment replace explicit solvent-solute interactions with a mean-field formalism. Implicit solvent offers computational advantages not just as a result of fewer explicit atomic interactions, but also by providing instantaneous averages over solvent degrees of freedom which are especially useful in scoring function applications. It is convenient to formulate an implicit solvent formalism by decomposing the solvation free energy of a given conformation into electrostatic, van der Waals, and entropy (cost of cavity) contributions [7-9].

The entropic cost of forming a cavity in solvent is approximated reasonably well as a simple linear function of the solvent-accessible surface area of the molecule [9, 10]. The van der Waals solute-solvent contribution has been included as part of the cavity term in previous formulations, however it is becoming clear that a separate treatment of solute-solvent van der Waals interactions is more accurate[9]. The dominant electrostatic contribution to the solvation free energy has received most of the attention. Most commonly, a continuum electrostatic description is used to derive an implicit term that accounts for solvent polarizability and the resulting polar interactions of a biomolecule with its environment[11]. In such a model a solvated biomolecule is represented as a set of partial charges that are embedded in a low dielectric (usually 1) and surrounded by a high dielectric (e.g. 80 for water) with the dielectric boundary located at the molecular surface [12]. The electrostatic potential and electrostatic solvation free energy for such a system can be calculated by solving the Poisson equation (or the Poisson-Boltzmann equation in the case of salt)[11]. While direct solutions to the Poisson equation are computationally expensive[13], more efficient approximations of the same underlying physical model based on the generalized Born (GB) formalism[14-20] have become popular over recent years[21].

In contrast to Poisson theory, Eq. 1 is limited to a simple two-dielectric model. As a result, applications have focused almost exclusively on biomolecules in aqueous solvent. An extension to more complex environments that require a heterogeneous dielectric environment such as integral membrane proteins interacting with lipid bilayers is not straightforward. First attempts to model membrane environments with GB models have focused largely on extending the solute cavity into the hydrophobic region of the lipid tails[22, 23]. Such models can approximate the effect of a hydrophobic layer in a high-dielectric environment fairly well[22, 24], however, they do not allow any polarization response within the lipid tail region ($\varepsilon > 1$) or intermediate dielectric regions. In the following we will describe an extension of the GB formalism that allows the representation of heterogeneous dielectric environments and is applied to the implicit modeling of dipalmitoyl-phosphatidylcholine (DPPC) bilayers.

2 THEORY

The electrostatic solvation energy in the original GB formalism is given as[15, 25]:

$$\Delta G_{elec} = -\frac{1}{2}\left(\frac{1}{\varepsilon_p} - \frac{1}{\varepsilon_w}\right)\sum_{i,j}\frac{q_i q_j}{\sqrt{r_{ij}^2 + \alpha_i \alpha_j \exp(-r_{ij}^2 / F\alpha_i\alpha_j)}} \tag{1}$$

where ε_p and ε_w are the internal and external dielectric constants, respectively, q_i are atomic partial charges, r_{ij} is the distance between two interacting atomic sites i and j, and α_i is the so-called GB radius of atom i. Eq. 1 by itself has been shown to approximate electrostatic solvation energies from Poisson theory very well[26]. However, the challenge in the successful application of the GB expression lies in the efficient calculation of GB radii α_i. According to the Born equation the GB radius of atom i is related to the electrostatic solvation energy for a system where all of the charges except i are set to zero[27]:

$$\alpha_i = -\frac{1}{2}\left(\frac{1}{\varepsilon_p} - \frac{1}{\varepsilon_w}\right)\frac{1}{G_{elec}^i} \tag{2}$$

While the G_{elec} can be calculated by solving the Poisson equation a more efficient approximation is possible with the Coulomb field approximation:

$$\frac{1}{\alpha_i} = \frac{1}{R_i} - \frac{1}{4\pi}\int_{solute,r>R_i}\frac{1}{r^4}dV \equiv A_4 \tag{3}$$

where R_i is an exclusion radius around atom i and the integral is carried out over the solute volume[14]. Recently, a correction to the Coulomb field approximation has been proposed and implemented in the GBMV method[16, 17]:

$$A_7 = \left(\frac{1}{4R_i^4} - \frac{1}{4\pi}\int_{solute,r>R_i}\frac{1}{r^7}dV\right)^{1/4} \tag{4}$$

GB radii are calculated from a combination of A_4 and A_7 as follows[16, 17]:

$$\alpha_i = \frac{1}{C_0 A_4 + C_1 A_7} + D \tag{5}$$

C_0, C_1, and D are parameters that were determined empirically in order to maximize the agreement with Born radii calculated from Eq. 2 based on results from Poisson theory. A comparison of Eq. 5 with the Kirkwood expression for the electrostatic solvation energy of a single off-center charge in a spherical cavity[28] suggests that GB radii α_i can be expressed as a function of internal and external dielectric constants by choosing[29]:

$$C_1 = C_1' \left(\frac{3\varepsilon_w}{3\varepsilon_w + 2\varepsilon_p} \right) \quad \text{and} \quad D = D' + \frac{E}{\varepsilon_w + 1} \tag{6}$$

The ability to calculate GB as a function of the external dielectric has prompted us to modify the original GB expression as follows[30]:

$$\Delta G_{elec} = -\frac{1}{2} \sum_{i,j} \left(\frac{1}{\varepsilon_p} - \frac{1}{\varepsilon_{ij}} \right) \frac{q_i q_j}{\sqrt{r_{ij}^2 + \alpha_i(\varepsilon_i)\alpha_j(\varepsilon_j) \exp[-r_{ij}^2 / F\alpha_i(\varepsilon_i)\alpha_j(\varepsilon_j)]}} \tag{7}$$

where ε_i is a local dielectric constant and ε_{ij} is defined as $(\varepsilon_i + \varepsilon_j)/2$. With Eq. 7 it is then possible to model heterogeneous dielectric environments. The value of ε_i at an atomic site i could be assigned simply from the dielectric constant at the closest dielectric boundary or in the case of a membrane slab geometry as a function of distance z from the membrane center perpendicular to the membrane. However, if ε_i is assigned in this manner in a system with multiple dielectric boundaries, the effect of additional polarization at each boundary would be neglected. In order to take polarization at multiple boundaries into account, we have introduced an effective dielectric function ε_{eff} that is determined by solving the Poisson equation for a probe ion at different locations in the heterogeneous dielectric environment[30]. In the case of a membrane environment this effective dielectric function only depends on z and needs to be calculated only once for a given dielectric profile. A continuous function $\varepsilon_{eff}(z)$ is then obtained by spline interpolation and used to assign $\varepsilon_i = \varepsilon_{eff}(z_i)$.

The new GB formalism for heterogeneous dielectric environments has been tested for the calculation of electrostatic solvation energies of a fixed conformation of the M2 channel-lining segment from nicotinic acetylcholine receptors in different orientations with respect to a membrane slab geometry[30]. Below we will present results from a more extensive set of test cases where electrostatic solvation energies according to Eq. 7 are compared to solvation energies obtained directly from Poisson theory.

When considering heterogeneous environments, the non-polar contribution to the solvation energy becomes more important than in a homogeneous environment. The cost of cavity formation decreases roughly with the dielectric constant of the solvent as the stronger interactions between polar solvent molecules incur a larger penalty for reorientation in order to accommodate a given biomolecule. The difference in the non-polar contribution to the solvation energy between the hydrophobic region of a membrane and water is usually much larger than the change in non-polar energy between different conformations of a given solute in water alone. In fact, it is the non-polar component that leads biomolecules to interact with the interior of a membrane since the electrostatic component is minimized where charge screening is largest, i.e. in the region with the highest dielectric constant. For a membrane bilayer the non-polar contribution to the solvation free energy can be estimated from explicit lipid/water simulations of O_2 insertion[31, 32]. The resulting free energy profile S(z), scaled to reach a value of one in bulk water, is then used to calculate the non-polar contribution to the solvation free energy for a given molecule with atoms i as follows:

$$\Delta G_{np} = \gamma \sum_i S(z_i) SA_i \qquad (8)$$

Where SA_i is the solvent-accessible surface area of atom i and γ is a factor set according to the cost of cavity formation in water. Based on previous tests[30] we chose $\gamma=15$ cal/mol/Å^2.

3 METHODS

All calculations presented here were carried out with the molecular mechanics program CHARMM[33], version c31b1, that was modified in order to incorporate extensions to the GBMV method to allow heterogeneous dielectric environments (called HDGB). The CHARMM force field for peptides, proteins, and amino acid side chain analogs[34] was employed throughout. No electrostatic cutoffs were applied in any of the calculations presented here. The PBEQ module[35] in CHARMM was used to solve the Poisson equation for membrane geometries.

Two test sets were used to test the agreement between the HDGB method and Poisson theory. The first and second test sets consist of representative conformations of bee venom melittin (50 structures) and *Halobacterium salinarium* bacteriorhodopsin (10 structures) sampled at 100 ps intervals from molecular dynamics simulations that were carried out with the HDGB method and have been described previously[30]. In addition we calculated electrostatic solvation energies for the crystal structures of three other integral membrane proteins: glycophorin A (PDB ID: 1AFO), outer membrane protein A (PDB ID: 1G90), and light-harvesting complex II (PDB ID: 1NKZ).

Simulations of explicit lipids in explicit water that were carried out in order to examine the true solvent-accessible volume of a water molecule in different regions of the bilayer employed the CHARMM force field for lipds[36] and were started from started from a pre-equilibrated system of 72 lipids with 2094 water molecules[37]. The initial box size was 47.5857 Å x 57.5857 Å x 66.8901 Å. The simulations were carried out in an NPT ensemble at a temperature of 50 °C and a pressure of 1 atm. Pressure and temperature were maintained with a Langevin piston[38] Electrostatic interactions were calculated with particle-mesh Ewald summation, with the direct sum and Lennard-Jones interactions truncated at 12 Å. The SHAKE algorithm was used to constrain heavy atom – hydrogen bonds along with an integration time step of 2 fs.

4 RESULTS AND DISCUSSION

Based the dielectric profile for a DPPC lipid bilayer that was calculated from explicit lipid/water molecular dynamics simulations[39] we use a dielectric profile of $\varepsilon=2$ (z=0-10 Å), $\varepsilon=7$ (z=10-15 Å), and $\varepsilon=80$ (z>15 Å)[30]. In addition, we will also investigate a slightly modified profile with $\varepsilon=20$ (z=15-20 Å) and $\varepsilon=80$ (z>20 Å) that reflects a more gradual increase of the dielectric constant as indicated in an earlier study[40]. The first profile will be denoted 2-7-80 and the second profile 2-7-20-80.

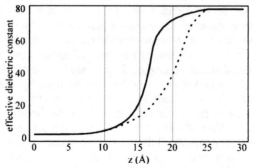

Figure 1   *Effective dielectric function for membrane slab geometry with dielectric profiles 2-7-80 (solid) and 2-7-20-80 (dashed) as a function of z in Å, the distance from the membrane center perpendicular to the membrane layer. A probe radius of 2 Å was used.*

The effective dielectric functions were calculated from Poisson theory and are shown in Fig. 1. The effective profiles were then used to calculate electrostatic solvation free energies with the HDGB formalism for the simulated melittin and bacteriorhodopsin conformations as well as selected experimental structures. The results are compared with Poisson theory in Fig. 2.

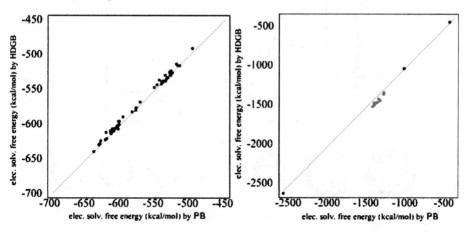

Figure 2   *Comparison of electrostatic solvation energies between HDGB model and Poisson theory for a 2-7-80 dielectric profile. Results for melittin conformations are shown on the left, results for bacteriorhodopsin conformations (red) and other membrane proteins are shown on the right.*

The results in Fig. 2 demonstrate that the HDGB formalism accurately reproduces the results from Poisson theory for different types of membrane proteins as well as different conformations of a given protein/peptide interacting with the membrane bilayer. These findings confirm and extend previous results where electrostatic solvation energies of the

M2 channel-lining segment from nicotinic acetylcholine receptors in different orientations with respect to the lipid bilayer were compared between HDGB and Poisson theory[30].

The HDGB model was tested further by calculating the membrane insertion free energy profile for a single water molecule. In this case, both the electrostatic and non-polar contributions are considered. It was noted previously[30] that the insertion profile for water is in qualitative but limited quantitative agreement with explicit lipid/water free energy calculations[31, 32, 41]. The agreement improves when the water oxygen radius that is defined to describe the dielectric boundary is increased towards the center of the lipid bilayer[30]. An effectively larger solvent-excluded volume in the lipid tail region is expected because the long lipid tails are unlikely to efficiently wrap around a small molecule like water. In order to quantify this point further we have carried out molecular dynamics simulations of an explicit lipid bilayer solvated in a box of water. In a series of simulations over 1 ns a single water molecule (TIP3P) was harmonically constrained to distances between $z=0$ (center of membrane) and $z=35$Å (bulk water). In each of these simulations the positions of surrounding solvent atoms (including their respective van der Waals radii) were counted on a three-dimensional grid. Regions of low density up to a given threshold were then integrated in order to obtain the effective solvent-excluded volume. The density threshold was chosen so that the solvent-excluded volume in bulk water matches $4/3 \pi r^3$ with r being the radius of oxygen (1.895 Å) where the continuum electrostatic solvation energy matches the explicit solvent charging free energy for the TIP3P water model (-8.5 kcal/mol, unpublished results). For each (irregular) volume the radius of an equivalent sphere with the same volume was calculated and is shown in Fig. 3 as a function of z.

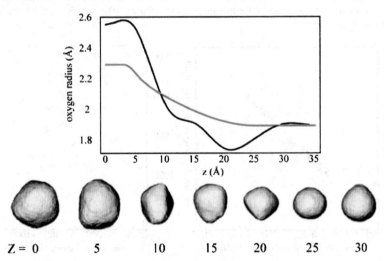

Figure 3   *Solvent excluded volume (bottom) and effective radii (top) of equivalent spheres for a water molecule constraint to different distances z from the membrane center (black). Also shown in red is the oxygen radius as a function of z fit to reproduce the explicit lipid/water energy profile (see Fig. 4).*

Fig. 3 shows that the effective radius increases indeed towards the center of the membrane. Fig. 3 (bottom) visualizes the three-dimensional shape of the solvent-excluded volume as a function of z. It can be seen that the solvent-excluded volume in the lipid tail region becomes not just larger but also deviates from a spherical shape as the water enters the lipid bilayer except for the very center of the membrane where a spherical shape is assumed again. Based on the change in excluded volume we have recalculated the total free energy profile for both the 2-7-80 and 2-7-20-80 dielectric profiles with oxygen radii that are modified accordingly. Fig. 4 compares the results with the data from explicit lipid simulations.

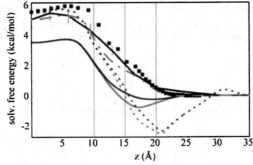

Figure 4   *Free energy in kcal/mol for the insertion of a single water molecule into a DPPC bilayer as a function of z in Å. Results from explicit lipid/solvent simulations[31, 32, 41](black) are compared with profiles from the HDGB method for the 2-7-80 (red) and 2-7-20-80 (blue) profiles as well as original (solid) modified radii from excluded volume (short dashes) and best-fit modified radii as shown in Fig. 3 (long dashes).*

It can be seen that the quantitative agreement improves inside the membrane with the modified radii. However, because the slight decrease in the effective radius around 20 Å leads to a pronounced minimum in the free energy profile in disagreement with the explicit lipid/solvent results. The discrepancy is due to a loss of translational and rotational entropy of the water molecule upon entry into the anisotropic lipid bilayer environment which is not taken into account in the implicit membrane model. The loss of entropy is strongest in the relatively ordered head group region where specific interactions with water greatly reduce the sampled space. While the effect on translational and rotational entropy as a result of an anisotropic environment cannot be readily incorporated into the implicit membrane model in a direct manner, we will target the explicit free energy profile directly by adjusting the water oxygen radii so that the entropic component is included implicitly. The resulting best-fit radii that optimize the agreement between the explicit and implicit water insertion profiles while maintaining a smooth function are shown in Fig. 3 with the resulting energy profile shown in Fig. 4. As Fig. 4 also shows there is only a very small difference between the 2-7-80 and 2-7-20-80 profiles.

In the next step, the HDGB model was applied to the calculation of membrane insertion energy profiles for amino acid side chain analogs. Explicit lipid simulations are available for comparison for methanol and acetamide[41]. Furthermore, the relative free energies from bulk to the center of the membrane can be compared to experimental transfer free energies between water and the lipid tail-mimic cyclohexane[42]. Insertion profiles were calculated for the 2-7-80 profile with the original radii and with all of the atomic radii scaled in an equivalent manner as the best-fit oxygen radii shown in Fig. 3.

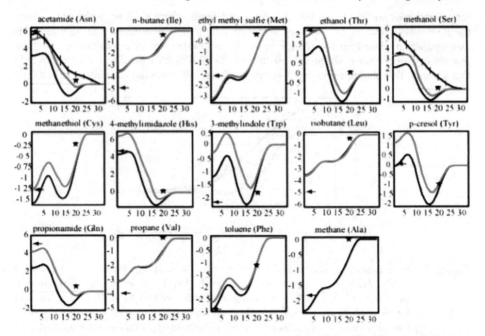

Figure 5   *Membrane insertion free energy profiles with HDGB method for uncharged amino acid side chain analogs with 2-7-80 profile. Free energies based on electrostatic solvation energies with unmodified radii (black) are compared with best-fit radii from Fig.3 (red). Explicit lipid/water curves are shown for methanol and acetamide. Experimental transfer free energies between water and cyclohexane[42] and whole-residue water-interface free energies[43] are indicated with arrows and stars (shown at z=20Å) respectively.*

We find that the energy profiles with the best-fit radii agree well with the explicit lipid/water profiles, but also with the experimental transfer free energies. The profiles also agree with previous established hydrophobicity scales[43] and correctly reproduce the preference of 3-methylindole and p-cresol to localize at the membrane/water interface[44]. We notice, however, that the transfer free energies between the membrane interior and bulk water systematically deviate for ring side chain analogs (Trp, Tyr, His) where calculated energies are too positive and for polar side chains (Gln, Asn, Thr, Ser) where the calculated energies are slightly too negative. Such deviations could be due to the approximations of the continuum model or the underlying force field, but it is also not clear how well the experimental data based on cyclohexane represents the actual membrane environment.

5 CONCLUSION

We have described the development of a new implicit model for the modeling of biological membranes based on a new, heterogeneous dielectric formulation of the generalized Born method. Detailed comparisons with explicit lipid/water and experimental data indicate good quantitative agreement that suggest suitability for a wide range of applications involving the interaction of peptides and proteins with lipid bilayers. As described elsewhere[30], we have begun to employ the HDGB model in molecular dynamics simulations of membrane-bound peptides and proteins. We will continue to refine the current model by including salt effects and allow the modeling of channels by the addition of explicit solvent molecules in the channel region.

Acknowledgements

We would like to thank Jonathan W. Essex for providing us with explicit lipid MD simulation data.

References

1. Allen, M.P. and D.J. Tildesley, *Computer Simulation of Liquids*. 1st ed. 1987, New York: Oxford University Press.
2. Ewald, P.P.. Annalen der Physik, 1921. **64**: p. 253-287.
3. Smith, P.E. and B.M. Pettitt. Journal of Chemical Physics, 1991. **95**: p. 8430.
4. Darden, T.A., D. York, and L.G. Pedersen. Journal of Chemical Physics, 1993. **98**: p. 10089-10092.
5. Toukmaji, A.Y. and J.A. Board. Computer Physics Communications, 1996. **95**: p. 73-92.
6. Mackerell, A.D.. Journal of Computational Chemistry, 2004. **25**(13): p. 1584-1604.
7. Fogolari, F., A. Brigo, and H. Molinari. Biophysical Journal, 2003. **85**: p. 159-166.
8. Kollman, P.A., et al.. Accounts of Chemical Research, 2000. **33**: p. 889-897.
9. Levy, R.M., et al.. Journal of the American Chemical Society, 2003. **125**: p. 9523-9530.
10. Sitkoff, D., K.A. Sharp, and B. Honig. Journal of Physical Chemistry, 1994. **98**: p. 1978-1988.
11. Sharp, K.A. and B. Honig. Annual Review of Biophysics and Biophysical Chemistry, 1990. **19**: p. 301-332.
12. Lee, B. and F.M. Richards. Journal of Molecular Biology, 1971. **55**: p. 379.
13. Feig, M., et al.. Journal of Computational Chemistry, 2004. **25**: p. 265-284.
14. Qiu, D., et al.. Journal of Physical Chemistry A, 1997. **101**: p. 3005-3014.
15. Still, W.C., et al.. Journal of the American Chemical Society, 1990. **112**: p. 6127-6129.
16. Lee, M.S., F.R. Salsbury, Jr., and C.L. Brooks, III. Journal of Chemical Physics, 2002. **116**(24): p. 10606-10614.
17. Lee, M.S., et al.. Journal of Computational Chemistry, 2003. **24**: p. 1348-1356.
18. Hawkins, G., C. Cramer, and D. Truhlar. Chemical Physics Letters, 1995. **246**: p. 122-129.
19. Ghosh, A., C.S. Rapp, and R.A. Friesner. Journal of Physical Chemistry B, 1998. **102**: p. 10983-10990.
20. Gallicchio, E. and R.M. Levy. Journal of Computational Chemistry, 2004. **25**: p. 479-499.
21. Feig, M. and C.L. Brooks III. Current Opinion in Structural Biology, 2004. **14**: p. 217-224.
22. Im, W., M. Feig, and C.L. Brooks III. Biophysical Journal, 2003. **85**: p. 2900-2918.
23. Spassov, V.Z., L. Yan, and S. Szalma. Journal of Physical Chemistry B, 2002. **106**: p. 8726-8738.

24. Im, W. and C.L. Brooks. Journal of Molecular Biology, 2004. **337**(3): p. 513-519.
25. Constanciel, R. and R. Contreras. Theoretica Chimica Acta, 1984. **65**: p. 1-11.
26. Onufriev, A., D.A. Case, and D. Bashford. Journal of Computational Chemistry, 2002. **23**: p. 1297-1304.
27. Born, M.. Zeitschrift für Physik, 1920. **1**: p. 45-48.
28. Kirkwood, J.G.. Journal of Chemical Physics, 1934. **2**: p. 351-361.
29. Feig, M., W. Im, and C.L. Brooks III. Journal of Chemical Physics, 2004. **120**.
30. Tanizaki, S. and M. Feig. Journal of Chemical Physics, 2005. **122**: p. in press.
31. Marrink, S.-J. and H.J.C. Berendsen. Journal of Physical Chemistry, 1994. **98**: p. 4155-4168.
32. Marrink, S.J. and H.J.C. Berendsen. Journal of Physical Chemistry, 1996. **100**: p. 16729-16738.
33. Brooks, B.R., et al.. Journal of Computational Chemistry, 1983. **4**: p. 187-217.
34. MacKerell, A.D., Jr., et al.. Journal of Physical Chemistry B, 1998. **102**: p. 3586-3616.
35. Roux, B.. Biophysical Journal, 1997. **73**: p. 2980-2989.
36. Feller, S.E. and A.D. MacKerell Jr.. Journal of Physical Chemistry B, 2000. **104**: p. 7510-7515.
37. Feller, S.E., R.M. Venable, and R.W. Pastor. Langmuir, 1997. **13**(24): p. 6555-6561.
38. Feller, S.E., et al.. Journal of Chemical Physics, 1995. **103**(11): p. 4613-4621.
39. Stern, H.A. and S.E. Feller. Journal of Chemical Physics, 2003. **118**: p. 3401-3412.
40. Zhou, F. and K. Schulten. Journal of Physical Chemistry, 1995. **99**: p. 2194-2207.
41. Bemporad, D., J.W. Essex, and C. Luttmann. Journal of Physical Chemistry B, 2004. **108**(15): p. 4875-4884.
42. Radzicka, A. and R. Wolfenden. Biochemistry, 1988. **27**(5): p. 1664-1670.
43. Wimley, W.C. and S.H. White. Nature Structural Biology, 1996. **3**(10): p. 842-848.
44. Wimley, W.C. and S.H. White. Biochemistry, 1993. **32**(35): p. 9262-9262.

ASSESSMENT AND TUNING OF A POISSON BOLTZMANN PROGRAM THAT UTILIZES THE SPECIALIZED COMPUTER CHIP MD-GRAPE-2 AND ANALYSIS OF THE EFFECT OF COUNTER IONS

S. Höfinger[1,2]

[1]Novartis Institutes for BioMedical Research, Vienna, IK@N, ISS, Brunner Straße 59, A-1235 Vienna, Austria. E-mail: siegfried.hoefinger@novartis.com

[2]present address: Michigan Technological University, Department of Physics, 1400 Townsend Drive, 49931 Houghton, MI, USA. E-mail: shoefing@mtu.edu

1 INTRODUCTION

Aqueous environments are crucial to all forms of life. In fact, most of the knowledge acquired today about the complex and fascinating interplay of biomolecular matter is bound to the solvated state, that is when biomolecules are dissolved in water. Consequently, the study of the physics governing the effect of solvation is of fundamental interest and has broad implications in many scientific disciplines.

Theoretical approaches have addressed the problem of solvation by the inclusion of either explicit[1] or implicit[2] solvation models. While explicit solvation models introduce a set of individual solvent molecules resolved to full atomic detail, implicit solvation models neglect any fine structure of the solvent completely. Thus the implicit solvation models simply approximate the solvent as a dielectric continuum, and the only thing that makes such a continuum specific to some particular solvent is the assignment of a characteristic dielectric constant (e.g., $\varepsilon = 80$ in the case of water, $\varepsilon = 10$ for octanol, etc.). The actual solvation effect, or better the actual polarization effect, is then commonly described from solutions to the Poisson Boltzmann equation[3] (PB). At this point it seems to be appropriate to recall that the net effect of solvation may be decomposed into major partial contributions[4],

$$\Delta G_{solv,H_2O} = \Delta G_{solv,H_2O}^{pol} + \Delta G_{solv,H_2O}^{cav} + \Delta G_{solv,H_2O}^{disp,rep} \qquad (1)$$

i.e., contributions due to polarization, cavitation, dispersion and repulsion. It is important to realize that PB will provide solutions only to the polarization term, which accounts for the dielectric response of the medium when a solute is embedded in the continuum. The remaining terms, frequently referred to as the non-polar contributions, are of equal importance and cover effects like the clearance cost of freeing the volume occupied by the solute from solvent molecules (cavitation), or the short-range type of interactions that typically occur at the solute-solvent interface (dispersion, repulsion).

Two variant schemes for solution of to the PB equation have become popular. These are the method of Finite Difference Poisson Boltzmann[5] (FDPB) and the Boundary Element Method[6] (BEM). FDPB usually solves the equation on a 3-dimensional grid while the BEM requires the construction of a 2-dimensional layer at the solute-solvent boundary, which is usually formed by the molecular surface constructed from simple geometric elements, i.e., flat or curved triangles, tesserae, etc. In the following text, we concentrate on several important aspects that have been found to play a critical role in the development of a PB/BEM program.

2 METHODS

2.1 Principle of the BEM

The fundamental relationship in the BEM has been derived by Zauhar and Morgan[6] from the avoidance of a discontinuity in the normal component of the electric displacement across the dielectric interface. Here the surface charge density σ_i that corresponds to a particular boundary element i is related to the electric field E_i field acting on that very boundary element via

$$\sigma_i = \frac{q_i}{A_i} = \frac{1}{4\pi} \left(\frac{\varepsilon_{int} - \varepsilon_{ext}}{\varepsilon_{int}} \right) \vec{E}_i . \vec{n}_i \tag{2}$$

where q_i is the partial charge (the unknown) located at the centre and A_i the partial area of the i-th boundary element. Furthermore, ε_{int} and ε_{ext} are the outer and inner dielectric constants (e.g., 80 and 1 for aqueous solutions), respectively, and the final term in eq. (2) is a scalar product between the outwards[7] pointing normal vector n_i and the net electric field vector E_i relevant to that particular boundary element. All the complexity in eq. (2) is hidden in the actual constituents of E_i. There are three main sources of that electric field,

$$\vec{E}_i = \vec{E}_i^{self} + \vec{E}_i^{solute} + \vec{E}_i^{\nabla BE} \tag{3}$$

which stem from
- self contribution (Gauss' Law)

$$\vec{E}_i^{self} = 2\pi \frac{q_i}{A_i} \tag{4}$$

- contribution due to all the solute atomic charges

$$\vec{E}_i^{solute} = \sum_k^{\#at} \frac{(\vec{r}_i - \vec{R}_k) Q_k}{|\vec{r}_i - \vec{R}_k|^3} \tag{5}$$

where Q_k and R_k are the partial charges and location vectors of the k-th atom of the solute and $\#at$ stands for the total number of atoms. Moreover, the centre of a particular boundary element is described as r_i.

- contribution due to all other charges on all the other boundary elements

$$\vec{E}_i^{\forall BE} = \sum_{j \neq i} \frac{(\vec{r}_i - \vec{r}_j) q_j}{|\vec{r}_i - \vec{r}_j|^3} \tag{6}$$

with q_j being the induced partial charge located at centre r_j of the j-th boundary element. It is clear that with growing solute size the number of boundary elements will increase considerably, hence this final constituent to the electric field as given in eq. (6) is the actual reason why solving the PB/BEM equation is also a computationally intensive task.

Upon substitution of eqs. (4) – (6) into eq. (3) and further insertion into eq. (2), a final system of equations may be derived, which may simply be written as

$$\mathbf{D} \cdot \vec{q} = \vec{q}_{ini} \tag{7}$$

where $\mathbf{D}$ is a matrix of dimension $\#BE \times \#BE$ and q as well as q_{ini} are vectors of dimension $\#BE$ where q represents all the unknowns and q_{ini} is a static vector closely related to the right hand side of eq. (5). The solution to eq. (7) may be obtained by direct matrix-inversion[6,8]. However, already for medium-sized molecules, the number of BEs forming the dielectric boundary becomes a limiting factor and conventional techniques for matrix-inversion cannot be used any longer[9]. Fortunately eq. (7) may be recast and an iterative approach may be pursued $q^m \rightarrow q^{m+l}$ (a detailed description of this iterative scheme will be the subject of a forthcoming report). The big advantage with this iterative approach is that one never has to store a matrix of dimension $\#BE \times \#BE$, but instead can get by with just vectors of dimension $\#BE$. In addition, the number of necessary iterations m to achieve a certain degree of accuracy may be kept small, and a well-known convergence acceleration scheme may be applied to increase the success rate of this iterative algorithm[10].

2.2 MD-GRAPE-2

A special type of computer aimed at large scale Molecular Dynamics (MD) simulations has been developed at the RIKEN Institute[11]. This machine, also known as the Molecular Dynamics Machine (MDM), consists of the specialized computer chips MD-GRAPE-2 and MD-WINE. Both of these computer chips may accomplish key-calculational steps of typical MD algorithms in hardware. A mini-version of the MDM is commercially available in the form of MD-GRAPE-2 boards, which are compatible with standard personal computers via a standard PCI-interface. All the present data has been obtained with such MD-GRAPE-2 PCI-cards. Although rather limited to a static and simplified functionality, when optimally employed, such specialized hardware can exhibit impressive performance. Speed up factors beyond 100-fold as opposed to current general-purpose CPUs may be reached. A comparable degree of optimization is difficult to achieve with conventional parallelism[12], i.e., MPI-based processing.

The iterative solution scheme was found to be particularly suited for an implementation involving the MD-GRAPE-2 architecture (again, a detailed description will be presented in a future article). It is mainly the Coulombic nature of eqs. (5) and (6) that allows a straightforward coupling to this specialized computer chip; it was a minor extension to the module for Coulomb force calculation that facilitated an effective deployment in the PB/BEM approach. Since this module is known to operate extremely efficiently, it was anticipated (and finally also shown) that MD-GRAPE-2 can improve the computational performance of the PB/BEM approach significantly.

2.3   PCM Calculations on a Set of Homodipeptide Conformations

In order to check the polarization results obtained with the PB/BEM solver, a set of reference data was needed. For this purpose a high level method such as the Polarizable Continuum Model[14] (PCM) should be employed. For example, a combination of the widely-accepted model B3LYP[15] from Density Functional Theory (DFT) together with a standard basis set like 6-31G*[16] could serve for the production of this reference data set. Since the ultimate goal for the novel PB/BEM solver was to handle large protein structures, the assessment of the predictive quality was most meaningfully done on structural data of peptides. Therefore a set of homodipeptides of all the 20 different amino acids was created. Substantial interest was also in the question of how dependent these calculations of the polarization term are on different conformations of a molecule. To address this problem the most prominent combinations of φ,ψ-angles of the peptide bond were extracted from a typical Ramachandran plot[13] and used for the construction of different conformations of the dipeptides. In so doing, a set of nine representative structures per species of dipeptide will emerge, which further led to an overall number of 180 individual structures to be calculated at the level of PCM/B3LYP/6-31G*. All structures were built as zwitterionic systems. All calculations were carried out with the GAUSSIAN-03 program[17].

2.4   Counter Ions

An additional problem in the PB/BEM approach is the treatment of background charge spread throughout the continuum. This is an important aspect in many biological systems that either exhibit a non-zero net charge or are found in aqueous solution of non-zero ionic strength. In fact it is the second capital letter "B" within the PB-abbreviation that refers to the proper theoretical mode of how to take into account such situations, which in particular is a suggestive link to what is called the Boltzmann term. Since a full implementation of the Boltzmann term within the BEM would require major additional programming[18], the question was if one could circumvent it by the inclusion of explicitly set counter ions. The easiest way of probing such a strategy was to compare electrostatic potentials derived from PB/BEM calculations on a test protein with and without consideration of explicitly included counter ions. Beside the possible consequences on spatial re-distribution of electrostatic potentials, the magnitude of alteration in the net polarization free energy upon inclusion of explicit counter ions as compared to their neglect was of particular interest.

    As a test case, Proteinase K (pdb code 1IC6) had been selected. This is a serine protease consisting of 279 residues. When using the program XLEAP from AMBER-6[19] for default parameter assignment, a net charge of +1 did emerge for that protein structure. XLEAP was used again to introduce a Cl⁻ ion at the most appropriate location according to XLEAP-internally computed electrostatic grids. This way the two systems were set up resulting in:
a) a neutralized form of Proteinase K including one Cl⁻ counter ion and
b) a positively charged form of Proteinase K that had no additional counter ions attached.
PB/BEM calculations were then performed on either system and electrostatic grid maps

computed. The electrostatic potentials were visualized as surfaces at ± 144 mV. A distinction was made between distant electrostatic potentials and proximate electrostatic potentials. The distant potentials did only consider locations in space that were at least 3 Å away from any atom of the protein and the sampling focused on iso-values, which means a collection of all those points that share the same potential value become connected to form a so-called iso-surface. Conversely, the proximate potentials did only concentrate on positions located directly on the molecular surface of the protein, which was represented as the Solvent Accessible Surface Area (SASA)[20]. Electrostatic potentials were computed for these points and common patches of ± 144 mV were visualized in an analogous fashion to what was described just above for the distant electrostatic potentials. For visualization purposes the program VMD[21] was used.

3   RESULTS

3.1   Comparison with PCM/B3LYP/6-31G*

A representative set of nine combinations for the φ,ψ-angles of the peptide bond was used to construct different conformations in a dipeptide. An example is shown in Figure 1 for the case of dialanine (AA). A similar process has been used to generate conformations for all the remaining 19 different types of homodipeptides. All conformations were set up as zwitterionic systems. PCM calculations and PB/BEM calculations were performed on this set of structures and the polarization free energies were read out and compared. What was realized immediately was the fact that one needs to scale up the default AMBER-6 van der Waals radii by a factor of 1.12 in order to operate at equal molecular volumes. For example, at default parameters, AA within the PCM model occupied a molecular volume of V_{PCM} = 230 Å^3, while within the PB/BEM approach, when using AMBER default van der Waals radii, the same structure of AA was assigned a volume of V_{AMBER} = 190 Å^3. The uniform, multiplicative application of this scaling factor of 1.12 to all AMBER default van der Waals radii could outbalance this inequality, and all volumes were thus brought approximately to comparable values of the PCM-volumes to provide a fair basis for proper comparison. In Figure 2, a comparison is shown between PCM results (triangles) and PB/BEM results (circles) for the AA system after the 1.12-scaling to all the AMBER default van der Waals radii has taken place. It is remarkable that the principal trend observed with varying molecular geometry is very well recovered from the PB/BEM approach when taking the high level PCM results as the baseline. This is not at all trivial considering the much simpler model chemistry applied in PB/BEM (just classic molecular mechanics employing atom-specific partial charges as well as atom-specific van der Waals radii). Another important point to recognize from Figure 2 is that there is a rather constant gap between PCM results and PB/BEM results. If we define a mean absolute deviation $\bar{\delta}$ like

$$\bar{\delta} = \sqrt{\frac{1}{\#configs} \sum_{i=1}^{\#configs} \left(\Delta G_i^{Pol,PCM} - \Delta G_i^{Pol,PB/BEM} \right)^2} \tag{8}$$

then for the case of AA we obtain a value of $\overline{\delta}^{PRE-SCALING}$ =16.7 kcal/mol. Furthermore, when looking at maximum unsigned deviations like

$$\delta_{max} = MAX \left(\left| \Delta G_i^{Pol,PCM} - \Delta G_i^{Pol,PB/BEM} \right| \right)_{i=1,2,...\#configs} \qquad (9)$$

then the initial conformation shows the greatest drift and we get a $\delta_{MAX}^{PRE-SCALING} = 19.9$ kcal/mol. Conversely, when looking at Figure 2 the question arises if one could find an appropriate scaling factor applied to AMBER default charges that could get the two graphs to match each other even better. Indeed, upon several trials it was found empirically that a multiplicative scaling factor of 0.9 applied to default AMBER partial charges could lead to very good agreement between PCM and PB/BEM results. A representation of the comparison after charge scaling is given in Figure 3. Improvement is seen from the mean deviation $\overline{\delta}^{POST-SCALING}$ = 2.9 kcal/mol and the greatest individual deviation $\delta_{MAX}^{POST-SCALING}$ = 6.4 kcal/mol.

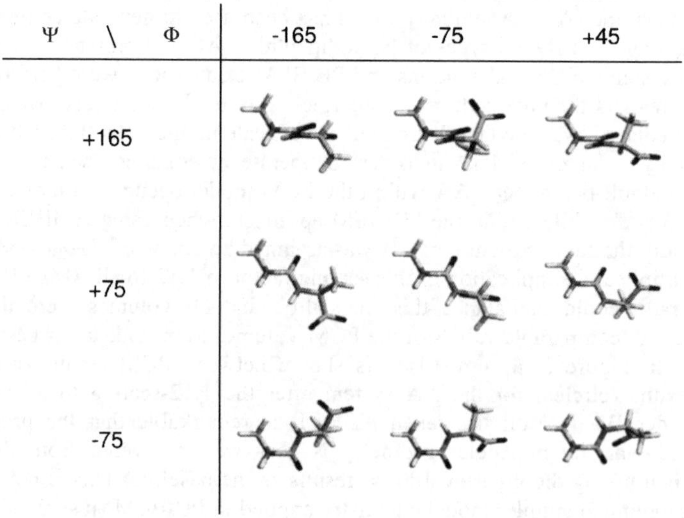

Ψ \ Φ	-165	-75	+45
+165			
+75			
-75			

Figure 1   *Representation of the employed torsional angles ψ,φ of the peptide bond to construct different conformers of the dialanine dipeptide. The actual values for ψ,φ have been extracted from the centres of the most densely populated islands in a Ramachandran plot[13].*

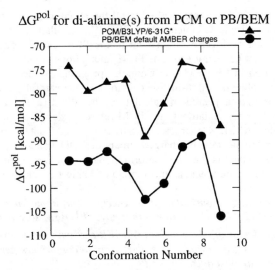

Figure 2 *Comparison of the polarization free energies for different conformations of the dialanine dipeptide computed from either PCM/B3LYP/6-31G* (triangles) or a novel PB/BEM solver based on default AMBER-6 charges[22] (circles).*

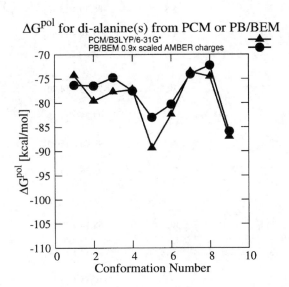

Figure 3 *Comparison of the polarization free energies for different conformations of the dialanine dipeptide computed from either PCM/B3LYP/6-31G* (triangles) or a novel PB/BEM solver employing AMBER-6 charges[22] uniformly scaled by a multiplicative factor of 0.9 (circles).*

The procedure described above was applied to all the remaining types of homodipeptides. A summary is given in Table 1. In general the PB/BEM calculations must be considered to reach a rather satisfactory level of agreement to the PCM calculations. However, a few conspicuous items must be mentioned. The PCM calculations failed for LL. The molecular surface algorithm used in PB/BEM (the Connolly program[23]) failed for FF and YY, and thus seems to have problems with 6-membered rings. When carefully examining the PCM-output, strange warnings appeared in calculations for PP, CC, RR and HH, which all belong to a group of severe violators to the PCM reference data. For the rest, the PB/BEM results due to scaled AMBER charges reconstruct the PCM data very well with minor deviations seen for the negatively charged dipeptides DD and EE. Another interesting point to note in this respect is that the conformational variability estimated from standard deviations included in square brackets in column 5 of Table 1 is surprisingly small.

Table 1 *Comparison of polarization free energies for homodipeptides computed at PCM/B3LYP/6-31G* level or with a novel PB/BEM solver after application of a scaling factor of 0.9 to default AMBER charges; shown are average ($\bar{\delta}$) and maximum (δ_{MAX}) deviations in a set of nine conformers per dipeptide species together with absolute average PCM energies.*

Dipeptide	$\bar{\delta}$ [a] [kcal/mol]	δ_{MAX} [b] [kcal/mol]	average $\Delta G^{Pol,PCM}$ [kcal/mol]	number of configurations usable to δ-calculations
AA	2.9	6.4	-79.55 ± [5.34]	9
GG	2.7	5.8	-85.30 ± [5.76]	9
VV	2.9	4.6	-72.49 ± [4.65]	9
LL	-	-	-	0
II	2.1	3.6	-70.45 ± [4.00]	6
PP	13.1	18.5	-77.22 ± [4.52]	2
SS	2.3	4.8	-99.76 ± [6.95]	9
TT	3.7	7.1	-93.36 ± [5.12]	6
YY	-	-	-	0
CC	23.8	30.2	-68.14 ± [9.38]	9
MM	3.3	7.1	-77.82 ± [5.04]	9
FF	-	-	-80.38 ± [6.15]	0
WW	8.7	12.3	-90.45 ± [10.45]	7
HH	21.0	29.1	-78.91 ± [5.12]	9

KK	14.7	18.7	-229.08 ± [8.46]	9
RR	88.1	108.8	-116.71 ± [16.52]	9
DD	32.8	36.2	-283.30 ± [11.57]	6
EE	29.1	33.4	-255.98 ± [7.13]	9
NN	4.3	6.0	-96.89 ± [4.69]	6
QQ	4.6	8.4	-106.05 ± [6.28]	9

[a] Defined in Equation (8). [b] Defined in Equation (9).

3.2 The Effect of Counter Ions

PB/BEM calculations were carried out on a medium-sized protein, the serine protease proteinase K (pdb-code 1IC6), which consists of 279 residues and exhibits a net charge of +1 when using AMBER-6/XLEAP residue assignment. Scaling of AMBER-6 non-bonded parameters was performed as described in the previous section; that is, van der Waals radii were scaled by a factor of 1.12 and atomic partial charges by a factor of 0.9 respectively. Two systems were studied for comparison. First, a +1 positively charged protein (the default structure) and second, a neutral form of proteinase K, which was obtained by adding a Cl⁻ counter ion appropriately to the structure, again a task to XLEAP. In Table 2 a confrontation of absolute values for the polarization free energies is given. As becomes clear from this comparison, the relative alteration of the absolute value for the polarization free energy upon introduction of an additional Cl⁻ ion is rather modest, -17.3 kcal/mol in particular or -0.8 % of the unmodified system (-2075.1 kcal/mol). Next, the effect of considering explicitly set counter ions was investigated. At first the periphery of the protein was studied by calculating the electrostatic potentials far outside the domain of the protein (at least 3 Å away from any protein atom). Iso-potential surfaces were computed from the PB/BEM results and representations are given in Figure 4. On the left hand side of Figure 4 the situation is given for an explicitly set counter ion Cl⁻ represented as a small sphere in the lower right quarter. The right panel in Figure 4 shows similar results obtained without consideration of any additional counter ions. Qualitatively there is no evidence for any modification of these distant electrostatic potentials upon introduction of an additional counter ion (color coded data not shown). In a further investigation, a similar analysis was done, but looking at local potentials distributed directly across the SASA of the protein. Corresponding results are shown in Figure 5. Visual representation similar to the previous experiment was carried out (color coded data not shown). In contrast to the previous example, local electrostatic potentials are considerably affected by an extra counter ion (see sector of the grey enclosed circle on the left hand side of Figure 5). Nevertheless, this effect seems to be bound to the closer proximity of the actual position of the additional counter ion and limited to a spherical extension of roughly 12 Å -15 Å.

Figure 4   *Iso-potential surfaces (dark-grey: +144 mV, light-grey: -144 mV) representing the distant electrostatic potential in the surrounding of proteinase K as obtained from PB/BEM calculations. Left panel: Inclusion of an explicit counter ion Cl⁻ represented as the small dark ball in the lower right corner; Right panel: Neglect of explicit counter ions.*

Figure 5   *Distribution of local electrostatic potentials across the SASA of proteinase K as obtained from PB/BEM calculations. Patches on the surface sharing potential values of +144 mV are shown in dark-grey, while regions of common potentials of -144 mV are given in light-grey. Left panel: Inclusion of an explicit counter ion Cl⁻; Right panel: Neglect of any counter ions.*

Table 2 *Comparison of PB/BEM results for the polarization free energies of proteinase K (pdb-code 1IC6) with or without the inclusion of an explicit counter ion Cl^-*

$\Delta G_{solv,H_2O}^{Polarization}$ including Cl^- [kcal/mol]	$\Delta G_{solv,H_2O}^{Polarization}$ neglecting Cl^- [kcal/mol]
-2092.4	-2075.1

4 CONCLUSION

A novel PB/BEM solver has been represented that can efficiently make use of the specialized hardware MD-GRAPE-2. The predictive quality of the calculations of free energies of polarization can be tuned to very high levels as shown for a comparison to a set of PCM/B3LYP/6-31G* reference data. When using AMBER-6 default parameterization concerning the van der Waals radii and atomic partial charges, scaling factors of 1.12 for the radii as well as 0.9 for the partial charges were found able to improve the quality of the results considerably. Explicitly set counter ions may serve as a first approximation to proteins where there is a net charge or a background charge. The introduced artefacts are small on an absolute scale, insignificant to distant electrostatic potentials and noticeable to local electrostatic potentials up to a distance of ± 15 Å.

Acknowledgements
The author is grateful to having been a post doctoral fellow in the Novartis Institutes for BioMedical Research Presidential Postdoctoral Fellowship program. Sincere thanks are given to Ing. Benjamin Almeida from NIBR Vienna and to Dr. Tetsu Narumi from RIKEN for very helpful technical assistance with the MD-GRAPE-2 architecture. Prof. Wendy Cornell, MERCK, and Prof. Thomas Simonson, Ecole Polytechnique, are acknowledged for many profitable discussions. The author wants to thank Dr. Michael Connolly for providing a test version of his popular Molecular Surface program and Susan C. DiClemente from NIBR Cambridge for help with the manuscript.

References

1   W.L. Jorgensen, J. Chandrasekhar, J.D. Madura, R.W. Impey and M.L. Klein, *J. Chem. Phys.*, 1983, **79** (2), 926.
2   C.J. Cramer, D.G. Truhlar, *Chem. Rev.*, 1999, **99** (8), 2161.
3   E.M. Lifshitz, L.D. Landau, L.P. Pitaevskii, *Electrodynamics of Continuous Media*, Publisher: Butterworth-Heinemann, 2 edition, ISBN: 0750626348, 1984.
4   J. Tomasi, M. Persico, *Chem. Rev.*, 1994, **94** (7), 2027.
5   R. Luo, L. David, M.K. Gilson, *J. Comp. Chem.*, 2002, **23** (13), 1244.
6   R.J. Zauhar, R.S. Morgan, *J. Mol. Biol.*, 1985, **186** (4), 815.
7   viewed from the core of the solute.
8   C.S. Pomelli, J. Tomasi, V. Barone, *Theor. Chem. Acc.*, 2001, **105**, 446.
9   S. Höfinger, T. Simonson, *J. Comp. Chem.*, 2001, **22** (3), 290.
10   P. Pulay, *Chem. Phys. Lett.*, 1980, **73** (2), 393.
11   http://atlas.riken.go.jp/mdm/mdgrape2.html

12   S. Höfinger, *Lect. Notes Comput. Sc.*, 2004, **3241**, 397.
13   N.J. Marshall, B.M. Grail, J.W. Payne, *J. Peptide Sci.*, 2000, **6**, 186.
14   J. Tomasi, R. Cammi, B. Mennucci, *Int. J. Quantum Chem.*, 1999, **75**, 767.
15   A.D. Becke, 1993, *J. Chem. Phys.*, **98**, 5648.
16   M.M. Francl, W.J. Petro, W.J. Hehre, J.S. Binkley, M.S. Gordon, D.J. DeFrees, J.A. Pople, 1982, *J. Chem. Phys.*, **77**, 3654.
17   Gaussian 03, Revision B.05, M.J. Frisch, G.W. Trucks, H.B. Schlegel, G.E. Scuseria, M.A. Robb, J.R. Cheeseman, J.A. Montgomery, Jr., T. Vreven, K.N. Kudin, J.C. Burant, J.M. Millam, S.S. Iyengar, J. Tomasi, V. Barone, B. Mennucci, M. Cossi, G. Scalmani, N. Rega, G.A. Petersson, H. Nakatsuji, M. Hada, M. Ehara, K. Toyota, R. Fukuda, J. Hasegawa, M. Ishida, T. Nakajima, Y. Honda, O. Kitao, H. Nakai, M. Klene, X. Li, J.E. Knox, H.P. Hratchian, J.B. Cross, V. Bakken, C. Adamo, J. Jaramillo, R. Gomperts, R.E. Stratmann, O. Yazyev, A.J. Austin, R. Cammi, C. Pomelli, J.W. Ochterski, P.Y. Ayala, K. Morokuma, G.A. Voth, P. Salvador, J.J. Dannenberg, V.G. Zakrzewski, S. Dapprich, A.D. Daniels, M.C. Strain, O. Farkas, D.K. Malick, A.D. Rabuck, K. Raghavachari, J.B. Foresman, J.V. Ortiz, Q. Cui, A.G. Baboul, S. Clifford, J. Cioslowski, B.B. Stefanov, G. Liu, A. Liashenko, P. Piskorz, I. Komaromi, R.L. Martin, D.J. Fox, T. Keith, M.A. Al-Laham, C.Y. Peng, A. Nanayakkara, M. Challacombe, P.M.W. Gill, B. Johnson, W. Chen, M.W. Wong, C. Gonzalez, and J.A. Pople, Gaussian, Inc., Wallingford CT, 2004.
18   A.H. Juffer, E.F.F. Botta, B.A.M. van Keulen, A. van der Ploeg, H.J.C. Berendsen, *J. Comput. Phys.*, 1991, **97**, 144.
19   D.A. Pearlman, D.A. Case, J.W. Caldwell, W.R. Ross, T.E. Cheatham, III, S. DeBolt, D. Ferguson, G. Seibel, P.A. Kollman, 1995, *Comp. Phys. Commun.*, **91**, 1.
20   http://www.netsci.org/Science/Compchem/feature14.html
21   W. Humphrey, A. Dalke, K. Schulten, 1996, *J. Mol. Graphics*, **14**, 33.
22   W.D. Cornell, P. Cieplak, C.I. Bayly, I.R. Gould, K.M. Merz Jr., D.M. Ferguson, D.C. Spellmeyer, T. Fox, J.W. Caldwell, P.A. Kollman, 1995, *J. Am. Chem. Soc.*, **117**, 5179.
23   M.L. Connolly, 1985, *J. Appl. Crystallogr.*, **18**, 499.

INTRINSIC ISOTOPE EFFECTS – THE HOLY GRAAL OF STUDIES OF ENZYME-CATALYZED REACTIONS

A. Dybała-Defratyka, R. A. Kwiecień, D. Sicińska and P. Paneth[*]

Institute of Applied Radiation Chemistry, Department of Chemistry, Technical University, Żeromskiego 116, 90-924 Łódź (Lodz), Poland

1 INTRODUCTION

The minimal geometrical information that defines the mechanism of a simple reaction is the structure of reactant(s), product(s), and the transition state. While numerous experimental techniques allow learning geometrical features of reactants and products, transition state is more elusive and not amenable to direct experimental scrutiny. Kinetic isotope effects (KIEs) are a very potent tool for inferring transition state structure. Apart from scientific curiosity, understanding details of the transition state structure is of great practical importance, e.g., in the rational drug design. The enzymatic processes, however, are always complex reactions involving more than one step. Even when the chemical step in an enzymatic reaction is the only isotope-sensitive step, the measured kinetic isotope effect (KIE_{exp}) may be different from the KIE for the chemical step of the reaction taking place in the enzyme active site. This is illustrated on the example of a simple enzymatic reaction that includes reversible binding steps. Such a reaction is characterized by five rate constants k_1, k_2, k_4, and k_5 corresponding to forward and reverse binding steps of the reactant S and the product P, and a rate constant k_3 corresponding to an irreversible chemical conversion step:

$$E + S \underset{k_2}{\overset{k_1}{\rightleftarrows}} ES \overset{k_3}{\longrightarrow} EP \underset{k_5}{\overset{k_4}{\rightleftarrows}} E + P \tag{1}$$

When KIEs are measured by competitive method, they provide mechanistic information only for steps up to and including the first irreversible step. If only the chemical step is isotope-sensitive, it is the first irreversible step, and the steady-state approximation is valid, the extent to which KIE_{exp} reflects KIE for the chemical step depends on all the rate constants in the mechanism up to this step as shown in equation:

$$KIE_{exp} = \frac{KIE_3 + k_3/k_2}{1 + k_3/k_2} = \frac{KIE_3 + C}{1 + C} \tag{2}$$

where KIE_3 is the kinetic isotope effect on the chemical rate constant k_3 and is called the **intrinsic kinetic isotope effect**. Parameter C is called **commitment to catalysis**. It can be seen from eq 2 that KIE_{exp} depends not only on the intrinsic KIE_3 but also on the parameter

C. When this parameter approaches zero, KIE_{exp} approaches KIE_3. However, when it is much larger than unity, KIE_{exp} approaches unity.

Frequent goal of studies of enzymatic reactions is to unravel the intrinsic KIE in order to elucidate the nature of the corresponding transition state. This is usually done by changing reaction conditions in a way that causes the parameter C to become negligible compared to unity and KIE_3. This can be achieved experimentally in several ways, which are briefly indicated below on the examples studied by us. In the case of a multisubstrate sequential reaction, the commitment to catalysis is described by a more complex expression that includes chemical concentration of the substrate that binds second. Thus, lowering the concentration of this reactant leads to lowering of the C value. This approach has been successfully applied in the case of the oxygen KIE on the phosphoenolpyruvate carboxylase-catalyzed reaction where the concentration of bicarbonate could be changed in a very wide range.[1] Since many enzymatic reactions involve shuffling of protons during the reaction, changes in pH can similarly lead to diminishing the value of the commitment to catalysis. This approach is nicely illustrated by the pH dependence of the carbon KIE on orotidine 5 -monophosphate decarboxylase.[2] At the physiological pH (7.4) KIE_{exp} is only 1.024. Lowering pH to about 3 causes chemical step to become mostly rate-limiting and busts the observed value of the kinetic isotope effect to over 1.05. Other parameters, such as temperature,[2] viscosity,[2,3] or changes of reactants can lead to changes of the parameter C magnitude and can lead to direct determination of the intrinsic isotope effect.

Changes of reactants can include solvent (e.g., use of D_2O instead of H_2O),[3] cofactor (metal),[3] the enzyme (e.g., mutations),[4,5] or the substrates.[4,5,6] Herein we will concentrate on this last option and consider cases when instead of physiological substrate another molecule that reacts more slowly is used. The philosophy behind this method is that for any substrates that react slowly, $k_3 \ll k_{-2}$, which causes that the chemical step becomes rate-determining as illustrated in Figure 1, and the commitment effectively approaches zero; thus KIE_{exp} approaches the intrinsic effect. A hidden assumption of this approach is that the intrinsic KIE for the physiological and slow substrates are identical. Although the slow-substrate hypotheses have been widely invoked,[7] the precise limits of its validity are unknown.

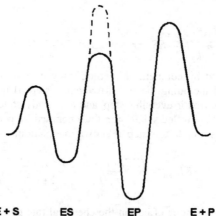

Figure 1 *Energy diagram for the reaction given in eq. 1 with the irreversible chemical step. Dashed line illustrates changes caused by the slow substrate*

Now-a-day computational techniques provide information that can frequently be equally useful as this gathered from the experiment. Thus it becomes possible to infer theoretically both quantities of the right-hand side of eq 2, i.e., the intrinsic value of KIE and the commitment to catalysis C. Herein we present examples from studies carried out in our laboratory that aim at the theoretical evaluation of both, the commitment factor and the kinetic isotope effects of the chemical step of the enzymatic catalysis. In addition we put into test the hypothesis that slow substrates are good models of the intrinsic KIEs for physiological substrates by comparing values calculated for both these substrates.

2 METHODS AND RESULTS

Theoretical modeling of individual steps within an enzyme has become practical only recently, due to the increase of computers power and the progress in QM/MM techniques. In this approach[8] different theory levels are applied to different parts of a system under investigation. Usually quantum mechanical (QM) levels are applied to reactants (and selected residues of the active site) while the rest of the system is treated by molecular mechanics (MM). Using these techniques magnitudes of both, the intrinsic KIE and commitment to catalysis C can be calculated theoretically. Below we present examples of such calculations carried out using different QM/MM schemes. We also test applicability of the slow-substrate hypothesis, as formulated in the Introduction.

2.1 Methylmalonyl-CoA mutase, MCM

Methylmalonyl-CoA mutase catalyzes reversible isomerization of methylmalonyl-CoA (MCA) to succinyl-CoA by radical mechanism. The key (pre-steady-state) steps in this mechanism are hydrogen atom transfer from the substrate to 5 -deoksyadenosyl (dAdo⁺) radical which is generated in the preceding homolytic cleavage of cobalt-carbon bond.[9] KIE_{exp} measured for this step is about 50 at 5 °C (see Table 1).[10] We have addressed the question whether the intrinsic KIE is fully expressed by modeling both reactions steps and the hydrogen KIE for the hydrogen atom transfer.

For modeling of the homolytic rupture of the Co-C bond the ONIOM protocol[11] has been employed. Based on crystallographic data we have built an enzyme model (Figure 2, structure on the left) describing the active site that includes all aminoacids within 15 Å from the cobalt atom of Cob(III)Ado. The N- and C-termini were capped with NHMe and C(O)Me moieties respectively, where protein chains were truncated. The quantum part, treated using spin unrestricted procedure with BP86 functional[12] and the def-SV(P) basis set,[13] comprised 71 atoms including the corrin ring without sidearm chains, ribose as the upper ligand and imidazole as the lower ligand. The remaining part of the cofactor, reactant, and the active site residues were included in the MM part of the model and treated using the Amber force field.[14] MCA was truncated at the 15 Å boundary of the model. Valences of the QM along the QM-MM border were complemented by hydrogen link-atoms.

For the hydrogen atom transfer step the potential energy surface was modeled by employing a combined generalized hybrid orbital quantum mechanical/molecular mechanical (GHO-QM/MM) method. GHO approach was used to describe the boundary between the quantum mechanical (QM) subsystem, which involved the adenosine moiety and methylmalonyl-SCH_2C part of the substrate (altogether 46 atoms) and was treated by the AM1[15] method. The remaining part of the active site, the protein and the solvent were treated at the molecular mechanics (MM) level by CHARMM22[16] and TIP3P[17] force fields. The radical species were treated with UHF method. The entire model built, on the basis of one of the MCM crystal structures, contained 14878 atoms (Figure 2 structure on the right). This includes 672 aminoacids, 1388 water molecules, and 299 atoms of the active site residues. A potential of mean force (PMF)[18] was computed to average the dynamical effects over the ensemble of possible conformations of transition states along the reaction coordinate, which was defined as the difference of distances between the transferred hydrogen and the donor and acceptor carbons. The PMF was determined with the use of stochastic boundary molecular dynamics method (SBMD)[19].

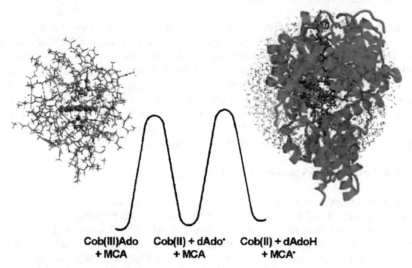

Cob(III)Ado	Cob(II) + dAdo·	Cob(II) + dAdoH
+ MCA	+ MCA	+ MCA·

Figure 2 *Energy diagram for the pre-steady-state steps of the MCM reaction with models used for homolysis and H-atom transfer steps*

The above model was then simplified to the "core" models, which included only fragments essential for the H atom transfer step of the mutase catalysis. Model M50 included 50 atoms and was analogous to the QM part in the QM/MM calculations. A smaller model (M37)[20] contained ribose ring instead of the adenosine moiety. The reactant in both cases was truncated by removing the CoA part, leaving only the methylmalonyl-SEt fragment. The rate constants within TST and CVT were carried out using MORATE program[21] for the electronic structure calculation using either AM1 or PM3 methods and POLYRATE[22] for the dynamics calculations. The transmission coefficients were computed using MEPSAG, CD-SCSAG, LCG3, COMT and μOMT methods. The calculations were limited to the experimentally explored temperature range, 5-20 °C.

Energetics obtained for both steps suggests strong stabilization of the intermediate radical pair by the active site; both activation barriers are nearly equal,[23,24] and close to the value of 18.8 kcal/mol observed experimentally. These results suggest that the commitment factor C (corresponding in the present case to the ratio of forward and reverse reactions of dAdo· radical) is close to unity. This implies that KIE$_{exp}$ corresponds to about half of the intrinsic KIE (cf., eq 1) which thus has a value of about 100. Such a large value suggests intervention of tunneling and agrees well with preliminary values obtained for small models[25] – see Table 1. Further studies of these steps are in progress.

Table 1 *CVT/μOMT D₃-KIEs obtained using semiempirical methods for electronic structure calculations*

	QM/model	AM1/M37	PM3[26]/M37	PM3/M50
T (°C)	*KIE_{exp}*			
5	49.9	127	113	96
20	35.6	89	94	80

The above example shows that QM/MM methods can be successfully applied to evaluations of the commitments to catalysis. In the following sections we will show applications of this methodology to the intrinsic KIEs on chemical steps of three enzymatic reactions for physiological and slow substrates.

2.2 Haloalkane dehalogenase, DhlA

Haloalkane dehalogenase from *Xanthobacter autotrophicus* GJ10 catalyzes dehalogenation of 1,2-dichloroethane to 2-chloroethanol. The mechanism of this reaction was studied extensively both theoretically[27,28] and experimentally.[20,29] The overall chemical catalysis proceeds in two S_N2 steps; in the first one carboxyl group of aspartate-124 acts as the nucleophile on the carbon of the substrate forming covalently bonded enzyme-intermediate complex and releasing chloride ion that is held in the active site by two strong hydrogen bonds from tryptophan 125 and 175 residues. In the subsequent step the enzyme-intermediate complex is hydrolyzed by a water molecule. At lower rates, the enzyme uses a number of other chlorinated alkanes.[30]

In an attempt to learn details of the transition state of the dehalogenation step we have studied theoretically and experimentally chloride KIEs for "physiological" substrate, dichloroethane, and for chlorobutane, the slow substrate.[6] After correcting for the presence of two equivalent positions of chlorine atom in 1,2-dichloroethane[31] we have found Cl-KIE_{exp} in both cases of about 1.0065.

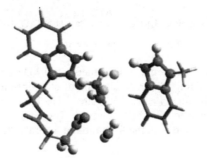

Figure 3 *Model of the transition state of the dehalogenation step within the active site of the haloalkane dehalogenase*

The chemical step of the enzymatic catalysis was modeled theoretically using ONIOM approach for both substrates. QM part,[a] treated at the B1LYP/6-31G(d) level of theory,[32] included the substrate, carboxylate part of the Asp124, and fragments of Trp125 and Trp175. The remaining part of the model that comprised the remaining parts of the aminoacids mentioned above and one water molecule was treated semiempirically using PM3 Hamiltonian (see Figure 3). Several tests that included changes in the size of the part of tryprophans included in QM calculations, inclusion of diffuse functions in the basis set, different semiempirical parametrization (AM1), and scalling of the frequencies were carried out to demonstrate that the model used is adequate. Isoeff98 program[33] was used for calculations of KIEs. The main results are summarized in Table 2

Table 2   *B1LYP/6-31G(d):PM3 Cl-KIEs on the dehalogenation*
step of the haloalkane dehalogenase-catalyzed reation

Trp model	"BuCl	1,2-dichloroethane
NH₃	1.0063	1.0065
pyrrole rings	1.0061	1.0063
indole rings	1.0065	not determined

Results collected in Table 2 indicate that theoretically predicted KIEs on the dehalogenation step for both substrates are very close to the experimentally determined value, suggesting that KIEs$_{exp}$ in these cases are practically equal to the intrinsic values, and the corresponding C parameter is effectively zero. Furthermore, intrinsic KIEs for both substrates are practically identical. Thus in the case of DhlA the slow-substrate hypothesis is applicable.

2.3 Aspartate transcarbamoylase, ATCase

ATCase catalyzes reaction of carbamyl phosphate and L-aspartate to yield carbamylaspartate and inorganic phosphate (Scheme 1) that is the first committed step in pyrimidine biosynthesis in many prokaryotes.

$$H_2N-C(O)-PO_3^{2-} + {}^-O_2C-CH_2-CH(N^+H_3)-CO_2^- \overset{\text{ATCase}}{\rightleftharpoons} H_2N-C(O)-NH-CH(CO_2^-)-CH_2-CO_2^- + P_i$$

Scheme 1

The enzyme is organized as two catalytic trimers (100 kDa each) and three regulatory dimers (34 kDa each). Binding of substrates promotes a conformational change from the low activity (T) state to the highly active (R) state.[34] The holoenzyme, but not the isolated catalytic trimer, is subjected to allosteric control.[35] The enzymatic reaction is thought to proceed by addition-elimination with the formation of a tetrahedral intermediate.[36]

---

[a] note that formally these calculations were carried at the QM/QM level where the smaller fragment was treated at the DFT level of theory while the whole model was treated semiempirically. For comparison with strict QM/MM formalism we refer here to the DFT part as the QM level.

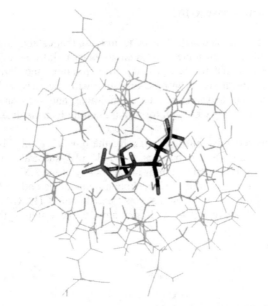

Figure 4   *Model of the ATCase active site used in the QM/MM calculations; reactants are rendered as tubes while remaining part of the model as wire-frame. Positions of MM atoms rendered in grey were constrained*

Calculations were carried out[37] on a model (Figure 4) that included 23 aminoacids which are either within the active site or are essential for the catalysis, and 11 molecules of water (233 heavy atoms and 257 hydrogens). The QM region included only reactants and was treated at the PM3 level. The remaining part of the system was treated using ESFF force field.[38] In the classical region only atoms within 3.8 Å from the reactants were subjected to geometry optimization while coordinates of the outer sphere were kept constrained. Intrinsic isotope effect has been modeled for this reaction for physiological reactant and the slow substrate, cysteine sulphate which differs from L-aspartate in that carbon of the carboxyl group is replaced by the sulphur atom. Theoretically obtained KIEs are compared with the experimental ones in Table 3. They indicate that for the slow substrate the commitment is practically wiped out and the KIE_{exp} corresponds directly to the intrinsic KIE. The two intrinsic values are slightly different but not to a point where using slow-substrate approach to intrinsic KIE would change conclusions. In the case of the physiological substrate sizable commitment masks about 30 % of the intrinsic KIE.

Table 3   *PM3/ESFF intrinsic carbon KIEs on the reaction catalyzed by ATCase*

substrate	KIE_{exp}	calculated KIE
L-aspartate	1.0240 ± 0.0005	1.0356
cysteine sulphate	1.0388 ± 0.0008	1.0383

2.4 Ornithine decarboxylase, ODC

ODC is a pyridoxal 5′-phosphate (PLP) dependent lyase that catalyzes the decarboxylation of ornithine to produce putrescine (1,4-diaminobutane), leading eventually to higher polyamines that are essential for cell growth, differentiation, and division.[39] The central steps of the catalysis are the reaction of L-ornithine with the enzyme-bound cofactor via transaldimination reaction to form an external aldimine, and subsequent cleavage of the carboxylate bond that releases CO_2, and forms a quinonoid intermediate.

We have studied theoretically[5] intrinsic KIEs for the physiological substrate, L-ornithine, and the slow substrate, lysine. In both cases whole PLP-substrate Schiff-base complex illustrated in Figure 5 was treated quantum-mechanically at the AM1 level. The remaining part of the system that consisted of 759 backbone aminoacids and 1986 water molecules resulting in a total of nearly 18000 atoms, was treated using CHARMM force field in the same way as described above for the hydrogen atom transfer step of the MCM-catalyzed reaction.

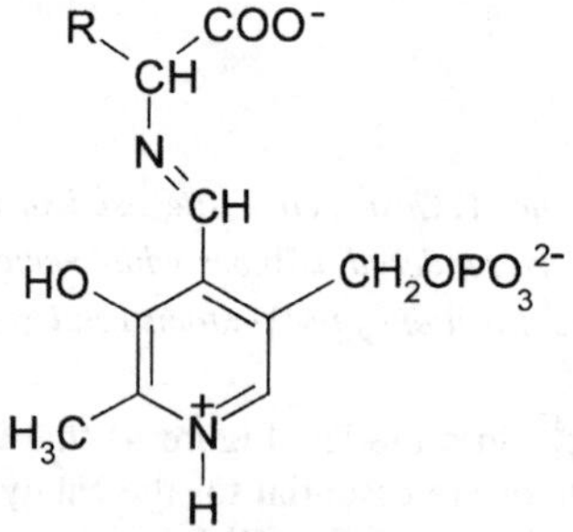

Figure 5 *QM part of the AM1/CHARMM model of the ODC reaction*

The results are summarized in Table 4. As can be seen from the comparison of the values reported in the last column, intrinsic KIEs for the two substrates are substantially different. The experimental C13-KIE for lysine is about twice as large as the one found for the physiological substrate.[40] This would, within slow-substrate hypothesis, imply a commitment of about unity. Presented results, however, show that this interpretation is not correct for the ODC reaction. The intrinsic KIE for L-ornithine is also quite close to the KIE_{exp} measured for this substrate indicating that also in this case chemical step is mostly rate-limiting. This is the first case documenting intrinsic KIEs for physiological and slow substrates being so different and providing first example when using KIE_{exp} for the slow substrate may lead to erroneous conclusions.

Table 4   *AM1/CHARMM intrinsic carbon KIEs on the reaction catalyzed by ODC*

substrate	KIE_{exp}	calculated KIE
L-ornithine	1.0325 ± 0.0002	1.0411
lysine	1.0633 ± 0.0003	1.0587

3 CONCLUSION

The examples shown above indicate that the current QM/MM techniques enable reliable theoretical predictions of the intrinsic KIEs and partitioning ratios that can be used in turn for analysis of mechanisms of enzymatic reactions and in particular the transition states of chemical steps in enzyme-catalyzed processes. Comparison of calculated intrinsic KIEs for physiological and slow substrates suggest that it is not always safe to draw conclusions regarding the structure of a transition state for the physiological substrate from kinetic isotope effects measured for slow substrates. It is, however, hard to predict *a priori* if such assumption is safe or not. Thus the results discussed herein argue that in the lieu of the modeled intrinsic KIEs for both substrates it is safer to use experimental methods of lowering the magnitude of the commitment to catalysis for the physiological substrate than studying KIEs for slow substrates.

Acknowledgements

Studies presented in this communication were supported by grants from the State Committee for Scientific Research (KBN, Poland) and National Institute of Health (FIRCA, USA). Supercomputer facilities at MSI (Minneapolis), ICM (Warsaw), Emory University (Atlanta), PCSS (Poznan), and Cyfronet (Cracow) were used.

References

1   M. H. O'Leary and P. Paneth, *Bioact. Mol.* 1987, **3**, 303.

2   J. A. Smiley, P. Paneth, M. H. O'Leary, J. B. Bell and M. E. Jones, *Biochemistry* 1991, **30**, 6216.

3   (a) P. Paneth and M. H. O'Leary, *Biochemistry* 1985, **24**, 5143. (b) P. Paneth and M. H. O'Leary, *J. Am. Chem. Soc.* 1985, **107**, 7381. (c) P. Paneth and M. H. O'Leary, *Biochemistry* 1987, **26**, 1728.

4   J. Pawlak, M. H. O'Leary and P. Paneth, *THEOCHEM* 1998, **454**, 69.

5   D. Sicińska, D. G. Truhlar and P. Paneth, *J. Am. Chem. Soc.* 2005, **127**, ASAP.

6   A. Lewandowicz, J. Rudziński, L. Tronstad, M. Widersten, P. Ryberg, O. Matsson and P. Paneth, *J. Am. Chem. Soc.* 2001, **123**, 4550.

7   (a) J. D. Hermes, P. A. Tipton, M. A. Fisher, M. H. O'Leary, J. F. Morrison and W. W. Cleland, *Biochemistry* 1984, **23**, 6263. (b) P. M. Weiss, C. Y. Chen, W. W. Cleland and P. F. Cook, *Biochemistry* 1988, **27**, 4814. (c) L. E. Parmentier and K. Smith, *J. Biochim. Biophys. Acta* 1998, **1382**, 333. (d) D. J. Merkler, P. C. Kline, P. Weiss and V. L. Schramm *Biochemistry* 1993, **32**, 12993. (e) J. L. Urbauer, D. E. Bradshaw and W. W. Cleland, *Biochemistry* 1998, **37**, 18026. (f) (g) A. C. Drohat, J. Jagadeesh, E. Ferguson and J. T. Stivers, *Biochemistry* 1999, **38**, 11866. (h) V. L. Schramm, *Methods, Enzymol.* 1999, **308**, 301. (i) M. Bruner and B. A. Horenstein, *Biochemistry* 2000, **39**, 2261. (j) J. Yang, S. Schenkman and B. A. Horenstein, *Biochemistry* 2000, **39**, 5902. (k) B. Gerratana, P. A. Frey and W. W. Cleland, *Biochemistry* 2001, **40**, 2972. (l) T. Swanson, H. B. Brooks, A. L. Osterman, M. H. O'Leary and M. A. Phillips, *Biochemistry* 1998, **37**, 14943. (m) M. J. Snider, L. Reinhardt, R. Wolfenden, R. And W. W. Cleland, *Biochemistry* 2002, **41**, 415. (n) R. Zheng, R. and J. S. Blanchard, *Biochemistry* 2003, **42**, 11289.

8   A. Warshel and M. Levitt, *J. Mol. Biol.* 1976, **103**, 227.

9   R. Banerjee, *Chem. Rev.* 2003, **103**, 2083 and references therein.

10   S. Chowdhury, R. Banerjee, *J. Am. Chem. Soc.* 2000, **122**, 5417.

11   S. Dapprich, I. Komáromi, K. S. Byun, K. Morokuma and M.-J. Frisch, *J. Mol. Struct.* *(THEOCHEM)* 1999, **461**, 1 and references therein.

12   (a) A. D. Becke, *Phys. Rev. A* 1988, **38**, 3098. (b) J. P. Perdew, *Phys. Rev. B* 1986, **33**, 8822.

13   A. Schaefer, H. Horn, and R. Ahlrichs, *J. Chem. Phys.* 1992, **97**, 2571.

14   (a) W. D. Cornell, P. Cieplak, C. I. Bayly, I. R. Gould, K. M. Merz Jr., D. M. Ferguson, D. C. Spellmeyer, T. Fox, J. W. Caldwell and P. A. Kollman, *J. Am. Chem. Soc.* 1995, **117**, 5179. (b) H. M. Marques, B. Ngoma, T. J. Egan and K. L. Brown, *J. Mol. Struct.* 2001, **561**, 71. (c) H. M. Marques and K. L. Brown, *THEOCHEM* 1995, **340**, 97.

15   (a) M. J. S. Dewar, E. G. Zoebisch, E. F. Healy and J. J. P. Stewart, *J. Am. Chem. Soc.* 1985, 107, 3902. (b) M. J. S. Dewar, C. Jie and G. Yu, *Tetrahedron*, 1993, **23**, 5003; (c) A. J. Holder, R. D. Dennington, C. Jie and G. Yu, *Tetrahedron*, 1994, **50**, 627.

16   (a) B. R. Brooks, R. E. Bruccoleri, B. D. Olafson, D. J. States, S. Swaminathan and M. Karplus, *J. Comp. Chem.* 1983, **4**, 187. (b) A. D. Mackerell, Jr., D. Bashford, M. Bellott, R. L. Dunbrack, J. D. Evanseck, M. J. Field, S. Fischer, J. Gao, H. Guo, S. Ha, D. Joseph-McCarthy, L. Kuchnir, K. Kuczera, F. T. K. Lau, C. Mattos, S. Michnick, T. Ngo, D. T. Nguyen, B. Prodhom, W. E. Reiher, III, B. Roux, M. Schlenkrich, J. C. Smith, R. Stote, J. Straub, M. Watanabe, J. Wiórkiewicz-Kuczera, D. Yin, M. Karplus, *J. Phys. Chem. B* 1998, **102**, 3586.

17   W. L. Jorgensen, J. Chandrasekhar, J. D. Madura, R. W. Impey, M. L. Klein, *J. Chem. Phys.* 1983, **79**, 926.

18   J. P. Valleau, G. M. Torrie, In *Statistical Mechanics*, Part A, Berne, B.J. (Ed.) Plenum, New York, 1977, p. 137.

19   S. L. III Brooks, A. Brünger, M. Karplus, *Biopolymers*, 1985, **24**, 843.

20   A. Dybała-Defratyka and P. Paneth, *J. Inorg. Biochem.* 2001, **86**, 681.

21   Y. -Y. Chuang, J. C. Corchado, P. L. Fast, W.-P. Hu, G. C. Lynch, Y.-P. Liu, D. G. Truhlar, MORATE –Version 8.5, University of Minnesota, Minneapolis, 2001.

22   Y.-Y. Chuang, J. C. Corchado, P. L. Fast, J. Villà, W. -P. Hu, Y.-P. Liu, G. C. Lynch, C. F. Jackels, K. A. Nguyen, M. Z. Gu, I. Rossi, E. L. Coitiño, S. Clayton, V. S. Melissas, B. J. Lynch, A. Fernandez-Ramos, R. B. Steckler, C. Garrett, A. D. Isaacson, D. G. Truhlar, POLYRATE–Version 8.5.1, University of Minnesota, Minneapolis, 2000.

23   R. A. Kwiecień, I. V. Khavrutskii, D. G. Musaev, K. Morokuma, R. Banerjee and P. Paneth, submitted.

24   A. Dybała-Defratyka, P. Paneth, R. Banerjee, D. G. Truhlar, in preparation.

25   R. Banerjee, D. G. Truhlar, A. Dybała-Defratyka and P. Paneth, "H-atom Transfers in B12 Enzymes" in "Handbook of Hydrogen Transfer" vol. 2, J. Klinman and R. L. Schowen (Eds.) "Biological Aspects of Hydrogen Transfer" Wiley-VCH, Germany, 2005, in press.

26   J. J. P. Stewart, *J. Comput. Chem.* 1989, **10**, 209 and 221.

27   A. Soriano, E. Silla, I. Tuñón and M. F. Ruiz-López, *J. Am. Chem. Soc.* 2005, **127**, 1946 and references therein.

28   P. Paneth, *Acc. Chem. Res.* 2003, **36**, 120.

29   T. Kurihara, N. Esaki and K. Soda, *J. Mol. Catalysis B* 2000, **10**, 57 and references therein.

30   (a) B. W. Dijkstra, J. Kingma and D. B. Janssen, *J. Biol. Chem.* 1996, **271**, 14747. (b) G. H. Krooshof, I. S. Ridder, A. W. J. W. Tepper, G. J. Vos, H. J. Rozeboom, K. H.

Kalk, B. W. Dijkstra and D. B. Janssen, *Biochemistry* 1998, **37**, 15013. (c) J. P. Schanstra and D. B. Janssen, D. B. *Biochemistry* 1996, 35, 5624.

31 P. Paneth, "Chlorine Isotope Effects in Biological Systems" in A. Kohen and H. H. Limbach (Eds.) "Isotope Effects in Chemistry and Biology" CRC Press, Inc, 2005, in press.

32 (a) D. A. Becke, *J. Chem. Phys.* 1996, **104**, 1040. (b) C. Lee, W. Yang and R. G. Parr, *Phys. Rev. B* 1988, **37**, 785. (c) P. C. Hariharan and J. A. Pople, *Theor. Chim. Acta* 1973, 28, 213. (d) M. M. Francl, W. J. Pietro, W. J. Hehre, J. S. Binkley, M. S. Gordon, D. J. DeFrees and J. A. Pople, *J. Chem. Phys.* 1982, 77, 3654.

33 V. Anisimov and P. Paneth, *J. Math. Chem.* 1999, **26**, 75.

34 (a) J. C. Gerhart and A. B. Pardee, *J. Biol. Chem.* 1962, **237**, 891. (b) M.R. Bethell, K. E. Smith, J. S. White and M. E. Jones, *Proc. Natl. Acad. Sci. U.S.A.* 1968, **60**, 1442.

35 J. C. Gerhart and H. K. Schachman, Biochemistry 1965, **4**, 1054.

36 G. L. Waldrop, J. L. Urbauer and W. W. Cleland, *J. Am. Chem. Soc.* 1992, **114**, 5941.

37 J. Pawlak, M. H.O'Leary and P. Paneth, *ACS Symp. Ser.* 1999, 721462.

38 S. Shi, L. Yan, Y. Yang, J. Fisher-Shaulsky and T. Thacher, *J. Comput. Chem.* 2003, **24**, 1059.

39 (a) L. K. Jackson, H. B. Brooks, A. L. Osterman, E. J. Goldsmith and M. A. Phillips, M. A. *Biochemistry* 2000, **39**, 11247. (b) A. L. Osterman, L. N. Kinch, N. V. Grishin and M. A. Phillips, *J. Biol. Chem.* 1995, **270**, 11797. (c) K. E. Tobias and C. Kahana, *Biochemistry* 1993, **32**, 5842.

40 T. Swanson, H. B. Brooks, A. L. Osterman, M. H. O'Leary and M. A. Phillips, *Biochemistry* 1998, **37**, 14943.

SUICIDE INACTIVATION IN THE COENZYME B$_{12}$-DEPENDENT ENZYME DIOL DEHYDRATASE

Gregory M. Sandala,[1,2] David M. Smith,[3] Michelle L. Coote,[2] and Leo Radom[1,2]

[1]School of Chemistry, University of Sydney, Sydney, NSW 2006, Australia. E-mail: radom@chem.usyd.edu.au
[2]Research School of Chemistry, Australian National University, Canberra, ACT 0200, Australia.
[3]Division of Organic Chemistry and Biochemistry, Rudjer Boskovic Institute, Zagreb, 10002, Croatia. E-mail: dsmith@irb.hr

1 INTRODUCTION

Diol dehydratase (DDH) is a coenzyme B$_{12}$-dependent enzyme that mediates the transformation of vicinal diols, e.g., ethane-1,2-diol (**1a**), into their corresponding aldehydes, e.g., acetaldehyde (**2a**), plus water:[1]

1a (R = H)   **2a** (R = H)
1b (R = CH$_3$)   **2b** (R = CH$_3$)

Like many other coenzyme B$_{12}$-dependent enzymes, this reaction is believed to operate via a radical mechanism, whereby the 5'-deoxyadenosyl radical (Ado–CH$_2\bullet$), derived from homolysis of the Co–C bond of adenosylcobalamin (AdoCbl), is intimately involved.[2]

Figure 1 depicts the generally accepted mechanism for the reaction catalyzed by DDH.[1] The presence of substrate (**1**) triggers the homolytic cleavage of the Co–C bond of AdoCbl yielding the cob(II)alamin and 5'-deoxyadenosyl (Ado–CH$_2\bullet$) radicals. Substrate catalysis begins with an initial hydrogen-atom abstraction by Ado–CH$_2\bullet$ from **1** to generate 5'-deoxyadenosine (Ado–CH$_3$) and a substrate-derived radical **3**, respectively (Step **A**). Then, DDH-assisted rearrangement of the substrate-derived radical **3** to the product-related radical **4** occurs (Step **B**). To complete the catalytic cycle, the product-related radical **4** re-abstracts a hydrogen atom from the unactivated methyl group of Ado–CH$_3$ to re-generate the 5'-deoxyadenosyl radical (Ado–CH$_2\bullet$) and form the closed-shell hydrated product **5** (Step **C**). Finally, enzyme-catalyzed dehydration of **5** occurs to yield the product aldehyde **2** plus water (Step **D**).

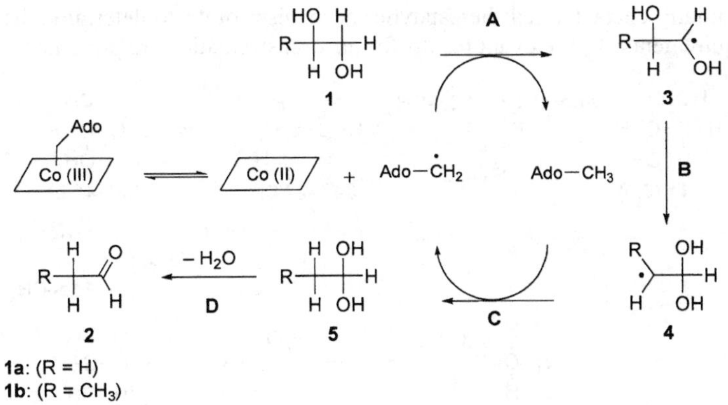

Figure 1 *Currently accepted mechanism for the reaction catalyzed by DDH*

The interplay of AdoCbl and apoenzyme DDH illustrates an elegant and efficient duality. Clearly, the contribution from each is critical to the successful catalysis of any substrate. At the same time, abnormal behaviour of either component will potentially result in some form of enzyme dysfunction. This latter fact has prompted researchers to develop clever methodologies for disrupting efficient catalysis in coenzyme B_{12}-dependent enzymes with the aim of probing their mechanisms.

Perhaps the most straightforward method for gaining insight into the nature of enzyme-catalyzed reactions is through the use of substrate analogues that result in mechanism-based inactivation. For DDH, such studies have been performed a considerable time ago.[3,4] However, the interpretation of these results has only recently been explored in detail.[5] To investigate these interpretations and to obtain a greater understanding of substrate-induced inactivation of DDH, we have carried out high-level ab initio calculations on appropriate small model systems.[6]

2 THEORETICAL METHODOLOGY

Standard ab initio[7] and density functional theory[8] calculations have been used in the present study. Geometries and scaled vibrational frequencies have been obtained at the MPW1K/6-31+G(d,p) level of theory.[9] Relative energies were derived with the high-level composite method G3(MP2)-RAD.[10] All energies refer to isolated molecules in the gas phase at 0 K. Ethanol was chosen as a model for 5'-deoxyadenosine (Ado–CH_3) in the abstraction reactions. All calculations were performed with the MOLPRO 2002.6[11] and Gaussian 03[12] programs.

3 RESULTS AND DISCUSSION

3.1 Mechanism for the Catalytic Substrate Ethane-1,2-diol

The ability of DDH to facilitate the transformation of vicinal diols into their corresponding aldehydes is exemplified by its reaction with ethane-1,2-diol (**1a**).[1] In order to gain insight into the nature of substrate-analogue-induced suicide inactivation of DDH, we have initially characterized the catalytic mechanism of **1a** to determine the associated energy requirements. The relevant results for this transformation are presented in Figure 2.

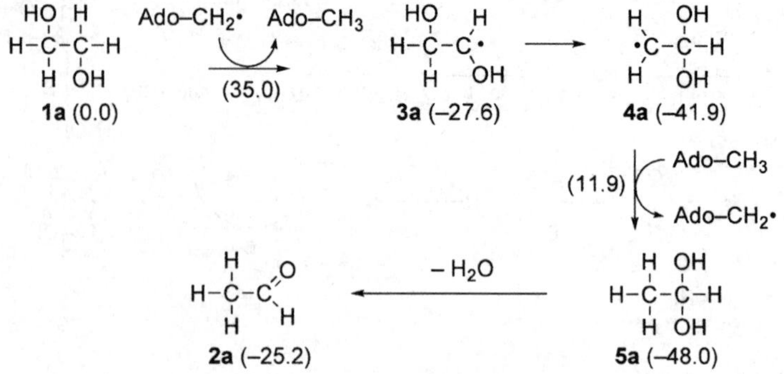

Figure 2 *Calculated energy requirements for the DDH-catalyzed reaction of ethane-1,2-diol (relative energies in parentheses, kJ mol^{-1})*

Substrate catalysis begins with an initial hydrogen-atom abstraction from **1a** by 5'-deoxyadenosyl radical (Ado–CH$_2$•) to form a substrate-derived radical **3a** and 5'-deoxyadenosine (Ado–CH$_3$), respectively. The formation of **3a** is exothermic by 27.6 kJ mol^{-1} and the reaction must overcome a barrier of 35.0 kJ mol^{-1}. Once formed, **3** rearranges to the product-related radical (**4a**) in a process that is also calculated to be exothermic, in this case by 14.3 kJ mol^{-1}. The interested reader is referred to earlier theoretical work from our group for an account of the salient features of this rearrangement step including its facilitation by so-called push-pull catalysis.[13]

Catalysis continues with the re-abstraction of a hydrogen atom by the product-related radical **4a** from Ado–CH$_3$. The barrier for this step is calculated to be 53.8 kJ mol^{-1} with an associated exothermicity of 6.1 kJ mol^{-1}. This critical step serves two purposes. Firstly, the closed-shell hydrated product (**5a**) is formed. In addition, Ado–CH$_2$• is re-generated, allowing it to re-combine with the enzyme-bound cob(II)alamin radical and dissociate freely from DDH. This enables the catalytic cycle to continue. The final step in the catalytic transformation of **1a** to products is a 22.8 kJ mol^{-1} endothermic dehydration of **5a** to form water plus acetaldehyde (**2a**).

With the energetic requirements for DDH and its catalytic conversion of ethane-1,2-diol in hand, we now turn our attention to mechanisms that may induce suicide inactivation of DDH upon incubation with the substrate analogues glycolaldehyde (Section 3.2) and 2-chloroacetaldehyde (Section 3.3).

3.2 Mechanism for the Substrate Analogue Glycolaldehyde

Mechanism-based inactivation of DDH by glycolaldehyde (**6**) has been known to occur for some time.[3a] Originally, the aerobic products were identified as glyoxal and 5'-deoxyadenosine, though at that time a mechanism for the inactivation process was not proposed.[3a] Recently, however, Frey, Reed and co-workers have proposed such a mechanism and, on the basis of EPR spectroscopic results, identified the causative agent for inactivation as *cis*-ethanesemidione radical (**10**).[5a] To explore this proposal further, we have calculated relevant barriers and thermodynamics for the reaction of glycolaldehyde (**6**) with 5'-deoxyadenosyl radical (Ado–CH$_2$•) (Figure 3).

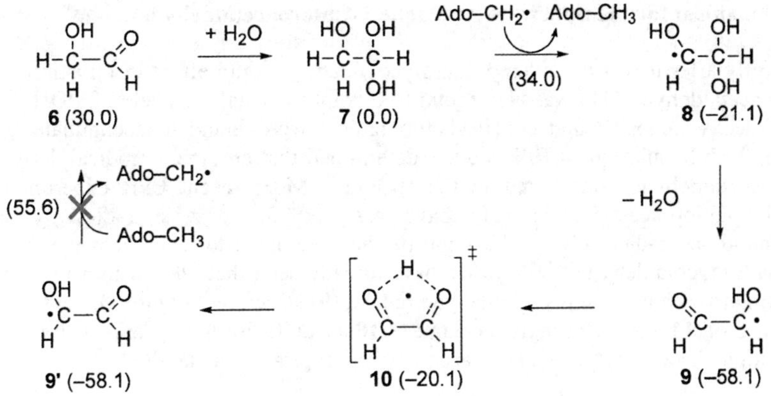

Figure 3 *Proposed mechanism for the DDH-catalyzed reaction of glycolaldehyde (relative energies in parentheses, kJ mol^{-1})*

The initial step is proposed to be a 30 kJ mol^{-1} exothermic hydration of glycolaldehyde (**6**) to form ethane-1,1,2-triol (**7**). Structurally, this hydrate (**7**) is very similar to the catalytic substrate ethane-1,2-diol (**1a**, Figure 2). On this basis alone, it is therefore not too surprising that DDH can bind and process (at least partially) this species.

The partial turnover of the hydrate of glycolaldehyde (**7**) begins with a hydrogen-atom abstraction from C2 by 5'-deoxyadenosyl radical (Ado–CH$_2$•) to generate the hydrated radical (**8**). This abstraction step must overcome a barrier of 34.0 kJ mol^{-1} and is exothermic by 21.1 kJ mol^{-1}. These results are very similar to those from the catalytic mechanism (35.0 and 27.6 kJ mol^{-1}, respectively, Figure 2). It is in the next step, however, where the catalytic and inactivation mechanisms begin to diverge. In the catalytic mechanism, the hydrogen-abstraction step (**1a**→**3a**) is followed by a 1,2-hydroxyl shift (**3a**→**4a**). While this may well occur in the inactivation mechanism, it would merely serve to generate an identical structure (**8**→**8'**). Thus, it is proposed that the suicidal tendencies of DDH in this context begin with the exothermic (by 37.0 kJ mol^{-1}) loss of water from **8** (or **8'**) to form glycolaldehyde radical (**9**).[14] In slight contrast to the original interpretation of the EPR spectrum,[5a] we find that *cis*-ethanesemidione radical (**10**) is not actually a minimum on the potential energy surface. Rather, it serves as a transition structure for the degenerate rearrangement of **9**, which is predicted to be associated with a barrier of 38.0 kJ mol^{-1}. On this basis, it is therefore quite likely that the observed EPR spectra correspond to an equilibrium mixture of glycolaldehyde radicals (**9** and **9'**) strongly interacting with amino acid residues within the active site of DDH, as opposed to a stable *cis*-ethanesemidione radical (**10**).[6]

Regardless of the detailed nature of the connectivity of **9** and **10**, for a catalytic mechanism to continue, a hydrogen atom must be re-abstracted from Ado–CH$_3$. Our calculations predict that for glycolaldehyde radical (**9**) to accomplish this task, an energetic barrier of 113.7 kJ mol^{-1} must be surmounted! Combining this result with the associated endothermicity of 88.1 kJ mol^{-1} makes it clear that **9** is a very stable radical species, a fact that can be attributed to the captodative stabilization provided to the radical center.[15] It is a direct result of this stabilization, and the concomitant inability of **9** to accomplish the required regeneration of Ado–CH$_2$•, that the cofactor cannot be reformed and the catalytic cycle becomes internally terminated.

3.3 Mechanism for the Substrate Analogue 2-Chloroacetaldehyde

Shortly after it was discovered that glycolaldehyde could effect inactivation of DDH, 2-chloroacetaldehyde (11) was also found to disrupt the catalytic cycle of DDH.[3b] At that time, 5'-deoxyadenosine and cob(II)alamin radical were found to accumulate from the reaction,[3b] while subsequent EPR studies determined that an organic radical derived from 2-chloroacetaldehyde contributed to the spectra.[16] More recent EPR experiments with labelled 2-chloroacetaldehyde (11) have assigned the organic radical to be *cis*-ethanesemidione radical (10),[5b] the same product predicted to arise from the reaction of DDH with glycolaldehyde.[5a] While we have already seen that 10 is a transition structure connecting two equivalent glycolaldehyde radical (9) structures (Figure 3), the fact that the processing of 11 can seemingly generate 9/10 is quite intriguing and warrants further examination. Accordingly, we have carried out appropriate calculations.

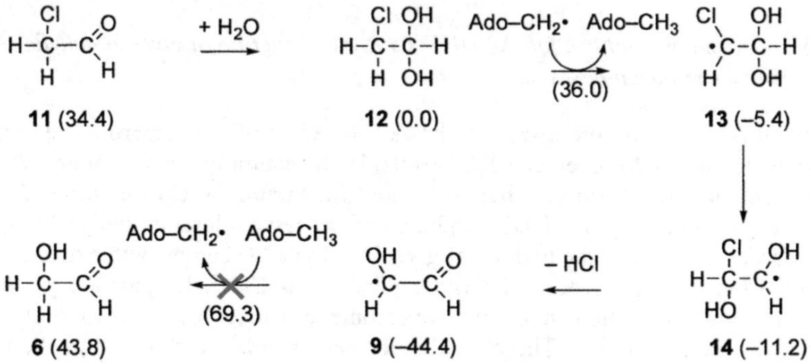

Figure 4 *Proposed mechanism for the DDH-catalyzed reaction of 2-chloroacetaldehyde (relative energies in parentheses, kJ mol^{-1})*

The similar EPR spectra[5] observed following the reactions of DDH with glycolaldehyde (6, Figure 3) and 2-chloroacetaldehyde (11, Figure 4) suggest that related reaction mechanisms are operative. Figure 4 depicts a proposed inactivation mechanism of DDH by 2-chloroacetaldehyde that is similar to that in Figure 3. The first step involves a 34.4 kJ mol^{-1} exothermic hydration of 11 to give 1,1-dihydroxy-2-chloroethane (12). 5'-deoxyadenosyl radical (Ado–CH$_2$•) then abstracts the C2 hydrogen atom from 12 to form the substrate-derived radical 13 and Ado–CH$_3$. This process is calculated to be mildly exothermic (by 5.4 kJ mol^{-1}) with a modest barrier (36.0 kJ mol^{-1}). As opposed to the situation shown in Figure 3, a 1,2-migration of a hydroxyl group at this point would lead to the distinct but isomeric radical 14. While this shift is calculated to be slightly exothermic (by 5.8 kJ mol^{-1}), it can be seen that the subsequent loss of HCl, which generates the glycolaldehyde radical (9), is even more exothermic (by 33.2 kJ mol^{-1}).

As we have seen previously, any attempt by 9 to re-abstract a hydrogen atom from Ado–CH$_3$ is prohibited by the very large endothermicity (88.1 kJ mol^{-1}) of the reaction. Therefore, in a manner identical to that operative in the inactivation of DDH by glycolaldehyde (6), the generation of the highly stabilized glycolaldehyde radical (9) from 2-chloroacetaldehyde (11) disrupts the ability of the cofactor to reform and therefore halts catalysis.

4 CONCLUSION

The ability of the substrate analogues glycolaldehyde (**6**) and 2-chloroacetaldehyde (**11**) to trigger self-inflicted suicide inactivation of holoenzyme DDH reflects the delicate balance that must be maintained for enzymes that operate with radical mechanisms. The turnover of both **6** and **11** proceeds along catalytic lines in a manner that ultimately leads to the highly stabilized glycolaldehyde radical (**9**). The enhanced stability of **9** manifests itself in the prohibitively large endothermicity associated with the hydrogen re-abstraction step from 5'-deoxyadenosine. The observed single turnover suicide inactivation of DDH is therefore a consequence of the lack of any feasible means to regenerate the 5'-deoxyadenosyl radical, which is thus unable to recombine with cob(II)alamin and dissociate from DDH. As a result, the catalytic cycle of the enzyme is terminated.

Acknowledgements

We gratefully acknowledge the award (to G.M.S.) of an Australian National University International Postgraduate Research Scholarship, the award (to M.L.C. and L.R.) of Australian Research Council Discovery grants, the appointment (of D.M.S.) as a Visiting Fellow in the School of Chemistry, University of Sydney, and generous allocations of supercomputer time from the Australian Partnership for Advanced Computing and the Australian National University Supercomputing Facility.

References

1   For recent reviews, see for example: (a) T. Toraya, *Chem. Rev.*, 2003, **103**, 2095. (b) P.A. Frey, *Chem. Rec.*, 2001, **1**, 277.
2   R. Banerjee, *Chemistry and Biochemistry of B$_{12}$*, Wiley: New York, 1999.
3   (a) O.W. Wagner, H.A. Lee, P.A. Frey and R.H. Abeles, *J. Biol. Chem.*, 1966, **249**, 1751. (b) T.H. Finlay, J. Valinsky, K. Sato, R.H. Abeles, *J. Biol. Chem.*, 1972, **247**, 4197.
4   W.W. Bachovchin, R.G. Eagar, Jr., K.W. Moore and J.H. Richards, *Biochemistry*, 1977, **16**, 1082.
5   (a) A. Abend, V. Bandarian, G.H. Reed, and P.A. Frey, *Biochemistry*, 2000, **39**, 6250. (b) P. Schwartz, R. LoBrutto, G.H. Reed and P.A. Frey, *Helv. Chim. Acta*, 2003, **86**, 3764.
6   For a preliminary account of part of this work, see G.M. Sandala, D.M. Smith, M.L. Coote and L. Radom, *J. Am. Chem. Soc.*, 2004, **126**, 12206.
7   (a) W.J. Hehre, L. Radom, P.v.R. Schleyer and J.A. Pople, *Ab Initio Molecular Orbital Theory*, Wiley, New York, 1986. (b) F. Jensen, *Introduction to Computational Chemistry*, Wiley, New York, 1999.
8   W. Koch and M.C. Holthausen, *A Chemist's Guide to Density Functional Theory*, Wiley-VCH, Weinheim, 2000.
9   (a) B.J. Lynch, P.L. Fast, M. Harris and D.G. Truhlar, *J. Phys. Chem. A.*, 2000, **104**, 4811. (b) B.J. Lynch and D.G. Truhlar, *J. Phys. Chem. A.*, 2001, **105**, 2936.
10   (a) D.J. Henry, M.B. Sullivan and L. Radom, *J. Chem. Phys.*, 2003, **118**, 4849. (b) D.J. Henry, C.J. Parkinson, P.M. Mayer and L. Radom, *J. Phys. Chem. A.*, 2001, **105**, 6750.
11   MOLPRO 2002.6 is a package of ab initio programs written by H.-J. Werner et al.
12   M.J. Frisch et al. Gaussian, Inc., Wallingford CT, 2004.

13 (a) D.M. Smith, B.T. Golding and L. Radom, *J. Am. Chem. Soc.*, 1999, **121**, 5700. (b) D.M. Smith, B.T. Golding and L. Radom, *J. Am. Chem. Soc.*, 2001, **123**, 1664.

14 It is worth noting that the analogous loss of water (from **4a**) could potentially occur in the mechanism shown in Figure 2. However, in that case, our calculations predict this to be 13.9 kJ mol^{-1} less advantageous than that (from **8**) shown in Figure 3.

15 H.-G. Viehe, Z. Janousek, R. Merényi and L. Stella, *Acc. Chem. Res.*, 1985, **18**, 148.

16 J.E. Valinsky, R.H. Abeles and A.S. Mildvan, *J. Biol. Chem.*, 1974, **249**, 2751.

SIMULATIONS OF PHOSPHORYL TRANSFER REACTIONS USING MULTI-SCALE QUANTUM MODELS

Brent A. Gregersen[1], Timothy J. Giese[1], Yun Liu[1], Evelyn Mayaan[1], Kwangho Nam[1], Kevin Range[1], and Darrin M. York[1,*]

[1]Department of Chemistry, University of Minnesota, 207 Pleasant St. SE, Minneapolis, MN 55455–0431
* E-mail: york@chem.umn.edu

1 INTRODUCTION

Computer simulation provides a tool of enormous potential impact in problems of biocatalysis.[1] From a theoretical perspective, ribozymes present several features that make them more difficult to model relative to most proteins. RNA is inherently highly charged and interacts strongly with monovalent and divalent metal ions and solvent. Careful treatment of long-range electrostatics is critical, as is the inclusion and equilibration of a sufficiently extensive solvation and ion atmosphere. The importance of quantum many-body effects such as polarization and charge transfer, which are neglected in conventional molecular mechanical force fields, is amplified in RNA simulations. Reactions catalyzed by RNA can exhibit fairly large scale conformational changes, such as those encountered in the hammerhead ribozyme, that either precede or occur in concert with reactive chemical steps. The chemistry of many RNA-catalyzed reactions, such as transphosphorylation and phosphate hydrolysis, involves ionic interactions, large polarization effects and transitions between tri-, tetra- and pentavalent phosphorus that require an accurate d-orbital quantum model for a proper description. Reliable molecular simulations of phosphoryl transfer reactions, therefore, need to take into account accurate quantum models, complex macromolecular, ionic and solvent environments, and extensive conformational sampling. Consequently, making accurate predictions about the mechanism and rates of phosphoryl transfer reactions requires theoretical methods that are robust and reliable over a broad range of time and length scales.

The present work describes a multi-faceted theoretical approach toward the development of methods that allow simulations of phosphoryl transfer reactions to be performed with increased reliability and predictive capability. The focus is to outline recent progress in the design of new multi-scale quantum models to study phosphoryl transfer reactions in non-enzymatic and enzymatic environments. Here, "multi-scale" implies the integration of a hierarchy of methods that simultaneously span a broad range of spatial and temporal domains and work in concert to provide insight into complex problems.

2 METHODS

The present work describes an approach toward the design of multi-scale quantum models for phosphoryl transfer reactions. In this section, a brief overview is given of some of the general theoretical methods and computational details

2.1 Density-functional calculations

Density-functional calculations were performed using the B3LYP exchange-correlation functional[2, 3] with the 6–31++G(d,p) basis set for geometry and frequency calculations followed by single-point energy refinement with the 6–311++G(3df,2p) basis set in a manner analogous to recent studies of biological phosphates.[4-7] Energy minimum and transition state geometry optimizations were performed in redundant internal coordinates with default convergence criteria,[8] and stability conditions of the restricted closed shell Kohn-Sham determinant for each final structure were verified.[9, 10] Frequency calculations were performed to establish the nature of all stationary points and to allow evaluation of thermodynamic quantities. Solvent was treated using the PCM solvation model.[11,12] Due to instabilities in the geometry optimization of the DFT calculations on the solvated potential energy surface, solvent corrections were taken into account by single point calculations at the gas phase-optimized B3LYP/6–31++G(d,p) geometries, as in previous work.[4-7] All density-functional calculations were performed with the GAUSSIAN03[13] suite of programs.

2.2 Semiempirical calculations

All semiempirical/implicit solvation calculations were performed using a modified version of the MNDO97 program,[14] and often employ the combined MNDO/d Hamiltonian[15] and smooth COSMO method.[16] Details of the method, implementation and testing have been presented elsewhere.[17, 18] Convergence criteria for the SCF energy was 10^{-6} eV, and for geometry optimizations was 1.0 kcal/mol/Å on the gradient norm. Unless otherwise stated, solvation calculations used a discretization level of 110 points per sphere, smooth COSMO switching parameter $\gamma_s = 1.0$, and shift parameter $\alpha \approx 0.5$ as described in detail elsewhere.[17] The MNDO/d and smooth COSMO methods have been implemented into the MNDO97 code[14] and interfaced to the CHARMM molecular modeling package[19] and will be available in future MNDO and CHARMM releases.

2.3 Hybrid QM/MM calculations

Hybrid QM/MM calculations were performed using the MNDO97 program[14] interfaced with CHARMM.[19] Activated dynamics calculations were performed using either stochastic boundary molecular dynamics (SBMD)[20, 21] without cut-off, or else full cubic periodic boundary conditions using a recently introduced linear-scaling QM/MM-Ewald method[22] that employs the smooth particle mesh Ewald (PME) method.[23, 24] Unless otherwise stated, simulations were performed at constant temperature (298.15 K) with a 0.5 fs integration time step, and used the TIP3P water model[25] with SHAKE constraints[26] on the internal water geometries. Potential of mean force (PMF) profiles were determined using umbrella sampling[27] and the weighted histogram analysis method (WHAM).[28]

3 RESULTS AND DISCUSSION

3.1 Density-functional calculations of phosphoryl transfer reactions

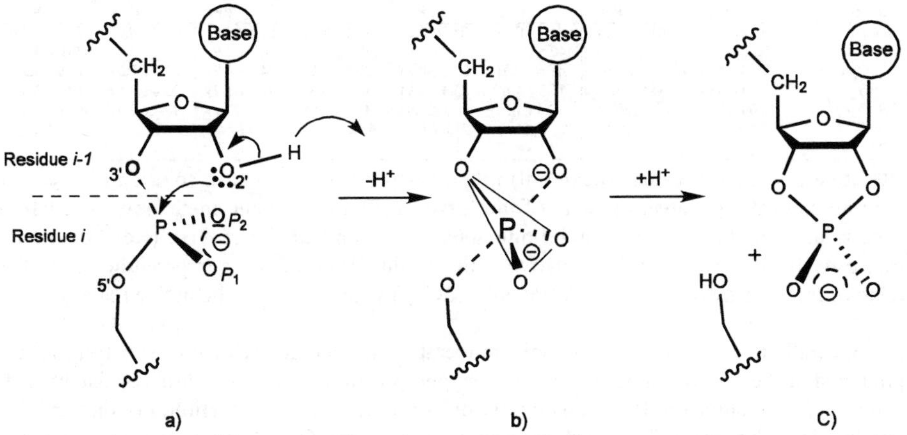

Scheme 1   *RNA transesterification*

Consider the cleavage transesterification reaction (Scheme 1) that occurs in RNA, and is catalyzed by the prototype RNA enzymes such as the hammerhead[29, 30] and hairpin[31, 32] ribozymes. One method to probe the mechanism of ribozymes is to introduce chemical modifications at specific sites, [33–38] and make mechanistic inferences from the measured change in reaction rate. A commonly applied modification involves the substitution of key phosphoryl oxygens with sulfur.[33, 35, 38–40] A subsequent change in the reaction rate is called a *thio effect*, and can provide insight into the specific role these positions play in the biological reaction. However, often multiple mechanistic pathways are able to fit the observed kinetics equally well.[41] Theoretical methods offer a potentially powerful tool to aid in the mechanistic interpretation of experimental kinetic data and provide additional atomic-level insight into the structural and chemical reaction dynamics.[1, 42–44]

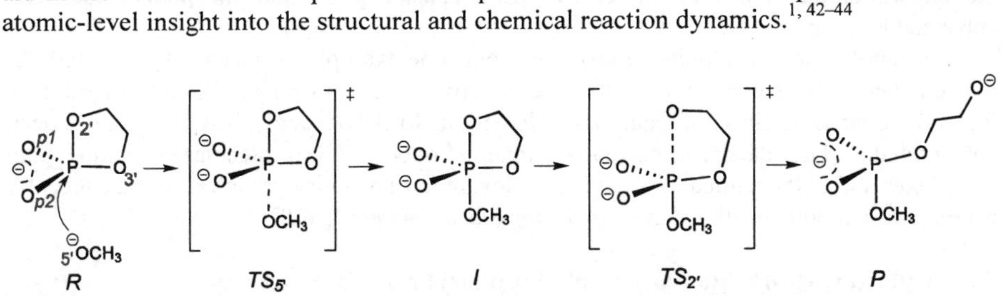

Scheme 2   *In-line dianionic mechanism of ethylene phosphate methanolysis (a reverse-reaction model for RNA phosphate transesterification)*

Consider the dianionic in-line mechanism of methanolysis of ethylene phosphate (Scheme 2) as reverse reaction model for RNA transesterification. Solvent has a large stabilization effect for all the dianionic reactions[45] (Table 1). In general, sulfur substitution tends to yield lower solvation free energies, due to the larger size and more delocalized concentration of negative charge of sulfur.

Table 1 *Forward and reverse rate-controlling free energy barriers for dianionic methanolysis reactions in the gas phase and in solution*[a]

	$\Delta G^{\ddagger}$		$\Delta G^{\ddagger}_{aq}$					
			PCM		CPCM		SM5	
Reaction	Fwd	Rev	Fwd	Rev	Fwd	Rev	Fwd	Rev
native	98.3 (0.0)	41.7 (0.0)	41.1 (0.0)	28.3 (0.0)	40.0 (0.0)	26.1 (0.0)	30.2 (0.0)	23.1 (0.0)
S: O_{P1}	93.0 (-5.3)	39.0 (-2.7)	40.6 (-0.5)	26.7 (-1.6)	39.0 (-1.0)	24.2 (-1.9)	27.7 (-2.5)	20.0 (-3.1)
S: O_{P1},O_{P2}	90.3 (-8.0)	38.1 (-3.6)	41.6 (0.5)	27.5 (-0.8)	39.8 (-0.2)	24.4 (-1.7)	27.9 (-2.3)	18.9 (-4.2)
S: $O_{3'}$	89.4 (-8.9)	35.9 (-5.8)	36.3 (-4.8)	23.4 (-4.9)	36.7 (-3.3)	22.4 (-3.7)	25.4 (-4.8)	19.7 (-3.4)
S: $O_{2'}$	88.0 (-10.3)	62.6 (20.9)	42.4 (1.3)	51.2 (22.9)	41.7 (1.7)	48.8 (22.7)	27.7 (-2.5)	49.4 (26.3)
S: $O_{5'}$	108.0 (9.7)	25.1 (-16.6)	55.9 (14.8)	21.8 (-6.5)	53.8 (13.8)	19.9 (-6.2)	51.5 (21.3)	15.1 (-8.0)

[a]Relative free energy values (kcal/mol) in the gas phase ($\Delta G^{\ddagger}$) and in solution ($\Delta G^{\ddagger}_{aq}$) for the rate-controlling transition state of the forward (Fwd) dianionic methanolysis reaction (i.e., relative to the reactants at infinite separation), and of the reverse (Rev) dianionic transesterification reaction (i.e., relative to the product state). Shown in parentheses are the relative free energy differences ($\Delta\Delta G^{\ddagger}$ and $\Delta\Delta G^{\ddagger}_{aq}$) with respect to the native reaction.

Overall, in aqueous solution only moderate rate changes with thio substitution are predicted at the non-bridging phosphoryl oxygen positions. This result is consistent with experimental results for RNA analogs with non-bridging thio substitutions that exhibit only modest thio effects.[46–48] Substitution at the bridging $O_{3'}$ position leads to a decrease in the activation barrier by 3.3–4.8 kcal/mol. This is consistent with the experimentally observed 200-fold[49] and 2000-fold[50] rate enhancements observed for base-catalyzed hydrolysis of 3′-thio modified RNA dinucleotides. Thio substitution at the $O_{2'}$ position has a moderate effect on the forward methanolysis activation barrier, and also a considerable increase in the reverse transesterification barrier that ranges from 22.7–26.3 kcal/mol. This is consistent with kinetic measurements for modified 2′-thio ribonucleotides[51] that indicate thiolate attack to the phosphate center is 10^7 times slower than the corresponding alkoxide. Thio substitution at the $O_{5'}$ position results in the largest forward activation barrier (51.5–55.9 kcal/mol). These results are consistent with the increased stability of the 5′ thiolate leaving group as indicated by lower theoretical and experimental pKa values of thiols with respect to the corresponding alcohols,[52, 53] and by the experimentally observed increase in rate constants for transphosphorylation with enhanced leaving groups.[39, 54–56]

The application highlighted above represents one example of the use of DFT methods to elucidate key factors that influence reactivity in phosphoryl transfer reactions. Recently, a large database of Quantum Calculations for RNA catalysis, QCRNA, has been constructed. The database currently contains over 1,700 molecular structures and complexes and 200 chemical mechanisms relevant to phosphoryl transfer, and has recently been made available on-line at *http://riesling.chem.umn.edu/QCRNA*.

3.2　Hybrid QM/MM simulations of phosphoryl transfer reactions

A particularly useful strategy to study chemical reactions in complex aqueous biological environments is to perform activated molecular dynamics simulations with hybrid QM/MM potentials[57–60] to derive reaction free energy profiles for different possible mechanistic pathways. A key factor in these simulations is to employ quantum methods that are sufficiently accurate and efficient to derive reliable free energy results. Detailed discussion of the QM/MM profiles, average geometric parameters and relative free energy values has been presented elsewhere[61] and the trends observed are qualitatively similar to those derived from DFT with implicit solvation discussed in the previous section. The focus here is to highlight the role of explicit solvation on the mechanism and barriers.

Solvation plays a key role in stabilizing the dianionic reactive intermediates. Figure 1 illustrates the change that occurs in the radial distribution function of water oxygens around the nucleophilic (2′) and leaving group (5′) positions as the reaction proceeds. In the transesterification reaction, as the activated 2′ nucleophile approaches phosphorus in the endocyclic bond formation step, it must become partially de-solvated. In the rate-controlling exocyclic cleavage step, the 5′ leaving group becomes solvated as the reaction proceeds as indicated by the appearance of an ordered solvation shell in Fig. 1. Thio substitution exhibits electronic effects as well as solvation effects, each of which is sensitive to the position of the thio substitution. For thio substitution at the non-bridging position, the solvation effect appears dominant, whereas at the bridging nucleophilic and leaving group positions, the electronic effects have the greatest influence.

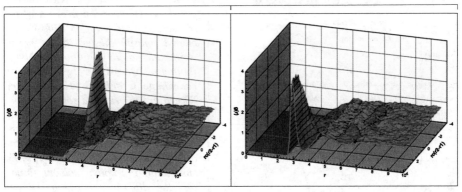

Figure 1 *Radial distribution functions of water oxygens (O_w) around the O2′ nucleophile (left) and O5′ leaving group (right) as a function of reaction coordinate (reactant state is in the background, product state is in the foreground) for the doubly thio-substituted reaction.*

3.3 Smooth implicit solvation models

There is growing interest in the development of implicit solvation models that can be used simultaneously with molecular simulation and linear-scaling electronic structure methods to address complicated biological problems. Certain solvation methods, particularly those of the "boundary element" type,[62] are among the most widely applied in electronic structure calculations.[11, 63] Incorporated into electronic structure or molecular mechanics methods, implicit solvation models provide a powerful tool to study solution-phase molecular properties, chemical reactions, and molecular dynamics. Consequently, it is important to ascertain the reliability of new and existing quantum and solvation models in order to design improved multi-scale quantum models for these reactions.

Recently, a smooth COSMO model[16] has been introduced that circumvents many of the problems encountered by conventional boundary-element methods through the use of Gaussian functions for expansion of the reaction field surface charges and a differentiable switching layer that allows new surface elements to appear and disappear smoothly with respect to the energy. The method has been implemented with analytic gradients[17] within a d-orbital semiempirical framework[15, 64] and applied to phosphoryl transfer reactions.[18]

Figure 2 compares PMF profiles derived from QM/MM simulation and from the MNDO/d-smooth COSMO (MNDO/d- SCOSMO) method for the transesterification of 3′-ribose, 5′-methyl phosphodiester and the doubly thio-substituted derivative. The two methods agree well for both profiles, each predicting similar transition states and intermediates. The MNDO/d- SCOSMO requires a fraction of the computational cost of the explicit QM/MM simulation. One of the key results that arise from this comparison is that the use of the present MNDO/d-SCOSMO model may afford an avenue for efficient parameterization of QM/MM van der Waals radii that are key to obtain reliable free energy profiles of reactions that involve ionic species.[65] A large number of radii can be examined rapidly using the MNDO/d-SCOSMO model, and the most promising sets of radii can then be applied in the more expensive QM/MM simulations.

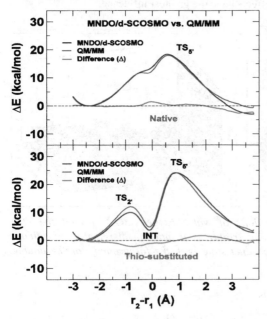

Figure 2 *MNDO/d-SCOSMO and QM/MM reaction profiles of the transesterification of 3′-ribose, 5′-methyl phosphodiester in the native (top) and doubly sulfur-substituted (bottom) forms. The reaction coordinate is defined to be $r_2 - r_1$. Shown are the relaxed solution-phase energy profiles optimized with MNDO/d-SCOSMO (black), the potential of mean force profile obtained from hybrid QM/MM simulation in explicit solvent (blue), and the difference between the two curves (red). All energies (kcal/mol) are with respect to the reactant.*

3.4   Linear-scaling electrostatic and solvation methods

3.4.1   Linear-scaling Ewald method for hybrid QM/MM calculations

In hybrid QM/MM models, the electrostatic environment affects the polarization of the solute[66] and plays a significant role in the stabilization of macromolecules in solution[67] and the rate enhancement of some enzymes.[68]   Consequently, it is critical to compute long-range electrostatic interactions accurately in QM/MM simulations of biochemical reactions.   Nonetheless, due to the lack of availability of algorithms that extend linear-scaling electrostatic methods to hybrid QM/MM potentials, a large percentage of QM/MM applications routinely employ electrostatic cut-offs.[69]

The present section highlights a recently developed linear-scaling Ewald method for efficient calculation of long-range electrostatic interactions in hybrid QM/MM simulations using semiempirical quantum models. Details of the method and simulations have been presented elsewhere.[22] To characterize the effects of treatment of long-range electrostatic interactions on QM/MM free energy profiles, the dissociative phosphoryl transfer reaction of acetyl phosphate (Figure 3) was examined using the semiempirical MNDO/d model.[15,27] Comparisons were made between periodic boundary molecular dynamics (PBMD) simulations using QM/MM-Ewald method and using spherical electrostatic cut-offs.

The effect of cut-off on the PMF profile for the dissociative phosphoryl transfer reaction of acetyl phosphate (Figure 3) is dramatic. The free energy of dissociation from current PMF profiles with QM/MM-Ewald method is 6.8 kcal/mol for acetyl phosphate, and the activation free energy barrier is 12.2 kcal/mol. The effect of cut-off in the PBMD simulations raised the activation free energy barrier to 13.8 kcal/mol (an increase of 1.6 kcal/mol, or 13%). The effect of cut-off for this reaction is to cause an artificial linear drift in the PMF profile at fairly large separation, and raise the activation free energy barrier at short distance. This artifact vanishes when the QM/MM-Ewald method is employed. The cut-off results are comparable to those of Bader and Chandler,[70] in which PMFs of like-charged anions increased with inter-ionic distance, while PMFs using Ewald sum become flat at long range. Despite the known problems associated with the use of electrostatic cut-offs, the majority of present day applications of QM/MM methods routinely employs cut-offs in simulations of biological reaction dynamics.[69] Consequently, the present method offers an important extension of linear-scaling Ewald techniques to hybrid QM/MM calculations of large biological systems.

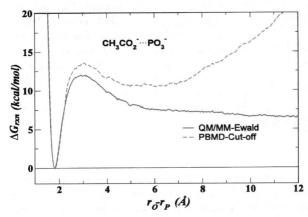

Figure 3 *Free energy profiles from PBMD simulations using hybrid QM/MM-Ewald sum potential (solid blue line) and with 11.5 A cut-off (dashed red line) for the dissociation of acetyl phosphate $(CH_3CO_2^- \cdots PO_3^-)$ in water.*

3.4.2 Variational electrostatic projection method for modeling the solvated macromolecular environment

Recently, a new approach has been introduced to efficiently model a complex solvated macromolecular environment in hybrid QM/MM simulations that does not require explicit consideration of a large simulation cell as is required for Ewald simulations. In this method a variational electrostatic projection (VEP) technique[71] is employed to determine a set of charges on a discretized surface surrounding, for example, a ribozyme active site, such that these charges accurately reproduce the electrostatic potential and forces inside the surface.

In the present section, a new VEP-charge scaling (VEP-cs) method is outlined where charges on a relatively small set of frozen atoms in a SBMD simulation can be re-scaled such that the electrostatic potential and forces inside the active dynamical region reflect the fully solvated macromolecular electrostatic environment. The VEP-charge scaling method has the advantage that the solvent effect due to the entire macromolecular charge distribution is represented, and the method does not require specification of fitting points.

Table 2 compares the average magnitude of the force errors associated with the VEP, VEP-RVM, and VEP-cs methods in the active dynamical region of the hammerhead ribozyme system.[71] In the VEP-charge scaling method, a charge scaling zone of size ΔR_{cs} is employed (Figure 4). Two key quantities regulate the accuracy of the methods: 1) the discretization level (number of points) of the γ variational projection surface (N_γ), and 2) the discretization level of the intermediate ω surface used in the RVM and charge scaling procedures (N_ω). Also shown for comparison is a hybrid method that uses the VEP-RVM procedure with an auxiliary set of charge scaling points (VEP-RVM+cs).

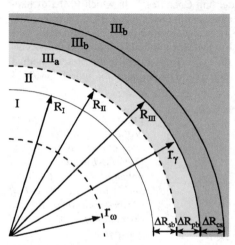

Figure 4   *Regions defined by the VEP charge scaling procedure. Regions I and II are the active dynamical regions, with the atoms of region I propagated using Newtonian dynamics, and those of region II propagated using Langevin dynamics. Region III (subdivided into IIIa and IIIb) comprises the external macromolecular and solvent environment. The electrostatic potential due to atoms of external environment (region IIIb), excluding those of the projection buffer (region IIIa), is replaced by the electrostatic potential of auxiliary charges placed at the centers of the atoms in the charge scaling region R_{cs}.*

The direct VEP method with ΔR_{cs}=2.0 Å leads to average relative force errors in the active dynamical region of 1.07, 0.30 and 0.056 % for N_γ values of 302, 590 and 1202, respectively. The VEP-RVM method offers considerable improvement, decreasing the relative errors by approximately 2 orders of magnitude, irrespective of the N_ω value. The VEP-cs method slightly outperforms the VEP-RVM method for N_γ discretization levels of 302 and 590 points. With the VEP-RVM+cs method, little improvement is gained with smaller N_ω and N_γ discretization levels, but considerably higher accuracy is obtained with N_ω=5810 and N_γ=1202.

Table 2 *Comparison of relative force errors in the active dynamical region of the hammerhead ribozyme (as described in reference 71) for the VEP-based methods (R_{II} = 20.0 Å, R_{pb} = 2.0 Å). The charge-scaling region contained a total of 1761 atoms (the ribozyme plus non-solute atoms in the charge-scaling buffer region, R_{cs} = 2.0 Å). The atoms in the charge-scaling region were not variationally projected.*

Method	N_ω	Relative RMS Force Error			
		$N_\gamma = 0$	$N_\gamma = 302$	$N_\gamma = 590$	$N_\gamma = 1202$
VEP	–		1.071E-02	2.959E-03	5.558E-04
VEP-RVM	1202		8.665E-04	6.035E-05	1.197E-06
	3890		8.409E-04	5.753E-05	1.200E-06
	5810		8.551E-04	5.779E-05	1.207E-06
VEP-cs	1202	1.408E-05			
	3890	1.446E-06			
	5810	2.157E-06			
VEP-RVM+cs	1202		6.380E-06	9.058E-06	1.270E-06
	3890		6.295E-07	2.902E-07	3.208E-07
	5810		1.032E-06	4.247E-07	7.441E-08

To put these results into perspective, the total hammerhead system with full solvation and counter table consists of over 20,000 atoms. In the case of the VEP-cs method, the interaction of the atoms of the active dynamical region with the 20,000 external frozen atoms is replaced by the interaction with around 2,000 atoms that make up the charge scaling zone. This represents almost an order of magnitude reduction of pairwise interactions with the external environment with a relative force error on the order of 0.001%. Further improvement can be gained using a hybrid VEP-RVM+cs approach where and additional set of N_γ surface elements are used to augment the set of scaled charges in the charge scaling buffer.

4 CONCLUSION

The present work describes the development and application of multi-scale quantum models for simulations of phosphoryl transfer reactions. The approach is multi-faceted, and contains components that include high-level density-functional calculations, semiempirical hybrid QM/MM quantum potentials, smooth implicit solvation models and linear-scaling methods for treatment of electrostatic interactions of the solvated macromolecular environment. These methods work together to provide deeper insight into the mechanisms of phosphoryl transfer reactions in complex environments and processes that simultaneously span a broad range of spatial and temporal domains. It is the hope that further development and application of multi-scale quantum models for phosphoryl transfer reactions may, together with experiment, ultimately unravel important questions in cell signaling, energy transfer and RNA catalysis.

5 ACKNOWLEDGMENT

D.Y. is grateful for financial support provided by the National Institutes of Health (grant GM62248), and the Army High Performance Computing Research Center (AHPCRC) under the auspices of the Department of the Army, Army Research Laboratory (ARL) under Cooperative Agreement number DAAD19-01-2-0014.

References

1   Warshel, A. *Computer Modeling of Chemical Reactions in Enzymes and Solutions.* John Wiley and Sons, New York, **1991**.
2   Becke, A. D. *J. Chem. Phys.* **1993**, *98*, 5648–5652.
3   Lee, C.; Yang, W.; Parr, R. G. *Phys. Rev. B.* **1988**, *37*, 785–789.
4   Range, K.; McGrath, M. J.; Lopez, X.; York, D. M. *J. Am. Chem. Soc.* **2004**, *126*, 1654–1665.
5   López, C. S.; Faza, O. N.; Gregersen, B. A.; Lopez, X.; de Lera, A. R.; York, D. M. *Chem. Phys. Chem.* **2004**, *5*, 1045–1049.
6   Mayaan, E.; Range, K.; York, D. M. *J. Biol. Inorg. Chem.* **2004**, *9*, 807–817.
7   López, C. S.; Faza, O. N.; R. de Lera, A.; York, D. M. *Chem. Eur. J.* **2005**, *11*, 2081–2093.
8   Peng, C.; Ayala, P. Y.; Schlegel, H. B.; Frisch, M. J. *J. Comput. Chem.* **1996**, *17*, 49–56.
9   Bauernschmitt, R.; Ahlrichs, R. *J. Chem. Phys.* **1996**, *104*, 9047–9052.
10   Seeger, R.; Pople, J. A. *J. Chem. Phys.* **1977**, *66*, 3045–3050.
11   Tomasi, J.; Persico, M. *Chem. Rev.* **1994**, *94*, 2027–2094.
12   Cossi, M.; Scalmani, G.; Rega, N.; Barone, V. *J. Chem. Phys.* **2002**, *117*, 43–54.
13   Frisch, M. J.; Trucks, G. W.; Schlegel, H. B.; Scuseria, G. E.; Robb, M. A.; Cheeseman, J. R.; Montgomery Jr., J. A.; Vreven, T.; Kudin, K. N.; Burant, J. C.; Millam, J. M.; Iyengar, S. S.; Tomasi, J.; Barone, V.; Mennucci, B.; Cossi, M.; Scalmani, G.; Rega, N.; Petersson, G. A.; Nakatsuji, H.; Hada, M.; Ehara, M.; Toyota, K.; Fukuda, R.; Hasegawa, J.; Ishida, M.; Nakajima, T.; Honda, Y.; Kitao, O.; Nakai, H.; Klene, M.; Li, X.; Knox, J. E.; Hratchian, H. P.; Cross, J. B.; Adamo, C.; Jaramillo, J.; Gomperts, R.; Stratmann, R. E.; Yazyev, O.; Austin, A. J.; Cammi, R.; Pomelli, C.; Ochterski, J. W.; Ayala, P. Y.; Morokuma, K.; Voth, G. A.; Salvador, P.; Dannenberg, J. J.; Zakrzewski, V. G.; Dapprich, S.; Daniels, A. D.; Strain, M. C.; Farkas, O.; Malick, D. K.; Rabuck, A. D.; Raghavachari, K.; Foresman, J. B.; Ortiz, J. V.; Cui, Q.; Baboul, A. G.; Clifford, S.; Cioslowski, J.; Stefanov, B. B.; Liu, G.; Liashenko, A.; Piskorz, P.; Komaromi, I.; Martin, R. L.; Fox, D. J.; Keith, T.; Al-Laham, M. A.; Peng, C. Y.; Nanayakkara, A.; Challacombe, M.; Gill, P. M. W.; Johnson, B.; Chen, W.; Wong, M. W.; Gonzalez, C.; Pople, J. A. Gaussian, Inc., Pittsburgh PA, 2003.
14   Thiel, W. *Program MNDO97*. University of Zurich, 1998.
15   Thiel, W.; Voityuk, A. A. *J. Phys. Chem.* **1996**, *100*, 616–626.
16   York, D. M.; Karplus, M. *J. Phys. Chem. A* **1999**, *103*, 11060–11079.
17   Khandogin, J.; Gregersen, B. A.; Thiel, W.; York, D. M. *J. Phys. Chem. B* **2005**, *109*, 9799-9809.
18   Gregersen, B. A.; Khandogin, J.; Thiel, W.; York, D. M. *J. Phys. Chem. B* **2005**, *109*, 9810-9817.
19   Brooks, B. R.; Bruccoleri, R. E.; Olafson, B. D.; States, D. J.; Swaminathan, S.; Karplus, M. *J. Comput. Chem.* **1983**, *4*, 187–217.

20   Brooks III, C. L.; Karplus, M. *J. Chem. Phys.* **1983**, *79*, 6312–6325.
21   Brooks III, C. L.; Brunger, A.; Karplus, M. *Biopolymers* **1985**, *24*, 843–865.
22   Nam, K.; Gao, J.; York, D. M. *J. Chem. Theory Comput.* **2005**, *1*, 2–13.
23   Essmann, U.; Perera, L.; Berkowitz, M. L.; Darden, T.; Hsing, L.; Pedersen, L. G. *J. Chem. Phys.* **1995**, *103*, 8577–8593.
24   Sagui, C.; Darden, T. A. *Annu. Rev. Biophys. Biomol. Struct.* **1999**, *28*, 155–179.
25   Jorgensen, W. L.; Chandrasekhar, J.; Madura, J. D.; Impey, R. W.; Klein, M. L. *J. Chem. Phys.* **1983**, *79*, 926–935.
26   Ryckaert, J. P.; Ciccotti, G.; Berendsen, H. J. C. *J. Comput. Phys.* **1977**, *23*, 327–341.
27   Torrie, G. M.; Valleau, J. P. *J. Comput. Phys.* **1977**, *23*, 187–199.
28   Kumar, S.; Bouzida, D.; Swendsen, R.; Kollman, P.; Rosenberg, J. *J. Comput. Chem.* **1992**, *13*, 1011–1021.
29   Scott, W. G.; Murray, J. B.; Arnold, J. R. P.; Stoddard, B. L.; Klug, A. *Science* **1996**, *274*, 2065–2069.
30   Scott, W. G. *Q. Rev. Biophys.* **1999**, *32*, 241–294.
31   Walter, N. G.; Burke, J. M. *Curr. Opin. Chem. Biol.* **1998**, *2*, 24–30.
32   Rupert, P. B.; Massey, A. P.; Sigurdsson, S. T.; Ferré-D'Amaré, A. R. *Science* **2002**, *298*, 1421–1424.
33   Herschlag, D.; Piccirilli, J. A.; Cech, T. R. *Biochemistry* **1991**, *30*, 4844–4854.
34   Usman, N.; Cedergren, R. *Trends Biochem. Sci.* **1992**, *17*, 334–339.
35   Catrina, I. E.; Hengge, A. C. *J. Am. Chem. Soc.* **1999**, *121*, 2156–2163.
36   Puerta-Fernández, E.; Romero-López, C.; Barroso-delJesus, A.; Berzal-Herranz, A. *FEMS Microbiol. Rev.* **2003**, *27*, 75–97.
37   Smith, J. S.; Nikonowicz, E. P. *Biochemistry* **2000**, *39*, 5642–5652.
38   Catrina, I. E.; Hengge, A. C. *J. Am. Chem. Soc.* **2003**, *125*, 7546–7552.
39   Zhou, D.-M.; Taira, K. *Chem. Rev.* **1998**, *98*, 991–1026.
40   Breslow, R.; Chapman, Jr., W. H. *Proc. Natl. Acad. Sci. USA* **1996**, *93*, 10018–10021.
41   Åqvist, J.; Kolmodin, K.; Florian, J.; Warshel, A. *Chem. Biol.* **1999**, *6*, R71–R80.
42   Karplus, M. *J. Phys. Chem. B* **2000**, *104*, 11–27.
43   Friesner, R. A.; Beachy, M. D. *Curr. Opin. Struct. Biol.* **1998**, *8*, 257–262.
44   Warshel, A. *Annu. Rev. Biophys. Biomol. Struct.* **2003**, *32*, 425–443.
45   Yun Liu, X. L.; York, D. M. *Chem. Comm.*, submitted.
46   Almer, H.; Strömberg, R. *Tetrahedron Lett.* **1991**, *32*, 3723–3726.
47   Oivanen, M.; Kuusela, S.; Lönnberg, H. *Chem. Rev.* **1998**, *98*, 961–990.
48   Ora, M.; Järvi, J.; Oivanen, M.; Lönnberg, H. *J. Org. Chem.* **2000**, *65*, 2651–2657.
49   Liu, X.; Reese, C. B. *Tetrahedron Lett.* **1996**, *37*, 925–928.
50   Weinstein, L. B.; Earnshaw, D. J.; Cosstick, R.; Cech, T. R. *J. Am. Chem. Soc.* **1996**, *118*, 10341–10350.
51   Dantzman, C. L.; Kiessling, L. L. *J. Am. Chem. Soc.* **1996**, *118*, 11715–11719.
52   Silva, C. O.; da Silva, E. C.; Nascimento, M. A. C. *J. Phys. Chem. A* **2000**, *104*, 2402–2409.
53   Lide, D. R., edt. *CRC handbook of chemistry and physics.* CRC Press LLC, Boca Raton, FL, 83 edition, **2003**.
54   Perreault, D. M.; Anslyn, E. V. *Angew. Chem. Int. Ed.* **1997**, *36*, 432–450.
55   Liu, X.; Reese, C. B. *Tetrahedron Lett.* **1995**, *36*, 3413–3416.
56   Thomson, J. B.; Patel, B. K.; Jiménez, V.; Eckart, K.; Eckstein, F. *J. Org. Chem.* **1996**, *61*, 6273–6281.
57   Warshel, A.; Levitt, M. *J. Mol. Biol.* **1976**, *103*, 227–249.

58 Åqvist, J.; Warshel, A. *Chem. Rev.* **1993**, *93*, 2523–2544.
59 Gao, J. *Rev. Comput. Chem.* **1995**, *7*, 119–185.
60 Gao, J.; Truhlar, D. G. *Annu. Rev. Phys. Chem.* **2002**, *53*, 467–505.
61 Gregersen, B. A.; Lopez, X.; York, D. M. *J. Am. Chem. Soc.* **2004**, *126*, 7504–7513.
62 Rashin, A. A. *J. Phys. Chem.* **1990**, *94*, 1725–1733.
63 Cramer, C. J.; Truhlar, D. G. *Chem. Rev.* **1999**, *99*, 2161–2200.
64 Thiel, W.; Voityuk, A. A. *Theor. Chim. Acta* **1992**, *81*, 391–404.
65 Riccardi, D.; Li, G.; Cui, Q. *J. Phys. Chem. B* **2004**, *108*, 6467–6478.
66 Gao, J.; Xia, X. *Science* **1992**, *258*, 631.
67 York, D. M.; Lee, T.-S.; Yang, W. *J. Am. Chem. Soc.* **1996**, *118*, 10940–10941.
68 Garcia-Viloca, M.; Truhlar, D. G.; Gao, J. *J. Mol. Biol.* **2003**, *327*, 549–560.
69 Garcia-Viloca, M.; Gao, J.; Karplus, M.; Truhlar, D. G. *Science* **2004**, *303*, 186–195.
70 Bader, J. S.; Chandler, D. *J. Phys. Chem.* **1992**, *96*, 6423–6427.
71 Gregersen, B. A.; York, D. M. *J. Phys. Chem. B* **2005**, *109*, 536–556.
72 Jackson, J. D. *Classical electrodynamics*. Wiley, New York, 3 edition, 1999.

SELECTIVITY AND AFFINITY OF MATRIX METALLOPROTEINASE INHIBITORS

V. Lukacova,[1] A. Khandelwal,[1] Y. Zhang,[1] D. Comez,[2] D.M. Kroll[3] and S. Balaz[1]

[1]Department of Pharmaceutical Sciences,
[2]Department of Mathematics,
[3]Department of Physics,
North Dakota State University, Fargo, ND 58105, USA. E-mail: stefan.balaz@ndsu.edu

1 INTRODUCTION

Matrix metalloproteinases (MMPs) degrade extracellular matrix (ECM) and many other extracellular proteins including proteases and their inhibitors, latent growth factors, growth factor binding proteins, cell surface receptors, and adhesion molecules mediating cell-cell and cell-matrix interactions.[1] The expression levels are cell-type dependent and generally low, yet prone to a rapid increase when tissues undergo remodeling, such as in inflammation, wound healing, and cancer. MMPs, synthesized as proenzymes, are mostly secreted and anchored to the cell surface or ECM proteins. This compartmentalization limits their activity radius to specific substrates within the pericellular space.[2] Further spatial control of MMP expression comes from signaling starting with integrin-substrate contact.[3] After proteolytic activation, the activities are controlled by endogenous tissue inhibitors and, in plasma, by α2-macroglobulin.[1] In various pathological conditions, e.g. in rheumatoid arthritis, cancer, and neural diseases, specific MMPs manifest imbalanced activities, which are targeted by drug development projects. This study intends to contribute to such projects by providing procedures to improve selectivity and affinity of developed inhibitors. A Similarity Analysis of MMP binding site structures is performed with the aim to ascertain which MMPs can be selectively inhibited, which features confer specificity on inhibitors, and which MMPs are to be tested to avoid redundancy. The selectivity clues are used to design inhibitor structures that need to have binding affinities predicted before synthesis by a reliable procedure.

MMPs contain two conserved sequences: one in the pro-domain and one in the catalytic domain. The pro-domain including the conserved sequence PRCGVPDV is removed during activation of MMPs that are synthesized as inactive pro-enzymes. The PRCGVPDV sequence is acting as a broad-spectrum inhibitor against various MMPs.[4] The C-terminal portion of the catalytic domain contains another conserved sequence HELGHXXGXXH, with the three histidines coordinating catalytic zinc atom.[5]

According to the domain arrangement and specificity to macromolecular substrates, human MMPs are divided into several subgroups: collagenases 1-3 (MMPs 1, 8, and 13); gelatinases A and B (MMPs 2 and 9); stromelysins 1-3 (MMPs 3, 10 and 11); minimal MMPs matrilysin (MMP 7) and endometase (MMP 26); membrane type MMPs (MMPs 14-17, 24 and 25); and other MMPs (MMPs 12, 18-21, 23, 25, 27 and 28).[6] All MMPs contain three basic domains: signal peptide, pro-peptide, and catalytic domain consisting of about 170 residues.[7] These are the only domains found in the minimal MMPs. Other domains include hemopexin-like domain attached to the catalytic domain through the hinge region (MMPs 1-3, 8-20, 24, 25, 27, 28) or directly (MMP 21), fibronectin-like domain inserted into the catalytic domain (MMPs 2 and 9), collagen-like domain between the catalytic domain and linker leading to hemopexin-like domain (MMP 9), and transmembrane domain in MMPs 14-17, 24, and 25.[5] MMPs 2 and 9 have catalytic domains split into two parts by insertion of fibronectin-like domain. However, this insert does not change the overall fold of the catalytic domain.[8] The removal of the fibronectin-like domain decreased the activity of MMP 2 against gelatine[9,10] but did not change the substrate specificity and activity against smaller substrates.[9] These facts suggest that the fibronectin-like domain is important for the gelatine binding but does not significantly affect the structure of the catalytic domain.

Catalytic domains of all MMPs exhibit sequence similarity ranging from 56 % to 64 %, which leads to significant structure similarity and makes the design of inhibitors that are selective for only certain MMP type difficult. The MMP binding site is composed of subsites S3, S2, S1, S1', S2', and S3', each binding one amino-acid residue of a hexapeptide stretch of a substrate; the ligand parts binding in the subsites are denoted as P3, P2, P1, P1', P2', and P3', resp.[11] (P1 is the amino acid side chain that has the peptide-bond carbonyl coordinated to catalytic zinc). The other subsites extend in both directions from S1 along the backbone of the substrate. The amino acids forming MMP subsites were either previously described or we assigned them as conserved residues (Table 1).

Table1 *Amino acids forming MMP subsites: references are given for described sequences.*

MMP	S1	S1'	S2	S2'	S3	S3'
1	GGNLAHAF[12]	LHRVAAHE[13]	HSLGLSH	ALMYPSY[14]	DNSPFD&HAF	GGNL&PSY[14]
2	DGLLAHAF[12]	LFLVAAHE[13]	HAMGLEH	ALMAPIY[14]	DGYPFD&HAF	DGLL&PIY[14]
3	GNVLAHAY[12]	LFLVAAHE	HSLGLFH	ALMYPLY[14]	DFYPFD&HAY	GNVL&PLY[14]
7	GNTLAHAF	FLYAATHE[13]	HSLGMGH	AVMYPTY[14]	DSYPFD&HAF	GNTL&PTY[14]
8	NGILAHAF[15]	LFLVAAHE[13]	HSLGLAH[7]	ALMYPNY[14]	special[7,16]	special[7,13,16]
9	DGLLAHAF[12]	LFLVAAHE[13]	HALGLDH	ALMYPMY[14]	DGYPFD&HAF	DGLL&PMY[14]
10	GHSLAHAY	LFLVAAHE[13]	HSLGLFH	ALMYPLY	DFYSFD&HAY	GHSL&PLY
11	GGILAHAF	LLQVAAHE[13]	HVLGLQH	ALMSAFY	DDLPFD&HAF	GGIL&AFY
12	GGILAHAF	LFLTAVHE[13]	HSLGLGH	AVMFPTY	DFHAFD&HAF	GGIL&PTY
13	SGILAHAF[15]	LFLVAAHE[13]	HSLGLDH	ALMFPIY	DFYPFD&HAF	SGIL&PIY[14]
14	GGFLAHAY[12]	IFLVAVHE[13]	HALGLEH	AIMAPFY[14]	DSTPFD&HAY	GGFL&PFY
15	GGFLAHAY	LFLVAVHE[13]	HALGLEH	AIMAPFY	DSSPFD&HAY	GGFL&PFY
16	GGFLAHAY	LFLVAVHE[13]	HALGLEH	AIMAPFY	DSSPFD&HAY	GGFL&PFY
17	RRHRAHAF	LFLVAVHE[13]	HAIGLSH	SIMRPYY	DGYPFD&HAF	RRHR&PYY
18	GRVLAHAD	LRIIAAHE	HALGLGH	ALMAPVY	CSNTFD&HAD	GRVL&PVY
19	GRVLAHAD	LRIIAAHE	HALGLGH	ALMAPVY	CSNTFD&HAD	GRVL&PVY
20	RGTLAHAF	LFTVAAHE	HALGLAH	ALMYPTY	DSYPFD&HAF	RGTL&PTY
21	GQEFAHAW	LLKVAVHE	HVLGLPH	SIMQPNY	CPRAFD&HAW	GQEF&PNY
23	TGELAHAF	LVHVAAHE	HALGLMH	ALMHLNA	LHHCFD&HAF	TGEL&LNA
24	GGFLAHAY	LFLVAVHE	HALGLEH	AIMAPFY	DSSPFD&HAY	GGFL&PFY
25	GGTLAHAF	LFAVAVHE	HALGLGH	SIMRPFY	DSYPFD&HAF	GGTL&PFY
26	GGILGHAF	LFLVATHE	HSLGLQH	SIMYPTY	DGWPFD&HAF	GGIL&PTY
27	LGVLGHAF	LFLVAAHE	HALGLSH	ALMFPNY	CPRYFD&HAF	LGVL&PNY
28	GGALAHAF	LFVVLAHE	HTLGLTH	ALMAPYY	LGNAFD&HAF	GGAL&PYY

Traditionally, S1' subsite has been labeled as the specificity pocket[7,17] due to significant differences in the size and the shape among various MMPs that can be observed in the rigid x-ray structures. Several studies focusing on modification of P1' part of the ligand demonstrated improved selectivity.[18,19] Unexpected results in a number of other cases where the design of the P1' substituents has been based solely on the differences in S1' pockets in unliganded enzymes[20,21] may be explained by observed changes in the structure of the pocket upon inhibitor binding.[22,23] The results showing the importance of other parts of the binding site, especially the unprimed side,[24,25] for the design of more potent and selective inhibitors have been reported as well. Occupation of multiple subsites may be crucial for designing truly specific inhibitors.[25]

Human MMPs are structurally well characterized: x-ray or NMR structures for MMPs 1-3, 7-10, and 12-14, in some cases with several ligands, are catalogued in the Protein Data Bank,[26] and comparative models based on the amino-acid sequences in the SWISS-PROT sequence data bank[27] are provided by Mobashery et al.[28] (MMPs 11, 15-18) and by us (MMPs 19-21, 23-28).[29] Our comparative models are based on the alignment shown in Figure 1. Conserved residues in both binding sites (yellow) and other regions of the catalytic domains (cyan) were used to fold the target sequences of MMPs 19-21 and 23-28.

The wealth of structural data makes MMPs a unique target group for development of selective and effective inhibitors. In spite of two decades of effort, the only FDA approved MMP inhibitor is doxycycline hydrate, a tetracycline analog, for treatment of periodontal disease. In clinical trials, several MMP inhibitors failed to increase survival rates of cancer patients and exhibited serious side effects.[30] In addition to complex regulation mechanisms,[31] the problems might have been associated with: (i) improper choice of the patient populations;[32] (ii) poor inhibitor selectivity for specific MMPs; (iii) low specificity of inhibitors towards zinc as compared to other metals, especially iron;[33] and (iv) disparity between the inhibition rate and pharmacokinetics that kept inhibitors for insufficient time around tumors and for too long in places, where metzincin inhibition was undesirable. This communication illustrates our usage of the structural information for the design of selective MMP inhibitors and prediction of binding affinities of new structures.

2 COMPARISON OF MMP BINDING SITES: SIMILARITY ANALYSIS

The possibility of induced fit makes determination of similarity of catalytic domains based solely on their specificity to macromolecular substrates, sequence similarity, or the similarity of rigid structures elusive. More insight is expected from examination of structural changes induced by ligand binding and the energies of the resulting complexes. The methods calculating the force-field interaction energies of probes in individual points of a grid positioned in the binding site have a potential to provide more general results than the use of concrete inhibitor structures,[34] especially if the binding site is treated as flexible to account for induced fit.[35] The previously used Principal Component Analysis (PCA) or Consensus PCA (CPCA)[34,35] work best if the repulsive interaction energies are disregarded. Since the repulsive energies are extremely important in definition of the binding site, we compared all interaction energies of all possible pairs of MMPs directly, by linear regression analysis. Although more time consuming than PCA and CPCA, the method precisely quantifies all similarities instead of characterizing only the most pronounced dissimilarities. This approach provides overall comparison of similarities of entire binding sites or individual subsites, the sets of MMPs that cannot be selectively inhibited, and the non-redundant sets of MMPs for rational testing, in addition to the clues for the design of selective inhibitors that are also available in the PCA and CPCA results.

```
MMP  1 :            PRWEQTHLTYRIENYTPD--LPRADVDHAIEKAFQLWSNVTPLTFTKV-------S
MMP  2 :            RKPKWDKNQITYRIIGYTPD--LDPETVDDAFARAFQVWSDVTPLRFSRI-------H
MMP  3 :        FRTFPGIPKWRKTHLTYRIVNYTPD--LPKDAVDSAVEKALKVWEEVTPLTFSRL-------Y
MMP  7 :        YSLFPNSPKWTSKVVTYRIVSYTRD--LPHITVDRLVSKALNMWGKEIPLHFRKV-------V
MMP  8 :         MLTPGNPKWERTNLTYRIRNYTPQ--LSEAEVERAIKDAFELWSVASPLIFTRI-------S
MMP  9 :           FEGDLKWHHHNITYWIQNYSED--LPRAVIDDAFARAFALWSAVTPLTFTRV-------Y
MMP 12 :            GPVWRKHYITYRINNYTPD--MNREDVDYAIRKAFQVWSNVTPLKFSKI-------N
MMP 13 :        YNVFPRTLKWSKMNLTYRIVNYTPD--MTHSEVEKAFKKAFKVWSDVTPLNFTRL-------H
MMP 14 :          IQGLKWQHNEITFCIQNYTPK--VGEYATYEAIRKAFRVWESATPLRFREVPYAYIREGH
MMP 19 :          YLLLGRWRKKHLTFRILNLPST--LPPHTARAALRQAFQDWSNVAPLTFQEV-----QAG-
MMP 20 :          YRLFPGEPKWKKNTLTYRISKYTPS--MSSVEVDKAVEMALQAWSSAVPLSFVRI-----NSG-
MMP 21 : DGGAAQAFSKRTLSW------RLLGEALSSQLSAADQRRIVALAFRMWSEVTPLDFREDLAA-P--G-
MMP 23 :        YTLTPARLRWDHFNLTYRILSFPRN-LLSPRETRRALAAAFRMWSDVSPFSFREV--A--PE-Q
MMP 24 :          YALTGQKWRQKHITYSIHNYTPK--VGELDTRKAIRQAFDVWQKVTPLTFEEVPYHEIKS-D
MMP 25 :         YALSGSVWKKRTLTWRVRSFPQSSQLSQETVRVLMSYALMAWGMESGLTFHEVD----SP-Q
MMP 26 :         ISPGRCKWNKHTLTYRIINYPHD--MKPSAVKDSIYNAVSIWSNVTPLIFQQV----QNG-
MMP 27 :      GQYGYTLPGWRKYNLTYRIINYTPD--MARAAVDEAIQEGLEVWSKVTPLKFTKI-----SKG-
MMP 28 :        FAKQGNKWYKQHLSYRLVNWPEH--LPEPAVRGAVRAAFQLWSNVSALEFWEAP-A---TG-
```

```
MMP  1 : EGQADIMISFVRGDHRDNSP------FDGPGGNLAHAFQPGPG-IGGD-AHFDEDERWTNN----FR--
MMP  2 : DGEADIMINFGRWEHGDGYP------FDGKDGLLAHAFAPGTG-VGGD-SHFDDDELWSLGKGV-----
MMP  3 : EGEADIMISFAVREHGDFYP------FDGPGNVLAHAYAPGPG-INGD-AHFDDDEQWTKD----TT--
MMP  7 : WGTADIMIGFARGAHGDSYP------FDGPGNTLAHAFAPGTG-LGGD-AHFDEDERWTDG----SS--
MMP  8 : QGEADINIAFYQRDHGDNSP------FDGPNGILAHAFQPGQG-IGGD-AHFDAEETWTNT----SA--
MMP  9 : SRDADIVIQFGVAEHGDGYP------FDGKDGLLAHAFPPGPG-IQGD-AHFDDDELWSLGKG------
MMP 12 : TGMADILVVFARGAHGDFHA------FDGKGGILAHAFGPGSG-IGGD-AHFDEDEFWTTH----SG--
MMP 13 : DGIADIMISFGIKEHGDFYP------FDGPSGLLAHAFPPGPN-YGGD-AHFDDDETWTSS----SK--
MMP 14 : EKQADIMIFFAEGFHGDSTP------FDGEGGFLAHAYFPGPN-IGGD-THFDSAEPWTVRNEDL----
MMP 19 : --AADIRLSFHGRQSSYCSNT-----FDGPGRVLAHADIPELG-----SVHFDEDEFWTEGTY----
MMP 20 : --EADIMISFENGDHGDSYP------FDGPRGTLAHAFAPGEG-LGGD-THFDNPEKWTMG--------
MMP 21 : -AAVDIKLGFGRGRHLGCPR-A----FDGSGQEFAHAW-------RLGD-IHFDDDEHFTPPTSD-----
MMP 23 : --PSDLRIGFYPINHTDCLVSALHHCFDGPTGELAHAFFP-P--HGG-I-HFDDSEYWVLGPTRYSWKK
MMP 24 : RKEADIMIFFASGFHGDSSP------FDGEGGFLAHAYFPGPG-IGGD-THFDSDEPWTLGNANH----
MMP 25 : GQEPDILIDFARAFHQDSYP------FDGLGGTLAHAFFPGEHPISGD-THFDDEETWTFGSKD-----
MMP 26 : --DADIKVSFWQWAHEDGWP------FDGPGGILGHAFLPNSG-NPGV-VHFDKNEHWSASDT------
MMP 27 : --IADIMIAFRTRVHGRCPRY-----FDGPLGVLGHAFFPGLG-LGGD-THFDEDENWTKD-------
MMP 28 : --PADIRLTFFQGDHNDGLGNA----FDGPGGALAHAFLPRR----G-EAHFDQDERWSLS--------
```

```
MMP  1 : ---EYNLHRVAAHELGHSLGLSHSTDIGALMYPSY-T-F--SGDVQLAQDDIDGIQAIYGRS
MMP  2 : ---GYSLFLVAAHEFGHAMGLEHSQDPGALMAPIYTY----TKNFRLSQDDIKGIQELYGASP
MMP  3 : ---GTNLFLVAAHEIGHSLGLFHSANTEALMPLY|--LTDLTRFRLSQDDINGIQSLYGPPPDSPET
MMP  7 : --LGINFLYAATHELGHSLGMGHSSDPNAVMYPTYG--NGDPQNFKLSQDDIKGIQKLYGK
MMP  8 : ---NYNLFLVAAHEFGHSLGLAHSSDPGALMYPNY-A-FRETSNYSLPQDDIDGIQAIYG
MMP  9 : --QGYSLFLVAAHEFGHALGLDHSSVPEALMYPMYRF----TEGPPLHKDDVNGIRHLY
MMP 12 : ---GTNLFLTAVHEIGHSLGLGHSSDPKAVMFPTYK--YVDINTFRLSADDIRGIQSLYG
MMP 13 : ---GYNLFLVAAHEFGHSLGLDHSKDPGALMFPIY-T-YT|---FMLPDDDVQGIQSLYGPGDE
MMP 14 : --NGNDIFLVAWHELGHALGLEHSSDPSAIMAPFYQ--WMDTENFVLPDDDRRGIQQLYGGES
MMP 19 : --RGVNLRIIAAHEVGHALGLGHSRYSQALMAPVYEGY---RPHFKLHPDDVAGIQALYGKK
MMP 20 : -TNGFNLFTVAAHEFGHALGLAHSTDPSALMYPTYK--YKNPYGFMLPKDDVKGIQALYGP
MMP 21 : -T-GISLLKVAVHEIGHVLGLPHTYRTGSIMQPNYIP---QEPAFELDWSDRKAIQKLYGS
MMP 23 : GVWLTDLVHVAAHEIGHALGLMHSQHGRALMHLNA-TLRGWKA----LSQDELWGLHRLYGCLDR
MMP 24 : --DGNDLFLVAVHELGHALGLEHSSDPSAIMAPFYQ--YMETHNFKLPQDDLQGIQKIYGPPAE
MMP 25 : -GEGTDLFAVAVHEFGHALGLGHSSAPNSIMRPFYQGPVGDPDKYRLSQDDRDGLQQLY
MMP 26 : ---GYNLFLVATHEIGHSLGLQHSGNQSSIMYPTY--WYHDPRTFQLSADDIQRIQHLYG
MMP 27 : -GAGFNLFLVAAHEFGHALGLSHSNDQTALMFPNYV--SLDPRKYPLSQDDINGIQSIYGGLPKE
MMP 28 : RRRGRNLFVVLAHEIGHTLGLTHSPAPRALMAP-Y-Y-KRLGRDALLSWDDVLAVQKLYGKPL
```

Figure 1   *Sequence alignment of catalytic domains of MMPs that was used for construction of comparative models of MMPs 19-21 and 23-28.[29,36] MMPs 1-3, 7-9, and 12-14 served as structural templates and were aligned previously.[37] The residues in individual subsites (Table 1) that served as the basis for comparative modeling are highlighted yellow. Other conserved residues that were also used for comparative modeling are highlighted cyan.*

2.1 Interaction Energy Fields

The MMP catalytic domains were superimposed[38] using the PDB file 1A85, one of the x-ray structures for MMP 8, as a template. A regular grid with spacing 1.5 Å was constructed to completely enclose the binding sites. The grid points extending beyond the residues lining the binding site from the protein side were removed to minimize the number of points where to calculate the interaction energies. The remaining 941 points were associated with the subsites (Figure 2). Five probes were selected to represent the atom types frequently occurring in inhibitors and small substrates: sp3 oxygen atom with charge 0, sp3 carbon atom with charge 0, sp3 nitrogen atom with charge +1, hydrogen atom with charge +1, and carbonyl oxygen atom with charge –1.5.

The probe was placed into each grid point, geometry of the protein-probe complex was optimized, and the total energy was calculated in Sybyl using the electrostatic and van der Waals terms of the Tripos force field.[39] For geometry optimization, the amino acids that had at least one atom within 2-Å distance from the probe were set as flexible; the rest of the protein and the position of the probe were fixed. The temperature factors in backbones and side chains in MMP structures are not substantially different (Figure 3), indicating no major differences in flexibility; therefore the backbone parts of amino acids were also optimized, in contrast to a previous study[35] using only flexible side chains. The steepest descent method with termination criterion of energy change less than 0.5 and maximum of 1000 iterations was used for geometry optimization. After the optimization, the probe was removed and the energy of the protein was calculated. The interaction energy of the protein with the probe in particular grid point was calculated as the difference between energies of optimized protein-probe complex and unliganded protein.

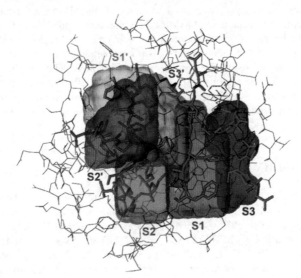

Figure 2   *Outer shape of the grid enclosing the binding site, with the parts corresponding to individual subsites highlighted. Catalytic zinc is shown in a spacefill representation. Catalytic domain of MMP-8 (PDB entry 1A85) is shown, with amino acids defining individual subsites depicted as tubes[36] (with permission of the American Society for Biochemistry and Molecular Biology).*

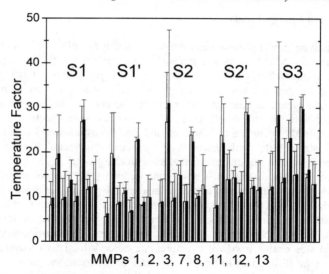

Figure 3   *Temperature factors for backbones (open bars) and side chains (full bars) in high-quality x-ray structures of MMPs catalogued in PDB. Each group describes one of subsites S1, S1', S2, S2' and S3. In each group, the backbone-side chain bar pairs are listed in the order of MMPs given in the inscription.*

Linear regression analysis was used to quantify the degree of similarity between corresponding regions of each pair of MMPs. The intercepts and slopes of the relationships between the probe-binding site interaction energies in compared regions did not differ from zero and one, resp.. Therefore, the similarity was completely characterized by the scatter of the points and quantified using the correlation coefficient R. The simple comparisons of interaction energies for the regions and probes of interest allowed evaluation of similarity for entire binding sites or subsites, and using all probes or individual probes. All results were succinctly catalogued[29,36] and are available for the design of selective ligands. Some more interesting conclusions are summarized below.

2.2   Similarities of MMP Binding Sites and Subsites

Among the five probes, the negative carbonyl oxygen probe was the least discriminating, especially in subsites S2, S3', and S1'. For entire binding sites, MMPs 3 and 7 were similar for all five probes at the level $R^2 \geq 0.9$. At a lower level of similarity, $R^2 \geq 0.8$, five groups of MMPs were identified: (i) MMPs 2, 3, 8, and 12; (ii) MMPs 3, 7, 8, and 12; (iii) MMPs 1, 8, and 13; (iv) MMPs 3, 7, 12, and 14; (v) MMPs 3, 8, 12, and 13. Interestingly, this similarity does not completely match the MMP classification based on macromolecular substrates: only the group (iii) is composed of collagenases, other groups combine MMPs from different subfamilies. Gelatinases (MMPs 2 and 9) and stromelysins (MMPs 3 and 10) do not appear together in any of the groups.

For individual subsites and all probes, the overall trend, in order of decreasing similarity of subsites, is: S1' > S2 > S3' > S1 ~ S3 > S2'. Subsite S1' exhibits the highest level of similarity, followed by S2 and S3' subsites. These results add to multiple experimental proofs of flexibility of S1' subsite,[22,23] help explain cases where desired specificity was not achieved upon optimization of groups fitting into S1' subsite,[20,21] and contradict the view of S1' subsite as the specificity pocket of MMPs that is based on variability of rigid x-ray structures.

An interesting picture is emerging from a comparison of subsite similarity with actual binding of 55 inhibitors for which the PDB files of complexes with MMPs are available (Figure 4). In this group of inhibitors, occupancies of individual subsites are: S1' – 49, S3' – 29, S1 – 20, S2' – 7, S2 –2, and S3 – 1. The most frequently occupied subsites S1' and S3' are among the most similar subsites; the least similar subsites S2' and S3 are seldom engaged in binding. Among the 55 inhibitors, 21 bind in subsites S1'/S3', 8 in S1/S1', 5 in S1', and groups of 1-4 inhibitors in other subsite combinations. Broadly speaking, if the MMP inhibitors in the PDB are indicative of overall synthetic trends, the development of MMP inhibitor structures apparently does not target the subsites that convey specificity.

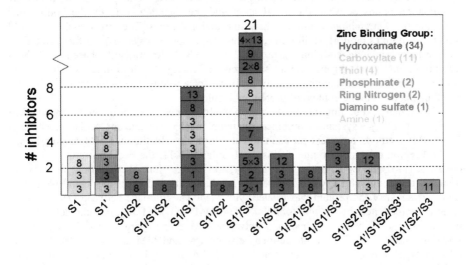

Figure 4   *Occupation of individual subsites S1, S1', S2, S2', S3 and S3' of the binding sites in MMP-ligand complexes catalogued in PDB. Colors denote the zinc binding groups of inhibitors. The (last) numbers in the boxes indicate the MMP. For instance, the leftmost column indicates that PDB contains three MMP-ligand complexes with ligands binding only in S1 subsite, two of them with MMP 3 and one with MMP 8. Column S1'/S3' is not scaled to proportion.*

It should be noted, however, that the overall subsite similarity trends are too fuzzy to be used for the design of concrete inhibitors and substrates. For this purpose, the subsites where the binding is anticipated and probes that occur in the studied compounds should be used. Published reports summarize the needed R^2 values for all possible combinations of the probes and subsites for individual MMPs[36] and for MMPs vs. tumor necrosis factor α converting enzyme (TACE).[29] The results provide clues for design of inhibitor structures with a potential for selectivity, using the methods of incremental growth or steered docking. Classical docking does not perform well for the large and flexible inhibitors that are required to exhibit selectivity in the set of the similar MMP binding sites.[40]

3 PREDICTION OF BINDING AFFINITIES

This task, in general, requires a sufficient conformation sampling of the protein-ligand complex. In the case of metalloproteinases and other metalloproteins, another key step is the structurally and energetically sound description of metal-ligand coordination bonds.

3.1 Ligand-Zinc Interactions

A description of the ligand interactions with zinc poses a challenge due to the possibility of multi-dentate coordination bonding. Simplified molecular mechanical approaches differ in the level of sophistication. In the nonbonded approach,[41,42] optimized electrostatics and van der Waals terms are used to model the metal interactions with ligands and proteins.[43,44] The coordination geometry was enforced using dummy cations placed around the metal atom.[45] The geometry enforcement is more stringent in the bonded model[46-48] that describes the coordination bonds between the metal atom and the ligand/enzyme by the bond terms including bond stretching, angle bending, and torsional terms. The directional force field YETI incorporates more flexibility in selection of appropriate valence.[49] Further force field enhancements include addition of polarizable bonds,[50] directionality based on orbital hybridization,[51] and ligand field stabilization energy.[52] Unfortunately, the more sophisticated force fields are not readily available for a routine use in modeling of ligand-metal interactions in the context of protein molecules.

Coordination bonding is most appropriately treated at the quantum mechanical level.[53-56] For large systems, combined quantum mechanical and molecular mechanical (QM/MM) methods represent an economical approach.[57] In the QM/MM calculations,[58] the QM region consisted of side chains of H405 and H411, the backbone atoms and side chains of H401 and E402, the entire inhibitor, and the zinc ion. The backbone atoms were included to obtain valid QM/MM cuts. The residues outside 5 Å of the ligand were kept rigid. The rest of the protein was treated with MM. The interface between QM and MM regions is mediated by frozen orbitals.[59] The QM and MM regions interact via two mechanisms: electrostatic interactions between MM point charges and the QM wave function, and van der Waals interactions between QM and MM atoms. For the MM and QM parts of the QM/MM calculations, OPLS-AA force field[60] and DFT functional B3LYP[61] were deployed, resp.. All charges in the MM region were treated using the OPLS-AA force field. The 6-31G* basis set was used in the interface region between the QM and MM regions. The LAV3P** basis set was employed for geometry optimization: for Zn, S and P atoms, this means the Los Alamos effective core potential (ECP)[62,63] with all the s functions and the last p and d gaussians uncontracted; for the remaining atoms, it implies the 6-31G** basis set.

3.2 Affinity Prediction Methods Utilizing Receptor and Ligand Structures

The armory of methods for calculations of free energies of binding contains several layers of approaches differing in the details of the treatment of ligand-receptor interactions. Free Energy Perturbation,[64] Thermodynamic Integration,[65] and similar techniques,[66] are the most sophisticated tools. Their routine use in drug design is precluded by limitation to close homologs and extensive sampling that results in extreme demands on computational resources. Simpler methods use the partitioning of the binding energy into individual contributions. The MM-PBSA[67] and MM-GBSA[68] methods form the parameter-free category while the Linear Response (LR) method[69] and its extended version (ELR)[70] represent the parametrized category. The contributions to the binding free energy are expressed as the differences Δ (between the solvated ligand in the bound and free states) in the ensemble averages (denoted by angle brackets) of respective quantities. In the ELR-type methods, binding free energy ΔG_b is calculated as the linear combination of the differences Δ of the molecular dynamics (MD) or Monte Carlo van der Waals and electrostatic energies,[69] and the solvent-accessible surface areas[70] (SASA):

$$\Delta G_b = \alpha \times \Delta \langle E_{vdW} \rangle + \beta \times \Delta \langle E_{el} \rangle + \gamma \times \Delta \langle SASA \rangle + \kappa \tag{1}$$

The protein-solvent and solvent-solvent interactions are absorbed in the adjustable parameters α, β, and γ. Further simplification can be achieved by the replacement of the ensemble averages by a single configuration, usually obtained by a direct geometry optimization of the receptor-ligand complex. This category is represented by the VALIDATE method,[71] the Free Energy Force Field approach,[72] COMBINE analysis,[73] and a single-structure version of the LR method using continuum electrostatics.[74] The simplest scoring functions are used to evaluate docking poses in high-throughput settings.

3.3 Affinity Prediction for MMP Inhibitors

We decided to use the ELR method because it seems to provide the right balance between the quality of the results and the cost. The force-field energies in Eq. 1 were replaced by the QM/MM energy, to correctly describe the zinc-ligand coordination:

$$\Delta G_b = \alpha \times \Delta \langle E_{QMMM} \rangle + \gamma \times \Delta \langle SASA \rangle + \kappa \tag{2}$$

For the final correlation, the QM/MM term was calculated for the time-averaged structures from MD simulations, as the difference between the energies of the complex and the energies of the protein and the ligand. The approach was implemented in a four-tier procedure consisting of: (1) docking with the selection of poses based upon appropriate metal binding; (2) QM/MM optimization of the best docked geometries; (3) MD simulation with the metal binding group of the ligand confined in the geometry from Step 2; and (4) QM/MM single point interaction energy calculation based on the time-averaged structures from Step 3. The QM/MM interaction energies are correlated in an ELR-type approach, along with SASA parametrizing desolvation, with experimental affinities (Eq. 2). The approach was tested using published data[75] on inhibition of MMP 9 by 28 hydroxamates. The studied compounds exhibit ~4000-fold difference in binding affinity, with the inhibition constants K_i ranging from 0.08 to 349 nM.

3.3.1 Step 1 - Docking of the inhibitors to the MMP-9 structure taken from the Protein Data Bank[26] (PDB file 1GKC) was performed using FlexX.[76,77] The ranking of poses was based upon the distance between catalytic zinc and hydroxamate oxygens in the interval 1.5 - 2.5 Å as the primary criterion and the FlexX score as the secondary criterion.[40] To reduce the QM/MM convergence time, for the top complexes of each ligand, the region within 5 Å of the ligand was briefly optimized by the conjugate gradient minimization using OPLS-AA force field.

3.3.2 Step 2 - QM/MM Geometry Optimization. The QM/MM approach for the whole binding domain is expected to provide a more realistic picture than the studies using reduced systems[53,55,78] that do not consider the protein surroundings of bound ligands. As compared to advanced force fields,[49,50,52] the QM/MM approach has the advantage of a more precise handling of electronic structures and thus of the issues like the zinc-ligand charge transfer[79] and tautomerism of hydroxamates.[51]

3.3.3 Step 3 - MD Simulations. The QM/MM optimized complexes were subjected to the MD simulation with the constrained bond lengths and angles between zinc and hydroxamate oxygens, to obtain conformational sampling for the rest of the complex. Inspection of the trajectories revealed that the secondary and tertiary structure of the ligand-receptor complex remained stable during the entire period of simulation. The ensemble averages of the van der Waals and electrostatic energies were calculated for the time-averaged structures obtained after 5 ps – 200 ps simulations. Interestingly, the 5-ps simulations provided the energy parameters that were as good as those for longer simulation times in fitting Eqs. 1 and 2 to experimental data (Table 2).

3.3.4 Step 4 - Calculation of QM/MM Interaction Energies. For the time-averaged structures resulting from the 5-ps MD simulations, single point QM/MM interaction energies were calculated. The time-averaged structures were used to preserve conformational sampling obtained in Step 3 that should result in more realistic descriptions than the use of the single optimized structures.[74] The QM/MM calculations properly treat the coordination bonds between zinc and the ligands; therefore, the QM/MM interaction energies are expected to provide better energy estimates for the correlations according to Eq. 2 than the MM-based force field simulations used in Eq. 1.

3.3.5 Correlations with Inhibitory Activities. The QM/MM energies, along with the SASA terms characterizing desolvation of the time-averaged structures upon binding, were correlated with experimental binding affinities using Eq. 2. The $\log K_i$ values were used directly instead of the ΔG_b values. The fits are summarized in Table 2. For a comparison, the minimization and MD results (steps 2 and 3, resp.) were also were correlated with biological data (lines 1 and 2, resp.). The adjustable parameters α (Eq. 2, line 1) and β (Eq. 1, line 2) were statistically insignificant.

The robustness of the regression equations and their predictive abilities were probed by cross-validation. For this purpose, the fits to the potency data are generated leaving out one or more inhibitors from the calibration process. The resulting equation for each fit is used to predict the potencies of the omitted compounds. We used the leave-several-out (LSO) approach, where 6 inhibitors were randomly omitted and the process was repeated 200 times. The correlations of the LSO predictions with the actual potencies were characterized by the root mean square errors (RMSE) that were equal to 0.751, 0.629 and 0.319 for the steps 2, 3 and 4, resp. (Table 2, lines 1, 2, and 3, resp.).

Table 2 *Correlations of inhibitory potencies with the energy and SASA terms obtained by different methods using Eqs. 1 and 2 for 28 inhibitors of MMP 9.*[75]

Method	Step	Eq.	$\alpha \times 10^{-3}$	$\gamma \times 10^{-3}$	κ	R^2	SD	F
QM/MM Minimization	2	2	-	11.550 ±2.260	-3.606 ±0.870	0.502	0.697	26.21
MD	3	1	4.630 ±4.060	8.680 ±1.150	-4.137 ±0.478	0.760	0.494	39.55
QM/MM	4	2	3.592 ±0.580	7.543 ±0.727	-2.623 ±0.394	0.900	0.318	112.8

4 CONCLUSIONS

Two components of the approach to the design of selective and potent MMP inhibitors have been presented. Similarity Analysis of the MMP binding sites provides clues for selectivity in terms of atom types that should be primarily used to bind in the given subsites. The structures complying with the Similarity Analysis results can have the affinities predicted by a computational approach combining docking, QM/MM calculations, and MD simulations. The application of the approach to the MMP-9 inhibition by 28 hydroxamates resulted in an excellent correlation ($R^2=0.900$) between experimental and calculated values for all tested compounds. Prediction ability of the correlation is characterized by RMSE ~ 0.3 for the $logK_i$ values, as compared to RMSE > 0.6 if the QM/MM term is not used.

Acknowledgements

This work was supported in part by the NIH NCRR grants 1P20RR15566 and 1P20RR16471, as well as by the access to resources of the Computational Chemistry and Biology Network and the Center for High Performance Computing, both at the North Dakota State University.

References

1   L.J. McCawley and L.M. Matrisian, *Curr. Opin. Cell Biol.,* 2001, **13**, 534.
2   S. Chakraborti, M. Mandal, S. Das, A. Mandal and T. Chakraborti, *Mol. Cell. Biochem.*, 2003, **253**, 269.
3   U.K. Saarialho-Kere, S.O. Kovacs, A.P. Pentland, J.E. Olerud, H.G. Welgus and W.C. Parks, *J. Clin. Invest.*, 1993, **92**, 2858.
4   A.J. Park, L.M. Matrisian, A.F. Kells, R. Pearson, Z. Yuan and M. Navre, *J. Biol. Chem.*, 1991, **266**, 1584.
5   Woessner, J. F., Nagase, H., *Matrix Metalloproteinases and TIMPs*, Oxford University Press, New York, 2000.
6   R. Lang, A. Kocourek, M. Braun, H. Tschesche, R. Huber, W. Bode and K. Maskos, *J. Mol. Biol.*, 2001, **312**, 731.
7   W. Bode, C.C. Fernandez, H. Tschesche, F. Grams, H. Nagase and K. Maskos, *Cell. Mol. Life Sci.*, 1999, **55**, 639.
8   I. Massova, R. Fridman and S. Mobashery, *J. Mol. Model.*, 1997, **3**, 17.
9   Q.Z. Ye, L.L. Johnson, A.E. Yu and D. Hupe, *Biochemistry*, 1995, **34**, 4702.
10   L. Banyai, H. Tordai and L. Patthy, *J. Biol. Chem.*, 1996, **271**, 12003.
11   A. Berger and I. Schechter, *Philos. T. Roy. Soc. B*, 1970, **257**, 249.
12   M. Yamamoto, H. Tsujishita, N. Hori, Y. Ohishi, S. Inoue, S. Ikeda and Y. Okada, *J. Med. Chem.*, 1998, **41**, 1209.
13   A.R. Welch, C.M. Holman, M. Huber, M.C. Brenner, M.F. Browner and H.E. van Wart, *Biochemistry*, 1996, **35**, 10103.
14   R.E. Babine and S.L. Bender, *Chem. Rev.*, 1997, **97**, 1359.
15   S. Hanessian, D.B. MacKay and N. Moitessier, *J. Med. Chem.*, 2001, **44**, 3074.
16   M. Aschi, D. Roccatano, A. di Nola, C. Gallina, E. Gavuzzo, G. Pochetti, M. Pieper, H. Tschesche and F. Mazza, *J. Comput. Aid. Mol. Des.*, 2002, **16**, 213.
17   G.E. Terp, I.T. Christensen and F.S. Jorgensen, *J. Biomol. Struct. Dyn.*, 2000, **17**, 933.

18 M.G. Natchus, R.G. Bookland, B. De, N.G. Almstead, S. Pikul, M.J. Janusz, S.A. Heitmeyer, E.B. Hookfin, L.C. Hsieh, M.E. Dowty, C.R. Dietsch, V.S. Patel, S.M. Garver, F. Gu, M.E. Pokross, G.E. Mieling, T.R. Baker, D.J. Foltz, S.X. Peng, D.M. Bornes, M.J. Strojnowski and Y.O. Taiwo, *J. Med. Chem.*, 2000, **43**, 4948.
19 A. Scozzafava and C.T. Supuran, *J. Med. Chem.*, 2000, **43**, 1858.
20 N.G. Almstead, R.S. Bradley, S. Pikul, B. De, M.G. Natchus, Y.O. Taiwo, F. Gu, L.E. Williams, B.A. Hynd, M.J. Janusz, C.M. Dunaway and G.E. Mieling, *J. Med. Chem.*, 1999, **42**, 4547.
21 S. Pikul, D.K. McDow, N.G. Almstead, B. De, M.G. Natchus, M.V. Anastasio, S.J. McPhail, C.E. Snider, Y.O. Taiwo, T. Rydel, C.M. Dunaway, F. Gu and G.E. Mieling, *J. Med. Chem.*, 1998, **41**, 3568.
22 X. Zhang, N.C. Gonnella, J. Koehn, N. Pathak, V. Ganu, R. Melton, D. Parker, S.I. Hu and K.Y. Nam, *J. Mol. Biol.*, 2000, **301**, 513.
23 S. Rowsell, P. Hawtin, C.A. Minshull, H. Jepson, S.M.V. Brockbank, D.G. Barratt, A.M. Slater, W.L. McPheat, D. Waterson, A.M. Henney and R.A. Pauptit, *J. Mol. Biol.*, 2002, **319**, 173.
24 J. Schroder, A. Henke, H. Wenzel, H. Brandstetter, H.G. Stammler, A. Stammler, W.D. Pfeiffer and H. Tschesche, *J. Med. Chem.*, 2001, **44**, 3231.
25 L.A. Reiter, P.G. Mitchell, G.J. Martinelli, L.L. Lopresti-Morrow, S.A. Yocum and J.D. Eskra, *Bioorg. Med. Chem Lett.*, 2003, **13**, 2331.
26 H.M. Berman, J. Westbrook, Z. Feng, G. Gilliland, T.N. Bhat, H. Weissig, I.N. Shindyalov and P.E. Bourne, *Nucleic Acids Res.*, 2000, **28**, 235.
27 B. Boeckmann, A. Bairoch, R. Apweiler, M.C. Blatter, A. Estreicher, E. Gasteiger, M.J. Martin, K. Michoud, C. O'Donovan, I. Phan, S. Pilbout and M. Schneider, *Nucleic Acids Res.*, 2003, **31**, 365.
28 I. Massova, L.P. Kotra and S. Mobashery, *Bioorg. Med. Chem Lett.*, 1998, **8**, 853.
29 V. Lukacova, Y. Zhang, D.M. Kroll, S. Raha, D. Comez and S. Balaz, *J. Med. Chem.*, 2005, **48**, 2361.
30 B. Fingleton, *Expert Opin. Ther. Targets*, 2003, **7**, 385.
31 M. Egeblad and Z. Werb, *Nat. Rev. Cancer*, 2002, **2**, 161.
32 L.M. Coussens, B. Fingleton and L.M. Matrisian, *Science*, 2002, **295**, 2387.
33 E. Breuer, J. Frant and R. Reich, *Expert Opin. Ther. Pat.*, 2005, **15**, 253.
34 H. Matter and W. Schwab, *J. Med. Chem.*, 1999, **42**, 4506.
35 G.E. Terp, G. Cruciani, I.T. Christensen and F.S. Jorgensen, *J. Med. Chem.*, 2002, **45**, 2675.
36 V. Lukacova, Y. Zhang, M. Mackov, P. Baricic, S. Raha, J.A. Calvo and S. Balaz, *J. Biol. Chem.*, 2004, **279**, 14194.
37 I. Massova, L.P. Kotra, R. Fridman and S. Mobashery, *FASEB J*, 1998, **12**, 1075.
38 Homology procedure, Biopolymer module, *Sybyl* 6.9. Tripos Inc., St. Louis, MO, 2003.
39 *Sybyl* 6.9, Tripos Inc., St. Louis, MO, 2003.
40 X.Hu, S.Balaz and W.H.Shelver, *J. Mol. Graphics Model.*, 2004, **22**, 293.
41 R.H. Stote and M. Karplus, *Proteins*, 1995, **23**, 12.
42 X. Hu and W.H. Shelver, *J. Mol. Graphics Model.*, 2003, **22**, 115.
43 O.A. Donini and P.A. Kollman, *J. Med. Chem.*, 2000, **43**, 4180.
44 T.J. Hou, S.L. Guo and X.J. Xu, *J. Phys. Chem. B*, 2002, **106**, 5527.
45 Y. Pang, K. Xu, J. Yazal and F.G. Prendergast, *Protein Sci.*, 2000, **9**, 1857.
46 S.C. Hoops, K.W. Anderson and K.M. Merz, Jr., *J. Am. Chem. Soc.*, 1991, **113**, 8262.
47 T.J. Hou, W. Zhang and X.J. Xu, *J. Phys. Chem. B*, 2001, **105**, 5304.
48 S. Toba, K.V. Damodaran and K.M. Merz, Jr., *J. Med. Chem.*, 1999, **42**, 1225.

49 A. Vedani, *J. Comput. Chem.*, 1988, **9**, 269.
50 J.P. Piquemal, B. Williams-Hubbard, N. Fey, R.J. Deeth, N. Gresh and C. Giessner-Prettre, *J. Comput. Chem.*, 2003, **24**, 1963.
51 J. El-Yazal and Y.P. Pang, *J. Phys. Chem. A*, 1999, **103**, 8346.
52 R.J. Deeth, *Coordin. Chem. Rev.*, 2001, **212**, 11.
53 U. Ryde, *Biophys. J.*, 1999, **77**, 2777.
54 D.W. Deerfield, C.W. Carter and L.G. Pedersen, *Int. J. Quantum Chem.*, 2001, **83**, 150.
55 M. Remko and V. Garaj, *Mol. Phys.*, 2003, **101**, 2357.
56 J. Koca, C.G. Zhan, R.C. Rittenhouse and R.L. Ornstein, *J. Comput. Chem.*, 2003, **24**, 368.
57 A. Warshel and M. Levitt, *J. Mol. Biol.*, 1976, **103**, 227.
58 *QSite*, Schrödinger LLC, Portland, OR, 2003.
59 R.B. Murphy, D.M. Philipp and R.A. Friesner, *J. Comput. Chem.*, 2000, **21**, 1442.
60 W.L. Jorgensen, D.S. Maxwell and J. Tirado-Rives, *J. Am. Chem. Soc.*, 1996, **118**, 11225.
61 A.D. Becke, *J. Chem. Phys.*, 1993, **98**, 5648.
62 P.J. Hay and W.R. Wadt, *J. Chem. Phys.*, 1985, **82**, 270.
63 W.R. Wadt and P.J. Hay, *J. Chem. Phys.*, 1985, **82**, 284.
64 P. Kollman, *Chem. Rev.*, 1993, **93**, 2395.
65 W.F. van Gunsteren, 'Methods for Calculation of Free Energies and Binding Constants: Success and Problems' in *Computer Simulation of Biomolecular Systems*, eds., W.F. van Gunsteren and P.K. Weiner, ESCOM, Leiden, 1989, pp. 27-59.
66 R.J. Radmer and P.A. Kollman, *J. Comput. Chem.*, 2003, **18**, 902.
67 J. Srinivasan, T.E. Cheatham III, P. Cieplak, P.A. Kollman and D.A. Case, *J. Am. Chem. Soc.*, 1998, **120**, 9401.
68 R.C. Rizzo, S. Toba and I.D. Kuntz, *J. Med. Chem.*, 2004, **47**, 3065.
69 J. Åqvist, C. Medina and J.E. Samuelsson, *Protein Eng*, 1994, **7**, 385.
70 H.A. Carlson and W.L. Jorgensen, *J. Phys. Chem.*, 1995, **99**, 10667.
71 R.D. Head, M.L. Smythe, T.I. Oprea, C.L. Waller, S.M. Green and G.R. Marshall, *J. Am. Chem. Soc.*, 1996, **118**, 3959.
72 J.S. Tokarski and A.J. Hopfinger, *J. Chem. Inf. Comp. Sci.*, 1997, **37**, 792.
73 A.R. Ortiz, M.T. Pisabarro, F. Gago and R.C. Wade, *J. Med. Chem.*, 1995, **38**, 2681.
74 D. Huang and A. Caflisch, *J. Med. Chem.*, 2004, **47**, 5791.
75 M. Sawa, T. Kiyoi, K. Kurokawa, H. Kumihara, M. Yamamoto, T. Miyasaka, Y. Ito, R. Hirayama, T. Inoue, Y. Kirii, E. Nishiwaki, H. Ohmoto, Y. Maeda, E. Ishibushi, Y. Inoue, K. Yoshino and H. Kondo, *J. Med. Chem.*, 2002, **45**, 919.
76 M. Rarey, B. Kramer, T. Lengauer and G. Klebe, *J. Mol. Biol.*, 1996, **261**, 470.
77 B. Kramer, M. Rarey and T. Lengauer, *Proteins*, 1999, **37**, 228.
78 D.P. Linder and K.R. Rodgers, *J. Phys. Chem. B*, 2004, **108**, 13839.
79 K. Raha and K.M. Merz, Jr., *J. Am. Chem. Soc.*, 2004, **125**, 1020.

INVESTIGATIONS OF CATALYTIC REACTION MECHANISMS OF BIOLOGICAL MACROMOLECULES BY USING FIRST PRINCIPLES AND COMBINED CLASSICAL MOLECULAR DYNAMICS METHODS

Mauro Boero[1] and Masaru Tateno[2]

[1] Institute of Physics, University of Tsukuba, 1-1-1 Tennodai, Tsukuba, Ibaraki 305-8571, Japan
[2] Center for Biological Resources and Bioinformatics, Tokyo Institute of Technology, 4259 Nagatsuda, Midori-ku, Yokohama 226-8501, Japan

1 INTRODUCTION

RNA enzyme molecules (ribozymes) have gained great interest not only in molecular biology, but also in medical science, since they are at the forefront research in cancer gene therapy[1,2]. Ribozymes operate via a cleaving reaction through the transesterification, for which the main steps are schematically shown in Figure 1 and consist in a nucleophilic attack from a deprotonated oxygen ($O^{2'}$), the formation of a trigonal bi-pyramidal (TBP) intermediate state, and a subsequent cut of a P-O bond (P-$O^{5'}$) of the RNA. In this process metal ions, indicated as M_1 and M_2 in the figure, play a crucial catalytic role in different stages of the reaction.

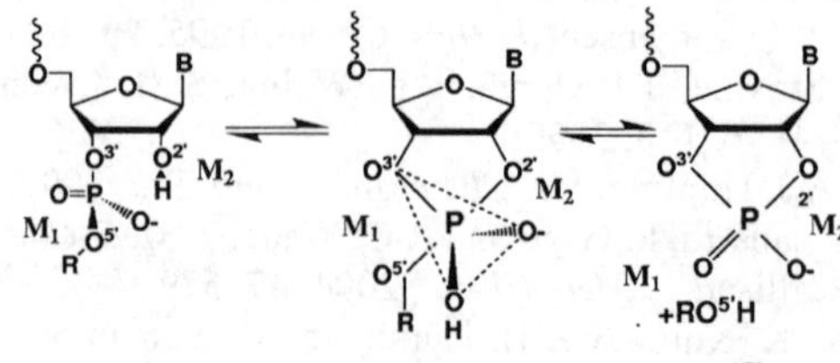

| (1) | (2) | (3) |

Figure 1 *Main steps of the cleavage reaction path for the transesterification of ribozymes promoted by two divalent metal ions* M_1 *and* M_2

Experimentally[3-6], ribozymes are metalloenzymes: it is well known that whenever Mg^{2+} ions are present, the reaction rate is enhanced and the reaction proceeds with a higher efficiency. Yet, unraveling the microscopic details of the reaction path and the exact role of metal catalysts is an issue that is still escaping experimental probes.

In this respect, accurate first principles molecular dynamics simulations can play the role of a useful complementary analysis tool. In our work, by using the best possible level of quantum simulations based on the Car-Parrinello scheme[7], coupled with the newly introduced metadynamics approach[8], we could show how it is possible to elucidate the details of the reaction mechanism at an atomic level and shed some light on the catalytic role of Mg^{2+} metal cations in withdrawing electrons from the cleavage site and in lowering the activation barrier.[9] In a first instance, we use a full quantum approach on a small model including one ribose ring, the phosphodiester backbone, and the metal ions. This model system was fully solvated in water (Figure 2 (a)). We then extended our study to a whole hammerhead ribozyme by using a hybrid quantum mechanics – molecular mechanics (QMMM) approach in which quantum the subsystem where the catalytic reaction occurs is coupled to the rest of the system treated classically (Figure 2 (b)).

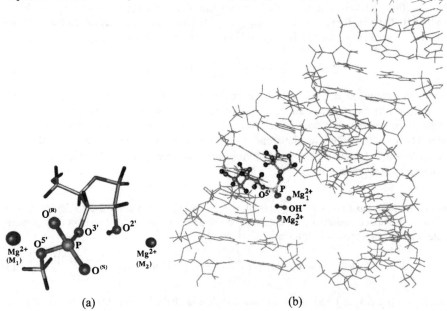

(a)   (b)

Figure 2   *The full-quantum model system (a) and the hybrid QMMM full hammerhead ribozyme (b) used in our simulations. In the QMMM system, quantum atoms are shown as tick sticks and balls, thin gray sticks represent the part treated classically. Although all of the water molecules are not shown for the sake of clarity, water belonging to the solvation shells of the two Mg^{2+} are treated quantum-mechanically. Main atoms are labeled according to the standard nomenclature.*

2 METHOD AND RESULTS

2.1 Computational details

First principles molecular dynamics simulations within the Car-Parrinello[7,10] (CP) scheme were adopted in the present work. The lagrangean potential functional is represented by a standard total energy as expressed in the density functional theory (DFT), with gradient corrections on the exchange and correlation after Hamprecht, Cohen, Tozer and Handy (HCTH)[11]. Only valence electrons are treated explicitly and their wavefunctions are expanded in a plane wave basis set with an energy cut-off of 70 Ry; Troullier-Martins[12] norm-conserving pseudopotentials account for the valence-core interaction.

The quantum systems consist of (N, V, T) ensembles containing one RNA anion model as shown in Figure 2 (a), amounting to 26 atoms, plus 60 water molecules and two metal cations or 61 H_2O molecules plus a single Mg^{2+} for the double- and single-metal-ion cases respectively. The temperature, set to 300 K, was controlled by a Nosé-Hoover thermostat chain[13-15]. An integration step of 4.0 a.u. (0.0967 fs) with a fictitious electron mass of 600 a.u. ensured good control of the conserved quantities.

The reaction path was sampled with the newly introduced the metadynamics approach[8], by adding to the Car-Parrinello Lagrangean L^{CP} the harmonic degrees of freedom of the collective variables s_α plus the history dependent *penalty* potential $V(s_\alpha, t)$

$$L = L^{CP} + \sum_\alpha \tfrac{1}{2} M_\alpha \dot{s}_\alpha^2 - \sum_\alpha \tfrac{1}{2} k_\alpha \left[s_\alpha(q) - s_\alpha \right]^2 - V(s_\alpha, t) \qquad (1)$$

where the argument q of $s_\alpha(q)$ can be any function of an arbitrary set of lagrangean coordinates. Specific details about the set of s_α adopted in our simulations will be given explicitly in the ongoing discussion. During the dynamics, the *penalty* contributions have been chose as Gaussian functions and a new Gaussian is added to the potential $V(s_\alpha, t)$ every $\Delta t = 0.012$ ps, amounting to 150 molecular dynamics generated configurations. The free energy $F(s_\alpha)$ results as

$$F(s_\alpha) = -\lim_{t \to \infty} V(s_\alpha, t) + const. \qquad (2)$$

where the limit has to be intended in the sense given in Ref. 8, i.e. until the selected portion of the phase space, spanned by the collective variables, is saturated. The history dependent potential has then the explicit form

$$V(s_\alpha, t) = \int_0^t dt' \left| \dot{\vec{s}}(t') \right| \delta \left[\frac{\dot{\vec{s}}(t')}{\left| \dot{\vec{s}}(t') \right|} (\vec{s} - \vec{s}(t')) \right] A(t') \exp\left[-\frac{(\vec{s} - \vec{s}(t'))^2}{2(\Delta s)^2} \right] \qquad (3)$$

where $\vec{s} = (s_1, ..., s_\alpha, ...)$ and the Gaussian amplitude $A(t')$, having the dimensions of an energy, is sampled in the interval $(0.02, 0.4)$ kcal/mol.

The estimated total and free energies are affected by an average error bar typical of first-principles approaches and amounting to about 2 kcal/mol.

2.1.1 The QMMM Approach. As far as the hybrid QMMM approach is concerned, we used as quantum main driver the same CPMD code. A hammerhead ribozyme system solvated in water was first pre-equilibrated with a standard Amber force field[16]. Then a QMMM simulation was performed including in the quantum subsystem one phosphate, the two ribose rings directly connected to it, the two metal ions and all the water molecules and/or OH- included in the solvation shell of the Mg^{2+} cations. The coupling between the CP quantum part and a classical Amber-like force filed was done in a fully Hamiltonian way[17-19] by extending the CP Hamiltonian H^{CP} as

$$H^{TOT} = H^{CP} + H^{MM} + E^{int}\left[\rho, \{\vec{r}_I\} \right] \qquad (4)$$

Due to the high computational cost implicit in the evaluation of the coupling functional

$$E^{int}\left[\rho, \{\vec{r}_I\} \right] = \sum_{I=1}^{MM} q_I \int d^3 x \frac{\rho(\vec{x})}{|\vec{x} - \vec{r}_I|} \qquad (5)$$

whose sum runs over all the MM classical atoms and the integration has to be performed of the whole space, we adopt a rescaled electrostatic potential (RESP) scheme according to

Ref. 19: in the region $r < r_1^{cut}$ very close to the quantum subsystem, the interacting functional (5) is computed "as is". We then identify an intermediate region, $r_1^{cut} < r < r_2^{cut}$, whose extension depends on the system under study and the magnitude of the interaction of the quantum part with the surrounding classical part, where the above exact expression can be well approximate by

$$E_{RESP}^{int} = \sum_{I=1}^{MM'} q_I \sum_{J=1}^{QM} \frac{q^{RESP}(\rho, \vec{r}_I)}{|\vec{r}_J - \vec{r}_I|} \tag{6}$$

and $MM' < MM$ is the subset of classical atoms inside this intermediate region. A multipolar expansion is instead used for the far region at $r > r_2^{cut}$. In equation (6), the quantity indicated as q^{RESP} is defined as an atomic point-like charge and is constructed by fitting its value to the electrostatic potential (ESP) due to the QM charge density ρ seen by the close MM atoms. A restrain penalty function (RESP) is included, since unphysical charge fluctuations have been observed in unrestrained ESP charges during dynamics[17-20]. In practice, to compute q^{RESP} we minimize on the fly, during the molecular dynamics run, the quantity

$$\chi = \sum_{I=1}^{MM''} \left(\sum_{J=1}^{QM} \frac{q_J^{RESP}}{|\vec{r}_I - \vec{r}_J|} - V_I \right)^2 + w_q \sum_{J=1}^{QM} \left(q_J^{RESP} - q_J^H \right)^2 \tag{7}$$

In equation (7), V_I is simply the Coulomb interaction modified at short range in order to avoid spurious over-polarization effects, w_q is a weight factor generally ranging from 0.1 to 0.25 and q_J^H is the Hirshfeld charge[20] given by

$$q_J^H = \int d^3x \rho(\vec{x}) \frac{\rho^{at}(|\vec{x} - \vec{r}_J|)}{\sum_K \rho^{at}(|\vec{x} - \vec{r}_K|)} - Z_J \tag{8}$$

being ρ the total charge density as expressed in our plane-wave DFT approach, ρ^{at} the atomic valence charge density and Z_J the bare valence charge of the Jth atom. In short, we request the RESP potential to reproduce the correct Coulomb interaction and, at the same time, the RESP charges to be as close as possible to the Hirshfeld charges. As it can be noticed, the expression (7) is just a least square procedure and, as such, it reduces to a simple matrix inversion operation that does not add significant computational cost to the simulation. Further details are given in the quoted literature.

2.2 Double-metal-ion vs. single-metal-ion mechanisms in the absence of OH⁻

The details of the reaction path were studied via a full quantum approach using the model described in the previous section and sketched in Figure 2 (a). After performing a standard CPMD for about 6 ps, in order to equilibrate the system, we inspected the reaction path using via metadynamics. In this case, as inferred from the accepted reaction mechanism[1,2], three collective variables were selected as reaction coordinates: the coordination number between P and $O^{2'}$ ($s_1(t) = N_{coord}(P-O^{2'})$), where a bond is expected to form as a consequence of the nucleophilic attack, the coordination number between P and $O^{5'}$ ($s_2(t) = N_{coord}(P-O^{5'})$), where the bond cleavage is expected to occur, and the coordination number between $O^{2'}$ and its bound H atom ($s_3(t) = N_{coord}(O^{2'}-H)$), since here a deprotonation is supposed to occur.

Here and in the following we adopt the definition of N_{coord}, i.e. the coordination of B around A

$$N_{coord}(A-B) = \frac{1}{N_A}\sum_{I=1}^{N_A}\sum_{J=1}^{N_B}\frac{1-\left(r_{IJ}/d_{AB}\right)^6}{1-\left(r_{IJ}/d_{AB}\right)^{12}} \tag{7}$$

where N_A and N_B are the number of atoms of species A and B, respectively, r_{IJ} is the interatomic distance and d_{AB} is the typical bond length between A and B. This particular choice is driven by the fact that we need a continuous derivable function $s_\alpha(t)$ representing N_{coord} and including a transition region in the vicinity of the equilibrium distance d_{AB} of intermediate weight[21]. Since in our case we are assigning this to a lagrangean variable, we need also a more stringent condition: a continuous derivable $s_\alpha(t)$ smooth enough to allow for a stable numerical integration and with an equally smooth derivative. Now, an expression like equation (7), i.e. $(1-(r/d_{AB}))^p/(1-(r/d_{AB}))^q$ has no singularity for $r = d_{AB}$ and assumes the value 1/2 if q = 2p. As far as p=6 is concerned, we simply tested various numbers from 1 to 8 and found that 6 is a sort of best compromise from too abrupt to too smooth in the case of H2O, Si (clusters) and hydrocarbon molecules.[8,22]

The fictitious effective masses of the collective variables in the Lagrangean (1) were set to $M_1 = M_2 = 30$ a.u. and $M_3 = 50$ a.u , i.e. slightly heavier for the case involving the lighter H atom; the harmonic coupling constants were $k_1 = k_2 = 0.6$ a.u. and $k_3 = 0.8$ a.u. . The result is summarized in Figure 3.

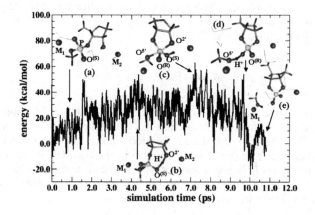

Figure 3 *Potential energy evolution during the metadynamics. (a) is the initial configuration, (b) the proton transfer from $O^{2'}$-H, (c) the TBP transition structure (d) the proton transfer to the departing group and (e) the final state. The arrows indicate approximately the location of the different configurations along the metadynamics trajectory.*

In order to disentangle the effects due exclusively to the metal catalysts from the effects coming from the presence of an OH- in the vicinity of M2,as postulated in the literature and inferred by experiments1,2, we first studied the case in which the reaction occurs in the presence of Mg^{2+} ions only. The system is then in solution in a neutral environment and starts its reaction path from the structure (a) of Figure 3; then it undergoes first a deprotonation of the -O$^{2'}$-H (Figure 3 (b)) with the transfer of the proton from O$^{2'}$ to the

$O^{(S)}$ oxygen - and not to *pro-R* oxygen $O^{(R)}$ – (Figure 3 (c)). During the dynamics, the system equilibrates with the *pro-R* oxygen coordinated to the metal ion M_1 and sharing the solvation shell of this Mg^{2+} with $O^{5'}$, as inferred from experiments[23,24]. $O^{2'}$, in turn, approaches $O^{(S)}$ and they take part in the solvation shell of M_2. In these particular conditions, an intramolecular hydrogen bond (H-bond) $-O^{2'}$-H...$O^{(S)}$ is formed, eventually resulting in a proton transfer.

This is more evident in Figure 4: the variable s_3 - the coordination number of $O^{2'}$ – is the first one to undergo a change. Namely, the proton is shared between $O^{2'}$ and $O^{(S)}$ from 3.4 to 6.2 ps before being stabilized on $O^{(S)}$. Let us remark that here and all the ongoing discussion about metadynamics simulations, the time has to be intended as the progress in the FES exploration and not as a "real" time, as explained in Refs. 8, thus it gives the energetically ordered sequence of the various steps of the reaction path.

The next step consists in the nucleophilic attack and the P-$O^{2'}$ bond formation: s_1 stabilizes to values around 1 and s_3 decreases to zero. This corresponds to the formation of the TBP intermediate[2] shown in the snapshot (c) of Figure 3. The energy barrier for the proton transfer, computed from this simulation, amounts to $\Delta E = 8.1 \pm 1.9$ kcal/mol and $\Delta F = 6.9 \pm 1.6$ kcal/mol for the total and free energies, respectively.

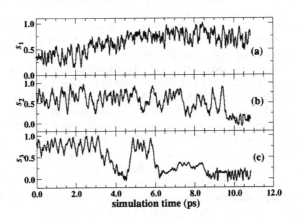

Figure 4 *Collective variables as a function of the metadynamics simulation time; s_1, s_2 and s_3 are the three coordination numbers as given in the text*

Going on in the exploration of the FES, the system reaches a configuration in which $O^{5'}$ and $O^{2'}$ occupy two opposite axial positions with $O^{2'}$, $O^{(R)}$ and $O^{(S)}$ at the three equatorial position of the TBP structure, as expected[1]. The P-$O^{5'}$ bond is finally cleaved and the departing R-$O^{5'}$ group migrates away together with the solvation shell of M_1, to which $O^{5'}$ belongs (Figure 3 (e)). This corresponds to the transition to zero of the collective variable s_2, as shown in panel (b) of Figure 4. After the cleavage of the P-$O^{5'}$ bond, the strengthened P-$O^{(S)}$ bond leads to destabilization of the $O^{(S)}$-H bond, the proton is released and is taken by the R- $O^{5'}$ group that reverts to R-$O^{5'}$-H (Figure 3 (d) and (e)).

No proton transfer to the *pro-R* oxygen was found, since it is hindered by the relatively large distance separating $O^{(R)}$ from H^+. On the other hand, the direct transfer of a proton to the departing group, mediated in our case by $O^{(S)}$, is not totally unexpected and may indeed occur[23], e.g. in the presence of NH_4^+. The computed activation barrier for the P-$O^{5'}$ cleavage reaction amounts to $\Delta E = 46.5 \pm 3.0$ kcal/mol and $\Delta F = 44.7 \pm 2.3$ kcal/mol for the total and free energies, respectively (Table 1). This is higher than the reported

Table 1. *Computed activation barriers (kcal/mol). The error bars reported here are computed as mean square dispersion of the free and total energies during the dynamical sampling of the stationary points.*

	No metal	One Mg^{2+}	Two Mg^{2+}	One Mg^{2+} with OH^-	Two Mg^{2+} with OH^-
ΔE	60.1 ± 3.1	55.2 ± 2.8^a 57.3 ± 2.7^b	46.5 ± 3.0	51.6 ± 2.9^b	43.8 ± 2.5
ΔF	58.5 ± 2.4	54.0 ± 2.5^a 55.5 ± 2.5^b	44.7 ± 2.3	49.2 ± 2.7^b	41.9 ± 2.2

^{a} One Mg^{2+} close to $O^{5'}$, ^{b} One Mg^{2+} close to $O^{2'}$.

experimental data[1], however we stress again the fact that here we are not considering the cooperating effect of OH^-. On the other hand, analogous calculations done on this same system in the absence of any metal catalyst turned out to be higher by 10-15 kcal/mol[9].

This is a clear indication that metal cations play a significant role in withdrawing electrons from the bonds that are expected to be cleaved, $O^{2'}$-H and P-$O^{5'}$, thus contributing to the lowering of the activation barriers and to the consequent enhancement of the reaction rate.

2.2.1 Single-Metal-Ion Reactions. To provide further support to this statement, we performed two auxiliary simulations: in the first one we eliminated from the system the metal cation corresponding to the catalytic site M_1 and replaced it with a water molecule. In the second case, instead, we suppressed M_2, substituted also in this case by a H_2O molecule. All the other parameters and details of the simulation were left unaltered. In both cases, and enhancement in the overall energy barrier was found.

In the fist case, the deprotonation of -$O^{2'}$-H is still promoted by the metal cation M_2, but the proton is released much closer to the transition state, while the breaking of the P-$O^{5'}$ bond is more problematic. Indeed, the TBP undergoes a *spontaneous* distortion that forces $O^{5'}$ and $O^{2'}$ to participate to the solvation shell of M_2., the only catalyst now available. In these conditions, the proton of -$O^{2'}$-H is directly transferred to $O^{5'}$ and the overall activation energies turns out to be $\Delta E = 57.3 \pm 2.7$ kcal/mol and $\Delta F = 55.5 \pm 2.5$ kcal/mol (Table 1), hence higher than the double-metal case. This indicates that a double-metal-ion mechanism should be favored with respect to a single-metal one.

Further confirmation comes from the second case studied. Upon suppression of M_2, the TBP transition state is formed when the P-$O^{2'}$ bond length reaches a value of 1.715 Å, shorter than the TBP formed in the presence of M_2 (P-$O^{2'}$ = 1.741 Å) and the release of the proton occurs simultaneously to the nucleophilic attack: only when $O^{2'}$ forms a chemical bond with P and the oxygen $O^{2'}$ becomes over-coordinated (3-fold) the proton is released. But this is in contrast with the accepted mechanism[1,2] that describes the nucleophilic attack as a consequence of the deprotonation[25,26] of $O^{2'}$. The energy barriers turns out to be $\Delta E = 55.2 \pm 2.8$ kcal/mol and $\Delta F = 54.0 \pm 2.5$ kcal/mol and the Mg^{2+} located at the M_1 site, occupies a dynamically stable catalytic site and does not participate to the deprotonation of $O^{2'}$. Instead, it helps in weakening the P-$O^{5'}$ bond, promoting its cleavage.

2.3 The effect of the presence of an OH⁻ in the reaction mechanism

The presence of a OH^- close to M_2 has long been postulated mostly on the basis of indirect experimental evidence[1,2,24]. In order to shed some light on the detailed role, we performed a metadynamics simulation with the two Mg^{2+} metal cations, but replacing one of the water molecules in the solvation shell of M_2 with an OH^- (Figure 5 (a)).

In this metadynamics investigation, we used the following collective variables: $s_1(t)=N_{coord}(\text{P-O}^{2'})$, coordination number of P with $\text{O}^{2'}$, $s_2(t)=N_{coord}(\text{P-O}^{5'})$, coordination number between P and $\text{O}^{5'}$, and $s_3(t)=N_{coord}(\text{H}^{2'}\text{-OH}^-)$, coordination number of the H atom of $\text{-O}^{2'}\text{-H}$ with the O atom of the hydroxyl anion. At variance with the former cases, this last choice id dictated from the fact that the detached proton is expected to form a water molecule with OH^-. The fictitious effective masses were set to $M_1 = M_2 = 30$ a.u. and $M_3 = 50$ a.u and the harmonic coupling constants were $k_1 = k_2 = 0.6$ a.u. and $k_3 = 0.8$ a.u. for the three variables, respectively.

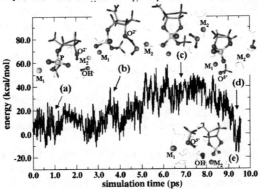

Figure 5 *Potential energy for the case of the OH$^-$ hydroxyl anion. (a) is the initial configuration, (b) the proton abstraction from $O^{2'}$-H, (c) the TBP (d) the P-O$^{5'}$ bond cleavage and (e) the final product.*

A total simulation (meta)time of about 9.7 ps allowed to explore the FES with a sufficient accuracy. In Figure 7 we report the most relevant steps of the reaction. Specific details will be given in a forthcoming publication[27]. They can be briefly summarized as follows: first, between 4.4 and 6.5 ps the proton of $\text{O}^{2'}$-H is shared between $\text{O}^{2'}$ OH$^-$ group (Figure 5 (b)), with M_2 that plays its catalytic role in the $\text{O}^{2'}$-H bond weakening. Then the proton is stabilized on the OH$^-$ that becomes a H_2O molecule and is released as an ordinary water molecule of the solvent.

This reactions proceeds by overcoming a rather modest activation barrier of $\Delta E = 6.7 \pm 2.0$ kcal/mol and $\Delta F = 5.4 \pm 1.6$ kcal/mol. Thus an hydroxyl anion allows for the proton release at an early stage of the transesterification reaction and the detached H$^+$ transferred to the ribozyme, in agreement with the experimental evidence[24], Then the nucleophilic attack occurs, leading to the formation of the TBP (Figure 5 (c)).

Finally, the P-O$^{5'}$ bond is cleaved (Figure 5 (d)), mediated by M_1. As soon as this bond is cleaved, a spontaneous water dissociation occurs close to the departing group (Figure 5 (e)), resulting in the protonation reaction $H_2O + \text{R-O}^{5'} \rightarrow \text{R-O}^{5'}\text{-H} + \text{OH}^-$. Thus, an OH$^-$ anion is released into the solvent, restoring the initial conditions. The activation barrier for the P-O$^{5'}$ bond cleavage is $\Delta E = 43.8 \pm 2.5$ kcal/mol and $\Delta F = 41.9 \pm 2.2$ kcal/mol, and we remark this corresponds to the lowest one among all the cases considered here (Table 1).

2.4 Further effects of the presence of an OH$^-$: a QMMM approach

Beside the crucial effect in the transesterification process itself, described in the previous paragraph, the role of an OH- has also been shown to be essential in smoothing the Coulomb repulsion between the two Mg^{2+} ion catalysts when they are too close to each other. Classical simulations have already attempted in the past to address this issue[28], however difficulties in modeling an appropriate model potential for Mg^{2+} metal cations, able to keep into account both polarization effects and their electron withdrawing properties, and OH⁻ anions make a pure force field approach a rather demanding task.

In order to overcome these problems, we performed hybrid QMMM simulations using a whole, fully hydrated hammerhead ribozyme, shown in Figure 2 (b). In the first simulation the OH⁻ was not included and we simply replaced it with a (quantum) water molecule. Then an unconstrained molecular dynamics simulation was performed to equilibrate the system. The system, indeed, equilibrated in about 1 ps as shown in the upper panel of Figure 6, where the value of the DFT total energy is shown as a function of the simulation time. Since the initial configuration is the one reported in Figure 2 (b), it can be noticed that in the presence of water only, one of the two Mg^{2+} cations, namely the one closer to the phosphate, indicated hereafter as Mg_1 (see Figure 2), and having in its solvation shell the pro-R oxygen $O^{(R)}$, reaches a stable position at an average distance of about 3.5 Å from the phosphorous atom and 2.1 Å from $O^{(R)}$. At the same time, the second metal ion, indicated hereafter as Mg_2, located beside Mg_1 in Figure 2 (b) but less close to the phosphate, departs rather quickly from its initial position, deduced from X-ray data, and displaces away from Mg_1.

This seems to indicate that in the absence of OH⁻ only one of the two metal cations can remain close enough to the ribozyme, thus the only possible reaction pathway would result in a single-metal-ion mechanism. On the other hand, if an OH⁻ is present as shown in Figure 2 (b), it finds a stable position in between the two Mg^{2+}, participating simultaneously to the solvation shell of both the metal cations and becoming basically a shared anion. The net effect results in a screening of the two positively charged metal ions, that reduces considerably the Coulomb repulsion between them and keeps both the ions close to the catalytic center of the ribozyme. Also in this case the equilibration is rather fast, as shown in Figure 7, and, as shown in the same figure, the OH⁻ is stabilized at an average distance of 2.1 Å from the two Mg^{2+} ions. The different behavior of the two metal cations in the two cases is summarized in Figure 8 and the conclusion that can be drawn from these simulations is that, beside the important catalytic properties described in the previous paragraph, a double-metal-ions mechanism includes as a fundamental ingredient also the participation of an OH⁻ anion.

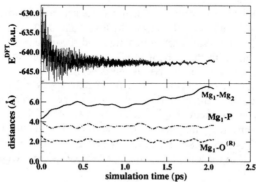

Figure 6 *DFT total energy (upper panel) and most relevant distances (lower panel) as a function of the simulation time in the absence of OH⁻ anions. The black line refers to the separation of the two Mg^{2+} ions, the dot-dashed line to the phosphorous-Mg_1^{2+} distance and the dashed line to the Mg_1^{2+}-$O^{(R)}$ distance.*

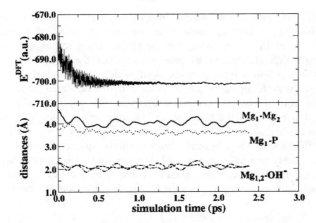

Figure 7 *DFT total energy (upper panel) and most relevant distances (lower panel) as a function of the simulation time for the system containing an OH⁻. The black line refers to the separation of the two Mg²⁺ ions, the dotted line to the phosphorous-Mg₁²⁺ distance and the dashed and dot-dashed lines to the distance of each of the two Mg²⁺ ions from the OH⁻ (see text for details).*

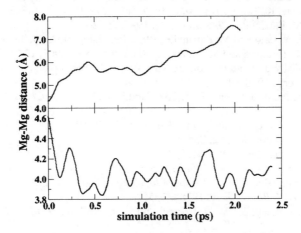

Figure 8 *Comparison of the evolution of the Mg₁²⁺-Mg₂²⁺ distance during the dynamics in the case of absence (upper panel) and presence (lower panel) of an OH⁻ hydroxyl anion.*

3 CONCLUSION

We have shown how the most advanced computational techniques allow to explore with reasonable computational cost the free and total energy landscape of a complex system and to inspect in great detail rather complicated reactions, such as the transesterification of RNA metalloenzymes. This offers a help both in the interpretation of the experiments and in unraveling all those aspects that are not directly accessible to experimental probes. In particular, we have been able to disentangle the separate roles of metal cations and hydroxyl anions, the main two ingredients in the self-cleavage reaction of ribozymes. The simulations clearly indicate that metal catalysts enhance the reaction rate with a high selectivity and that a double-ion mechanism, along with the presence of an OH⁻, represents the most efficient way to promote the phosphate hydrolysis in a large class of RNA enzymes, where one of the most representative is the widely studied hammerhead ribozyme. Last, but not least, the coupling of quantum calculations with classical force fields enables the simulation of a very realistic biological system, while keeping the computational effort within the affordable limits of the computers available nowadays. We then extended our study to a full hammerhead ribozyme in which the portion of the system that undergoes the catalytic reaction is still treated with the same accuracy of a full quantum approach, but no assumptions on the quantum (sub)model are introduced.

References

1   D. Zhou and K. Taira, *Chem. Rev.* 1998, **98**, 9919.
2   D. M. Perreault and E. V. Anslyn, *Angew. Chem. Int. Ed. Engl.* 1997, **36**, 433.
3   M. J. Fedor, *J. Mol. Biol.* 2000, **297**, 269.
4   A. M. Pyle, *Science* 1993, **261**, 709.
5   T. A. Steitz and J. A. Steitz, *Proc. Natl. Acad. Sci. USA* 1993, **90**, 6498.
6   D. M. Zhou and K. Taira, *Proc. Natl. Acad. Sci. USA* 1997, **94**, 14343.
7   R. Car and M. Parrinello, *Phys. Rev. Lett.* 1985, **55**, 2471.
8   M. Iannuzzi, A. Laio and M. Parrinello, *Phys. Rev. Lett.* 2003, **90**, 238302.
9   M. Boero, K. Terakura and M. Tateno, *J. Am. Chem. Soc.* 2002, **124**, 8949.
10   CPMD code by J. Hutter *et al.* at MPI für Festkörperforschung and IBM Zurich Research Laboratory (1990-2004).
11   F. A. Hamprecht, A. J. Cohen, D. J. Tozer and N. C. Handy, *J. Chem. Phys.* 1998, **109**, 6264.
12   N. Troullier and J. L. Martins, *Phys. Rev. B* 1991, **43**, 1993.
13   S. Nosé, *Mol. Phys.* 1984, **52**, 255.
14   S. Nosé, *J. Chem. Phys.* 1984, **81**, 511.
15   W. G. Hoover, *Phys. Rev. A* 1985, **31**, 1695.
16   T. E. Cheatham III and M. A. Young, *Biopolymers* 2001, **56**, 232.
17   F. J. Momany, *J. Phys. Chem.* 1978, **82**, 592.
18   C. I. Bayly, P. Cieplak, W. D. Cornell and P. A. Kollman *J. Phys. Chem.* 1993, **97**, 10269.
19   A. Laio, J. Vande Vondele and U. Roethlisberger, *J. Phys. Chem. B* 2002, **106**, 7300.
20   F. L. Hirshfeld, *Theo. Chim. Acta* 1977, **44**, 129.
21   M. Sprik, *Faraday Discuss.* 1998, **110**, 437.
22   M. Boero, T. Ikeshoji, K. Terakura, C. C. Liew and M. Parrinello, *J. Am. Chem. Soc.* 2004, **126**, 6280.
23   Y. Takagi and K. Taira, *J. Am. Chem. Soc.* 2002, **124**, 3850.
24   S. Sawata, M. Komiyama and K. Taira, *J. Am. Chem. Soc.* 1995, **117**, 2357.
25   L. B. Weinstein, B. C. Jones, R. Cosstick R. And T. R. Cech, *Nature* 1997, **388**, 805.
26   M. Oivanen, S. Kuusela and H. Lönnberg, *Chem. Rev.* 1998, **98**, 961.
27   M. Boero, M. Tateno and K. Terakura, submitted.
28   T. Hermann, P. Auffinger, W. G. Scott and E. Westhof, *Nucleic Acids Res.* 1997, **25**, 3421.

Toward Drug Discovery

CHANGING PARADIGMS IN DRUG DISCOVERY

Hugo Kubinyi

University of Heidelberg, Germany

The strategies of drug research did not change too much from the late 19th century till the seventies of the 20th century. New compounds were synthesized and tested in animals or organ preparations, following some chemical or biological hypotheses. Although synthetic output was relatively low, the real bottleneck were the biological test models. Pharmacological experiments, using dozens of animals for every new compound, most often needed more time for biological characterization than for chemical synthesis.

This situation started to change about thirty years ago. Slowly rational approaches developed, like QSAR and molecular modeling. The consequence was a lower output in such projects, when certain chemical structures had to be synthesized that were proposed by these methods. On the other hand, *in vitro* test systems like enzyme inhibition or the displacement of radio-labeled ligands in membrane preparations enabled a much faster investigation of new analogs. Now chemistry was the bottleneck. About ten to fifteen years ago, another significant shift in drug discovery paradigms happened: combinatorial chemistry suddenly flooded the biology laboratories with an overwhelming number of new compounds. It has been commented that combinatorial chemistry was the "revenge of the chemists" to the development of fast *in vitro* test models, with their large output of data within relatively short time. However, biologists were able to compete: ten thousands of compounds, later even more, could be investigated in just one week by automated high-throughput screening (HTS) systems.

In the past, wrong or misleading results were obtained too often just because of the use of animal models. Gene technology made an important contribution to drug discovery: the possibility to produce almost any protein in sufficient quantities enabled biologists to test new compounds at human targets. Genetically modified animals indicate whether a certain principle could work in therapy. The action of an enzyme inhibitor can be simulated before any compounds are synthesized and tested, by a knock-out of the corresponding enzyme; the action of drugs can be investigated in animals bearing a human protein. In addition, the production of larger quantities of a protein of therapeutic relevance allows the determination of its three-dimensional (3D) structure at atomic resolution by protein crystallography, alternatively by multidimensional NMR methods. As a consequence, methods developed for the structure-based design of ligands, by modeling or experimental determination of the 3D structures of protein-ligand complexes. Unfortunately, a new bottleneck resulted! Early combinatorial chemistry was guided by synthetic accessibility and the hype for large numbers. Due to this wrong focus, a huge amount of greasy, high-molecular weight compounds resulted, with all their problems in bioavailability and pharmacokinetics. Biological testing did not produce any valuable hits or supposed hits later failed in preclinical or clinical development. Whereas this situation fortunately changed in the last years, due to the maturation of combinatorial chemistry to an automated parallel synthesis of designed libraries, there was still a need for the fast measurement or prediction of ADME (absorption, distribution, metabolism, excretion) properties. Indeed, ADME became the new bottleneck.

Nowadays, we have the information on the sequence of the human genome; our combinatorial chemistry approaches are under control by medicinal chemists and biologists; in addition to structure-based design we apply computer-aided methods for data mining, virtual screening, docking and scoring, to predict valuable leads and optimized candidates; HTS models have developed to ultra-HTS models, with up to a million test points per 24 hours; we even have fast experimental models and prediction tools for some ADME parameters. Is there a new bottleneck? Yes, unfortunately, or better saying: yes, of course. Target validation, the proof that the modulation of a certain target by a small molecule will indeed work in therapy, is one of the new bottlenecks. The other one, even more problematic, is the fact that only for some targets small molecules can be discovered which modulate the protein or a certain protein-protein interaction in the desired manner; "druggable" defines the property of a certain target to be accessible by small molecule intervention.

In the past, serendipity played a big role in the discovery of new drugs [1-3]. Some other projects confirmed that the search for new drugs may be more efficient by establishing biological or structure-activity hypotheses and/or selecting certain scaffolds and substituents in the design of new drug candidates. The ratio of 10,000 compounds to produce one new drug is still very often cited. This applies to the situation where research starts from an endogenous ligand or any other lead structure. The "irrational approach", to test huge numbers of in-house compounds, commercially available compounds or chemistry-driven combinatorial libraries in HTS, did not deliver to the expected amount [4]. Hundred thousands to millions of compounds have to be investigated if such a search starts from scratch, without any knowledge of an active lead, and even then there is no guarantee for success.

The "Druggable Genome"

The human genome project has provided the information on all our genes. However, the situation is the same as the one with Egyptian hieroglyphs before the discovery of the Rosetta Stone. We read the text but we understand only a minor part. There are about 30,000 genes in the human genome but we do not know how many of them are disease-related and how many of the gene products will be druggable. It has been estimated that about 600 to 1,500 druggable, disease-related targets exist, if one assumes about 10% disease-related genes on the one hand and about 10% druggable gene products on the other hand [5]. However, this number has to be questioned because only the number of genes was considered [6]. First of all, much more proteins or protein variants (estimated to be in the range of some 100,000s) are produced by alternative splicing and/or posttranslational modification than there are genes in the genome. Second, proteins can form a multitude of heteromeric complexes that are made up from only a small number of different proteins, e.g. GABA and nicotinic acetylcholine (nACh) receptors, integrins, and heterodimeric G protein-coupled receptors (GPCR). Third, some proteins are involved in more than one signaling chain, interacting with different proteins to modulate certain effects. And, last but not least, many therapeutically used drugs do not interact with just one target but have a balanced effect on several different targets. A striking example for such a promiscuous drug is the atypical neuroleptic olanzapine, which interacts as a nanomolar ligand with many different GPCRs [7-9].

Thus, we should neither discuss a druggable genome, nor a druggable proteome, nor a "druggable targetome", but a "druggable physiome" [6]. Our problem is that we do not yet know how to define and design a drug with the right balance of different target affinities, e.g. for depression, schizophrenia and other CNS diseases.

Virtual Screening

Several new strategies have been developed for the structure-based and computer-aided design of active compounds. Drug research has often been compared with the search for a needle in a haystack. If neither active leads are known nor the 3D structure of the biological target, HTS seems to be the only reasonable approach. But much useful information can be derived from virtual screening [10], which reduces the size of the haystack. First of all, reactive compounds and other compounds with undesirable groups can be eliminated by so-called "garbage filters" [11]. In a next step, the Lipinski (Pfizer) rule of five may be applied to estimate the potential for oral bioavailability; this set of four rules demands that the molecular weight of a molecule should be lower than 500; the lipophilicity, expressed by log P (P = calculated octanol/water partition coefficient), should be smaller than 5; the number of hydrogen bond donors should not be larger than 5; the sum of oxygen and nitrogen atoms in the molecule (as a rough approximation of the number of hydrogen bond acceptors) should not exceed 10 [12]. A high risk of insufficient oral bioavailability is assumed if more than one of these conditions is violated. Often rule of five-compatible molecules are erroneously called "drug-like" [6]. However, most of the compounds of the ACD (Available Compounds Directory) [13] would get this label if only the Lipinski rules are applied. "Druglike" or "non-druglike" character can only be attributed by neural nets that have been trained with drugs and chemicals [14-16]. In this context it is important to notice that filters are valuable and efficient in the enrichment of interesting candidates out of large libraries. Single compounds should not be evaluated by such filters because the relatively large error rate of about 20% false positives and 20% false negatives would too often provide misleading results.

The situation in drug discovery is much better if already a certain number of active and inactive ligands of a target is known. If a chemical series belongs to a common scaffold or to some related scaffolds, 2D or 3D similarity methods, QSAR and 3D QSAR approaches, and pharmacophore approaches can be applied to derive structure-activity hypotheses (some problems of pharmacophore generation will be discussed in the next section). The results of such analyses are proposals for new syntheses or selections of compounds from a library. A highly valuable tool in this respect are feature tree similarity comparisons [17,18], where the molecules are coded as strings with nodes, to which the pharmacophoric properties of the corresponding functional group, ring or linker are attributed. Due to this simple representation of the molecules, similarity searches can be performed extremely fast. In this manner, screening hits can be compared in their similarity to a whole in-house compound library, to libraries of commercially available compounds, e.g. the MDL Screening Compounds Directory [19], and to even larger virtual libraries.

If the 3D structure of a new target is known from experimental determination or from reliable homology modeling, the situation seems to be better but in reality it isn't. There remains a high degree of uncertainty about the 3D structure of the protein in the bound state if no information on protein-ligand complex 3D structures is available. Relatively often the protein itself and its ligand complexes have significantly different 3D structures, the most prominent example being HIV protease. In addition, the relatively low resolution of most protein 3D structures does not allow to differentiate between the side chain rotamers of asparagine, glutamine, threonine and histidine; the protonation state of histidine remains unclear; water molecules being important for the binding of a ligand are sometimes neglected in protein 3D structures.

All these problems exist only to a minor extent if several protein-ligand complexes can be inspected, which leads to the fourth and best situation in ligand design: not only the protein but also some protein-ligand complexes are known. Molecular modeling and docking aids in the design of new ligands with hopefully improved binding affinity and/or selectivity with respect to other targets. It should be emphasized that structure-based design can result in a high-affinity ligand but affinity is only a necessary property of a drug, not a sufficient one. In addition, a drug has to be bioavailable, it must have a proper biological half-life time and is must not be toxic, among some other important properties.

Pharmacophores

The definition of a pharmacophore is simple [20]. A 3D pharmacophore corresponds to an arrangement of hydrogen bond donor and acceptor, lipophilic and aromatic groups in space, in such a manner that these moieties can interact with a binding site at the target protein; in addition, steric exclusion volumes can be defined. However, the identification of a pharmacophore within a congeneric group of compounds is far from being trivial. Although there are computer programs for the automated derivation of pharmacophores from series of active and less active analogs [21], a better and more reliable method seems to be a "construction by hand" [22]. Four independent problems have to be considered:
- the different pharmacophoric properties of oxygen atoms,
- the protonation and deprotonation of ionizable groups,
- the consideration of tautomeric forms, and
- the superposition of flexible molecules.

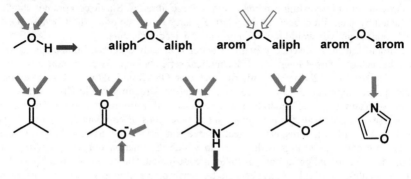

Figure 1 *The oxygen atoms of alcohols, aliphatic ethers, aldehydes, ketones, carboxylates and amides are strong hydrogen bond acceptors. The oxygen atoms of mixed aliphatic-aromatic ethers are weaker acceptors and the oxygen atoms of aromatic ethers and heterocycles are more or less without acceptor properties. The same applies to the sp_3 atom of an ester group, because of the electron-withdrawing effect of the carbonyl group, which itself is a strong hydrogen bond acceptor, and to oxygen in aromatic systems.*

Oxygen atoms are strong hydrogen bond acceptors, as long as they are connected to a carbon atom by a double bond (e.g. in aldehydes, ketones, carboxylic acids, carbonyl group of esters) or if they are substituted by hydrogen and/or aliphatic residues (water, aliphatic alcohols and aliphatic ethers). They are weak or even no acceptors at all (e.g. the sp_3 oxygen atom of an ester group) if the direct neighbor atoms are connected to another atom by a double bond or if they are part of an aromatic system, as in oxazoles and isoxazoles (Figure 1) [23,24].

Ionizable groups must be recognized and defined in the right manner to end up with correct pharmacophores. As this is still a mainly unsolved problem for many compounds that are not simple acids, phenols or anilines (at least considering the speed that is needed in the virtual screening of large libraries), a rule-based system has recently been proposed [25]. In this set of rules, all carboxylic acids, the strongly basic amidines and guanidines, and quaternary ammonium compounds are permanently charged. Neutral and protonated forms are generated and investigated in parallel for amines, imidazoles, pyridines and other nitrogen-containing heterocycles. For tetrazoles, thiols, hydroxamic acids, and activated sulfonamides, neutral and deprotonated forms should be investigated in parallel. Certain rules restrict the number of generated species, to avoid combinatorial explosions: there are definitions of the maximum number of charges in a molecule and no identical charges are allowed in adjacent positions of the molecule. Although this approach is definitely better than using all molecules in their neutral form, refined prediction models are urgently needed. An even more difficult problem arises from the fact that ionizable amino acid side chains in proteins may significantly change their pK_a value in dependence of their environment [26,27].

Protomers and tautomers constitute another serious problem in virtual screening and docking (Figure 2) [28,29]. 1,3-Diketones, acetoacetic esters, hydroxypyridines, oxygen-substituted pyrimidines and purines, and many other compounds may exist in several tautomeric forms that have to be recognized and considered.

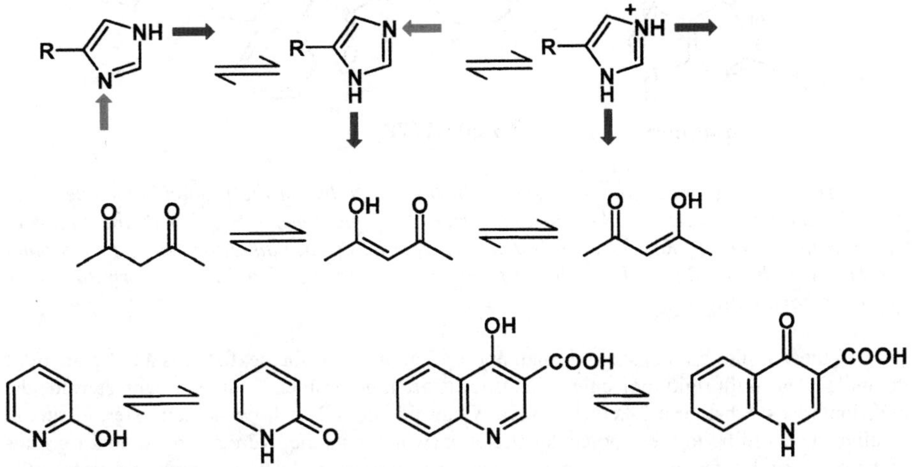

Figure 2 *The two different protomers of imidazole (upper left) present their donor and acceptor moieties in different positions; as imidazole has a pK_a value around 7, also the charged form with two donor functions has to be considered (upper right). The other compounds are typical examples of tautomeric forms of molecules, where donor and acceptor functions change their position.*

For the purine base guanine, 15 different tautomers can be formulated [29]. In this context it is interesting to note that for long time Watson and Crick had problems to derive the correct 3D structure of DNA because they only considered the enol tautomers of guanine and thymine, instead of the keto forms (Figure 3). When their colleague Donohue corrected this error, they immediately arrived at the correct base pairing [30]. Sometimes enol forms of a ligand are induced by the binding site, as is the case for the binding of the barbiturate Ro 200-1770 to a matrix metalloprotease (Figure 3) [31]. Computer programs for the generation of all possible tautomers have been described [29,32].

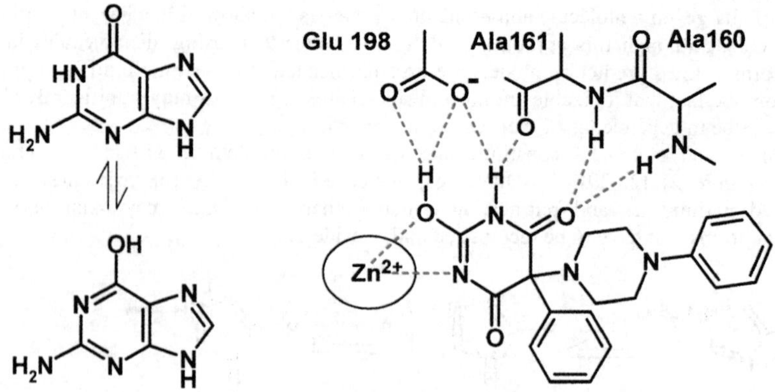

guanine   Ro 200-1770

Figure 3 *The guanine tautomer shown in the upper left is the predominant one of 15 possible tautomers, whereas Watson and Crick, for long time, worked with the tautomer shown in the lower left. Ro 200-1770 is a matrix metalloprotease inhibitor. Only one tautomer can bind with high affinity; the carbonyl form or the other enol form will not form favorable hydrogen bond networks.*

If all these topics are considered in an appropriate manner, the next step is an alignment of the molecules. With rigid molecules, this is most often no problem. But even with steroids the question arises, whether a 3-keto,17-hydroxy-steroid and a 3-hydroxy-17-keto-steroid should be aligned according to their molecular skeleton (which puts the hydrogen bond donor groups of both molecules far apart) or whether a head-to-tail superposition is more favorable; the steric superposition is about as good in this latter case as in the conventional superposition [33]. For flexible molecules, the most rigid, active species should be used as template onto which, step by step, the other molecules are superimposed. This can be done by visual inspection or by field fit methods, like SEAL [34,35]; most often several different conformations have to be considered. FlexS is a computer program for flexible superposition; one molecule is used as a rigid template and all other molecules are superimposed in a flexible manner onto this template [36,37]. 3D QSAR methods, like CoMFA (comparative molecular field analysis) [38,39] or CoMSIA (comparative molecular similarity index analysis) [40], surprisingly do not depend on a knowledge of the bioactive conformation. If all conformations are "wrong" to the same extent, the result of an analysis may nevertheless be useful. Another difficulty in the alignment of molecules arises from different binding modes of seemingly similar molecules; there are no general rules how to recognize such situations.

Once a pharmacophore hypothesis has been derived, 3D searches can be performed, using commercial software [21,41]. However, it must be emphasized that 3D searches are only meaningful if all structures of a database were defined according to their correct pharmacophoric properties; otherwise such searches are just useless.

Structure-based Ligand Design

In the seventies of the last century, the first structure-based design of ligands was performed. The 3D structure of the 2,3-diphosphoglycerate (2,3-DPG) hemoglobin complex was used to derive simple aromatic dialdehydes which mimicked the function of 2,3-DPG as an allosteric effector molecule. Another early example was the structure-based design of trimethoprim analogs with significantly improved affinities to dihydrofolate reductase.

However, neither the hemoglobin ligands nor the trimethoprim analogs could be optimized to become drugs for human therapy [42,43]. The first real success story was the structure-based design of the antihypertensive drug captopril, an angiotensin-converting enzyme (ACE) inhibitor. Its structure was derived in a rational manner from a binding site model, using the 3D information of an inhibitor complex of the closely related zinc protease carboxypeptidase A [44].

With the ongoing progress in protein crystallography and multidimensional NMR techniques, the 3D structures of many important proteins, especially enzymes, have been determined. This information led to the structure-based design of several therapeutically useful enzyme inhibitors, most of them still being in preclinical or clinical development. Marketed drugs that resulted from structure-based design are e.g. the antiglaucoma drug dorzolamide (Merck) and the newer HIV protease inhibitors nelfinavir (Agouron Pharmaceuticals, now Pfizer) and amprenavir (Vertex Pharmaceuticals; developed and marketed by GSK).

Neuraminidase is an interesting target for the structure-based design of anti-influenza drugs. In a very elegant study, Mark von Itzstein used the computer program GRID to estimate interaction energies of the neuraminic acid binding site of this enzyme with different probe atoms or small groups [45]. He realized that the introduction of basic groups, like $-NH_2$, $-C(=NH)NH_2$ or $-NH-C(=NH)NH_2$, into the relatively weak inhibitor neu5ac2en should significantly improve inhibitory activities. This is indeed the case: the neuraminidase inhibitor zanamivir is about 4 orders of magnitude more active than its 4-hydroxy-analog neu5ac2en. Due to its polar character, zanamivir (Relenza®, GSK) is orally inactive; it must be applied by inhalation [46].

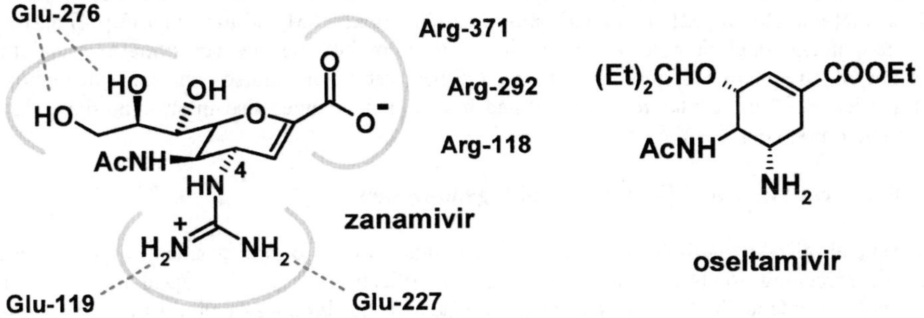

Figure 4 *Schematic presentation of the binding mode of the 4-desoxy-4-guanidino-analog of neu5ac2en, zanamivir, to neuraminidase (left); chemical structure of the orally available prodrug oseltamivir (right).*

Scientists at Gilead Sciences started from the observation that the glycerol side chain of certain zanamivir analogs does not contribute to affinity. In a series of carbocyclic analogs, strongest inhibitor activity was observed for a pent-3-yl ether. Its ethyl ester prodrug oseltamivir (GS 4104, Tamiflu®, Hoffmann-La Roche; Figure 4) shows good oral bioavailability [46,47]. Several other success stories of structure-based design have been published [43,48-51].

Computer-aided Ligand Design

Whereas structure-based design can be regarded as the predominant strategy of the last two decades, several computer-assisted methods were developed more recently. If thousands of candidates and even larger structural databases shall be tested whether they are suited to be ligands of a certain binding site, this cannot any longer be performed by visual inspection. The design process has to be automated with the help of the computer.

The very first computer-based approaches, to search for ligands of a certain binding site, were the programs DOCK [52] and GROW. The *de novo* design program LUDI, developed by Hans-Joachim Böhm at BASF in the early 1990's, was a significant improvement over these early prototypes [53,54]. After the definition of a binding site region by the user, the program automatically identifies all hydrogen bond donor and acceptor sites, as well as aliphatic and aromatic hydrophobic areas of this part of the protein surface. From the program-implemented information on the geometry of interaction of such groups with a ligand, the program creates vectors and regions in space, where complementary groups of a ligand should be located. In the next step, LUDI searches any database of 3D structures of small and medium-sized molecules for potential ligands. Every candidate is tested in a multitude of different orientations and interaction modes, optionally also in different conformations. After a rough evaluation, by counting the number of favorable interactions and by checking for unfavorable van der Waals overlap between the ligand and the protein, the remaining candidates are prioritized by a simple but efficient scoring function [55]. This scoring function estimates interaction energies on the basis of charged and neutral hydrogen bonding energies, hydrophobic contact areas, and the number of rotatable bonds of the ligand. In a last step, the program is capable to attach groups, fragments and/or rings to a hit or to an already existing lead structure. A flexible docking of ligands onto a rigid binding site can be achieved by the programs DOCK 4.0 [56], GOLD [57], FlexX [58,59], and the public domain program AutoDock [60,61], to mention just the most prominent ones; more docking programs and several success stories of computer-assisted drug design have been reviewed by Schneider and Böhm [62]. The FlexX modifications FlexE [63] and Flex-Pharm [64] allow a flexible ligand docking into an ensemble of different binding site conformations and the definition of pharmacophore constraints, respectively. Of course, the pharmacophoric properties of all molecules must be defined in a correct manner also in structure-based and computer-aided design.

Fragment-based and Combinatorial Ligand Design

Several other methods for the design of new ligands have been described in the past, e.g. needle screening [65,66], which starts from a collection of small drug-like ligands and attempts to extend the best ones to larger ligands. In the binding of biotin to avidin, some molecular fragments have only micromolar affinities, whereas biotin itself binds with femtomolar affinity [67]. This principle has recently been used in the rational design of a nanomolar enzyme inhibitor from two low-affinity natural products which bind to different sites of the protein [68]. The SAR by NMR method [69-71] searches for small, low-affinity ligands of proteins which bind to adjacent areas of the binding site. A linker combines both molecules to a nanomolar ligand. Some other NMR-based techniques for ligand discovery have been developed [72-78].

Fragment-based ligand design has been applied for combinatorial techniques [79]. Up to 10,000 low-molecular weight ligands can be tagged to a gold-coated glass surface [80]; binding of any protein to these microarrays of immobilized ligands is detected by surface plasmon resonance, in this manner avoiding the development of a specific screening method for a new protein. The dynamic assembly of ligands [81-84] generates new molecules from fragments which reversibly react with each other in the presence of a protein. Molecules that fit the binding site are preferentially formed and afterwards trapped by a reaction which freezes the equilibrium. Some other approaches for the combinatorial design of new leads have been described [85-89].

There are also several computer-assisted techniques for the combinatorial combination of fragments to new leads. The program CombiGen [90] designs libraries with a high percentage of drug-like compounds by assembling privileged and/or user-defined fragments and optionally modifying the resulting structures; virtual screening procedures eliminate molecules with undesired properties. TOPAS [91,92] dissects lead structures into fragments and assembles new molecules by re-combining a chemically similar scaffold with fragments that are similar to the original ones; cleavage and assembly of the molecules follow chemical reactions that are defined by a RECAP-like procedure [93]. In this manner, new chemistry is generated by "scaffold hopping" [94]. In principle, the docking program FlexX [58,59], which performs an incremental construction of a ligand within the binding site, could also arrive at new analogs, if many different building blocks are used instead of the original building blocks; no virtual library of millions of potential candidates needs to be constructed, only favorable intermediate solutions and final candidates are generated. The only unsolved problem in this respect is the lack of reliability of the scoring functions [95].

Summary and Conclusions

Virtual screening and fragment-based approaches are powerful techniques in the search for new ligands [10,62,96]; promising candidates can be enriched in compound collections and virtual libraries. The integration of protein crystallography, NMR techniques, and virtual screening will "significantly enhance the pace of the discovery process and the quality of compounds selected for further development" [97].

The similarity principle, that similar compounds should exert similar biological activities, has always been a most successful approach in drug research, despite many exceptions to this general concept [98,99]. Chemogenomics is a new term for the dedicated investigation of certain compound classes in target families, like the G protein-coupled receptors (GPCR), kinases, phosphodiesterases, serine proteases, ion channels, etc. [100-104]. An analogous concept, the "selective optimization of side activities" (SOSA approach), attempts to develop new drugs in the direction of a side effect of a certain drug [105,106]. Historical examples for the validity of this approach are e.g. antitussive and constipating morphine analogs, diuretic and antidiabetic sulfonamides, and many others [107]; some very recent examples have been reviewed [106,107].

As already mentioned, good ligands are not necessarily good drugs. High-throughput screening of in-house libraries, which originally contained large numbers of reactive, degraded, colored, fluorescent, and highly lipophilic molecules, and screening of combinatorial libraries of large, lipophilic molecules produced hits that could not be optimized to drug candidates. The awareness for the real problems came only after Lipinski had defined his set of rules [12]. On the other hand, the massive increase of screening failures due to such inappropriate compound collections or libraries turned this awareness of ADME problems into a hype. Prior statistics of 40% failure in clinical investigation (the most expensive phase of drug development), due to ADME problems [108,109] are cited in literature, again and again; a closer inspection of the data shows that the ADME-related failure can be neglected if antiinfectives are removed from the original sample [6,109]. This is an indication that medicinal chemists considered, all the time, the importance of ADME properties. Only HTS and early combinatorial chemistry generated so many problems in this direction. In addition to the Lipinski rules, several *in vitro* and *in silico* techniques are now available for the estimation of ADME properties [110].

With respect to biological testing, Horrobin has raised the question whether we are already living in Castalia, the famous virtual land of Hermann Hesse's novel "The Glass Bead Game", where the masters organize and play the most sophisticated, complex and brilliant games - without any context to reality [111]. Sometimes, this is also the case in modeling and drug design [6]. The new tools of drug research are extremely powerful but they will be successful only if the most important factors, some of them reviewed here, are considered in the right manner. In the fascinating search for better and safer drugs, the new paradigms of drug discovery have to merge with traditional medicinal chemistry experience [2,112,113].

Acknowledgements

This text is a reprint of: H. Kubinyi, Changing Paradigms in Drug Discovery, in: The Chemical Theatre of Biological Systems (Proceedings of the International Beilstein Workshop, May 2004, Bozen, Italy), Eds. M. G. Hicks and C. Kettner, Logos-Verlag, Berlin, 2005, pp. 51-72; with kind permission of Beilstein-Institut, Frankfurt, Germany.

References

[1]   Roberts, R. M., Serendipity. Accidental Discoveries in Science. John Wiley & Sons, New York, **1989**.

[2]   Sneader, W., Drug Prototypes and Their Exploitation, John Wiley & Sons, Chichester, **1996**.

[3]   Kubinyi, H., J. Receptor Signal Transduct. Res. **1999**, *19*, 15-39.

[4]   Lahana, R., Drug Discovery today **1999**, *4*, 447-448.

[5]   Hopkins, A. L., and Groom, C. R., Nature Rev. Drug Discov. **2002**, *1*, 727-730.

[6]   Kubinyi, H., Nature Rev. Drug Discov. **2003**, *2*, 665-668.

[7]   Schaus, J. M., and Bymaster, F. P., Ann. Rep. Med. Chem. **1998**, *33*, 1-10.

[8]   Bymaster, F. P., Calligaro, D. O., Falcone, J. F., Marsh, R. D., Moore, N. A., Tye, N. C., Seeman P., and Wong, D. T., Neuropsychopharmacology **1996**, *14*, 87-96.

[9]   Bymaster, F. P., Nelson, D. L., DeLapp, N. W., Falcone, J. F., Eckols, K., Truex, L. L., Foreman, M. M., Lucaites V. L., and Calligaro, D. O., Schizophr. Res. **1999**, *37*, 107-22.

[10]   Böhm, H.-J., and Schneider, G., Eds., Virtual Screening for Bioactive Molecules (volume 10 of Methods and Principles in Medicinal Chemistry, Mannhold, R., Kubinyi, H., and Timmerman, H., Eds.), Wiley-VCH, Weinheim, **2000**.

[11]   a) Rishton, G. M., Drug Discov. today **1997**, *2*, 382-384; b) Rishton, G. M., Drug Discov. today **2003**, *8*, 86-96.

[12]   Lipinski, C. A., Lombardo F., Dominy, B. W., and Feeney, P. J., Adv. Drug Del. Revs. **1997**, *23*, 3-25.

[13]   MDL Available Compounds Directory, MDL Information Systems Inc., San Leandro, CA, U.S.A.; www.mdl.com.

[14]   Ajay, Walters, W. P., and Murcko, M. A., J. Med. Chem. **1998**, *41*, 3314-3324.

[15]   Sadowski, J., and Kubinyi, H., J. Med. Chem. **1998**, *41*, 3325-3329.

[16]   Walters, W. P., and Murcko, M. A., Adv. Drug Deliv. Rev. **2002**, *54*, 255-271.

[17]   Rarey, M., and Dixon, J. S., J. Comput.-Aided Mol. Design **1998**, *12*, 471-490; www.biosolveit.de.

[18]   Rarey, M., and Stahl, M., J. Comput.-Aided Mol. Design **2001**, *15*, 497-520.

[19]   MDL Screening Compounds Directory, MDL Information Systems Inc., San Leandro, CA, U.S.A.; www.mdl.com.

[20]   Güner, O. F., Ed., Pharmacophore Perception, Development and Use in Drug Design, International University Line, La Jolla, CA, **2000**.

[21]   Catalyst, Accelrys Inc., San Diego, CA, U.S.A.; www.accelrys.com.

[22]   Höltje, H.-D., Sippl, W., Rognan, D., and Folkers, G., Molecular Modelling. Basic Principles and Applications, 2nd Edition, Wiley-VCH, Weinheim, **2003**.

[23]   Lommerse, J. P. M., Price, S. L., and Taylor, R., Comput. Chem. **1997**, *18*, 757-774.

[24] Böhm, H.-J., Brode, S., Hesse, U., and Klebe, G., Chem. Eur. J. **1996**, *2*, 1509-1513.

[25] Sadowski, J., Lecture presented at the ACS Meeting, Boston, August **2002**.

[26] Nielsen, J. E., and Vriend, G., Proteins **2001**, *43*, 403-12.

[27] Nielsen, J. E., and McCammon, J. A., Protein Sci. **2003**, *12*, 1894-901.

[28] Pospisil, P., Ballmer, P., Scapozza. L., and Folkers, G., J. Recept. Signal Transduct. Res. **2003**, *23*, 361-71.

[29] Trepalin, S. V., Skorenko, A. V., Balakin, K. V., Nasonov, A. F., Lang, S. A., Ivashchenko, A. A., and Savchuk, N. P., J. Chem. Inf. Comput. Sci. **2003**, *43*, 852-860.

[30] a) Watson, J. D., The Double Helix: A Personal Account of the Discovery of the Structure of DNA, Atheneum Press, New York, **1968**; b) Watson, J. D., with Berry, A., DNA. The Secret of Life, William Heinemann, London, **2003**; c) www.phy.cam.ac.uk/camphy/dna/dna13_1.htm and ... /dna14_1.htm.

[31] Brandstetter, H., Grams, F., Glitz, D., Lang, A., Huber, R., Bode, W., Krell, H. W., and Engh, R. A., J. Biol. Chem. **2001**, *276*, 17405-17412.

[32] Pospisil, P., and Ballmer, P., ETH Zurich, Switzerland; www.pharma.ethz.ch/pc/Agent2/.

[33] Kubinyi, H., Hamprecht, F. A., and Mietzner, T., J. Med. Chem. **1998**, *41*, 2553-2564.

[34] Kearsley, S. K., and Smith, G. M., Tetrahedron Comp. Methodol. **1990**, *3*, 615-633.

[35] Klebe, G., Mietzner, T., and Weber, F., J. Comput.-Aided Mol. Design **1994**, *8*, 751-778.

[36] Lemmen, C., and Lengauer, T., J. Comput.-Aided Mol. Design **1997**, *11*, 357-368; www.biosolveit.de; www.tripos.com.

[37] Lemmen, C., and Lengauer, T., J. Comput.-Aided Mol. Design **2000**, *14*, 215-232; www.biosolveit.de; www.tripos.com.

[38] Cramer III, R. D., Patterson, D. E., and Bunce, J. D., J. Am. Chem. Soc. **1988**, *110*, 5959-5967.

[39] a) Kubinyi, H., Ed., 3D QSAR in Drug Design. Theory, Methods and Applications, ESCOM Science Publishers B.V.: Leiden, **1993**. b) Kubinyi, H., Folkers, G., and Martin, Y. C., Eds., 3D QSAR in Drug Design. Volume 2. Ligand-Protein Komplexes and Molecular Similarity, Kluwer/ESCOM, Dordrecht, **1998**; also published as Persp. Drug Discov. Design **1998**, *9-11*, 1-416. c) Kubinyi, H., Folkers, G., and Martin, Y. C., Eds., 3D QSAR in Drug Design. Volume 3. Recent Advances, Kluwer/ESCOM, Dordrecht, **1998**; also published as Persp. Drug Discov. Design **1998**, *12-14*, 1-352.

[40] Klebe, G., Abraham, U., and Mietzner, T., J. Med. Chem. **1994**, *37*, 4130-4146.

[41] Sybyl / Unity, Tripos Inc., St. Louis, MO, U.S.A.; www.tripos.com.

[42] Goodford, P. J., J. Med. Chem. **1984**, *27*, 557-564.

[43] Kubinyi, H., Curr. Opin. Drug Discov. Dev. **1998**, *1*, 4-15.

[44] Redshaw, S., in: Medicinal Chemistry. The Role of Organic Chemistry in Drug Research, 2nd Ed., Ganellin C. R., and Roberts, S. M., Eds., Academic Press, London, **1993**, pp. 163-185.

[45] von Itzstein, M., Wu, W. Y., Kok, G. B., Pegg, M. S., Dyason, J. C., Jin, B., Phan, T. V., Smythe, M. L., White, H. F., Oliver, S. W., Colman, P. M., Varghese, J. N., Ryan, D. M., Woods, J. M., Bethell, R. C., Hotham, V. J., Cameron, J. M., and Penn, C. R., Nature **1993**, *363*, 418-423.

[46] Abdel-Magid, A. F., Maryanoff, C. A., and Mehrman, S. J., Curr. Opin. Drug Discov. Devel. **2001**, *4*, 776-91.

[47]   Kim, C. U., Lew, W., Williams, M. A., Liu, H., Zhang, L., Swaminathan, S., Bischofberger, N., Chen, M. S., Mendel, D. B., Tai, C. Y., Laver, W. G., and Stevens, R. C., J. Am. Chem. Soc. **1997**, *119*, 681-690

[48]   Veerapandian, P., Ed., Structure-Based Drug Design. Marcel Dekker, New York, **1997**.

[49]   Babine, R. E., and Bender, S. L., Chem. Rev. **1997**, *97*, 1359-1472.

[50]   Gubernator, K., and Böhm, H.-J., Eds., Structure-Based Ligand Design (volume 6 of Methods and Principles in Medicinal Chemistry, Mannhold, R., Kubinyi, H., and Timmerman, H., Eds.), Wiley-VCH, Weinheim, **1998**.

[51]   Babine R. E., and Abdel-Meguid, S. S., Protein Crystallography in Drug Discovery (volume 20 of Methods and Principles in Medicinal Chemistry, Mannhold, R., Kubinyi, H., and Folkers, G., Eds.), Wiley-VCH, Weinheim, **2004**.

[52]   DesJarlais, R. L., Sheridan, R. P., Seibel, G. L., Dixon, J. S., Kuntz, I. D., and Venkataraghavan, R., J. Med. Chem. **1988**, *31*, 722-729.

[53]   Böhm, H.-J., J. Comput.-Aided Mol. Design **1992**, *6*, 61-78.

[54]   Böhm, H.-J., J. Comput.-Aided Mol. Design **1992**, *6*, 593-606.

[55]   Böhm, H.-J., J. Comput.-Aided Mol. Design **1994**, *8*, 243-256.

[56]   Ewing, T. J., Makino, S., Skillman, A. G., and Kuntz, I. D., J. Comput.-Aided Mol. Design **2001**, *15*, 411-428.

[57]   Jones, G., Willett, P., Glen, R. C., Leach, A. R., and Taylor R., J. Mol. Biol. **1997**, *267*, 727-748.

[58]   Rarey, M., Kramer, B., Lengauer, T., and Klebe, G., J. Mol. Biol. **1996**, *261*, 470-489; www.biosolveit.de; www.tripos.com.

[59]   Lengauer, T., and Rarey, M., Curr. Opin. Struct. Biol. **1996**, *6*, 402-406.

[60]   Goodsell, D. S., and Olson, A. J., Proteins **1990**, *8*, 195-202.

[61]   Morris, G. M., Goodsell, D. S., Halliday, R. S., Huey, R., Hart, W. E., Belew, R. K., and Olson, A. J., J. Comput. Chem. **1998**, *19*, 1639-1662.

[62]   Schneider, G., and Böhm, H.-J., Drug Discov. today **2002**, *7*, 64-70.

[63]   Claussen, H., Buning, C., Rarey, M., and Lengauer, T., J. Mol. Biol. **2001**, *308*, 377-395; www.biosolveit.de; www.tripos.de.

[64]   Hindle, S. A., Rarey, M., Buning, C., and Lengauer, T., J. Comput.-Aided Mol. Design **2002**, *16*, 129-149; www.biosolveit.de; www.tripos.de.

[65]   Hilpert, K., Ackermann, J., Banner, D. W., Gast, A., Gubernator, K., Hadvary, P., Labler, L., Müller, K., Schmid, G., Tschopp, T. B., and van de Waterbeemd, H., J. Med. Chem. **1994**, *37*, 3889-3901.

[66]   Boehm, H.-J., Boehringer, M., Bur, D., Gmuender, H., Huber, W., Klaus, W., Kostrewa, D., Kuehne, H., Luebbers, T., Meunier-Keller, N., and Mueller, F., J. Med. Chem. **2000**, *43*, 2664-2674.

[67]   Green, N. M., Adv. Protein Chem. **1975**, *29*, 85-133.

[68]   Hanessian, S., Lu, P.-P., Sanceau, J.-Y., Chemla, P., Gohda, K., Fonne-Pfister, R., Prade, L., and Cowan-Jacob, S. W., Angew. Chem. Int. Ed. Engl. **1999**, *38*, 3159-3162.

[69]   Shuker, S. B., Hajduk, P. J., Meadows, R. P., and Fesik, S. W., Science **1996**, *274*, 1531-1534.

[70]   Hajduk, P. J., Meadows, R. P., and Fesik, S. W., Science **1997**, *278*, 497-499.

[71]   Hajduk, P. J., Sheppard, G., Nettesheim, D. G., Olejniczak, E. T., Shuker, S. B., Meadows, R. P., Steinman, D. H., Carrera Jr., G. M., Marcotte, P. A., Severin, J., Walter, K., Smith, H., Gubbins, E., Simmer, R., Holzman, T. F., Morgan, D. W., Davidsen, S. K., Summers, J. B., and Fesik, S. W., J. Am. Chem. Soc. **1997**, *119*, 5818-5827.

[72]   Zerbe, O., Ed., BioNMR in Drug Research, (volume 16 of Methods and Principles in Medicinal Chemistry, Mannhold, R., Kubinyi, H., and Folkers, G., Eds.), Wiley-VCH, Weinheim, **2003**.

[73]   Fejzo, J., Lepre, C. A., Peng, J. W., Bemis, G. W., Ajay, Murcko, M. A., and Moore, J. M., Chem. Biol. **1999**, *6*, 755-769.

[74]   Lepre, C. A., Peng, J., Fejzo, J., Abdul-Manan, N., Pocas, J., Jacobs, M., Xie, X., and Moore, J. M., Comb. Chem. High Throughput Screen. **2002**, *5*, 583-590.

[75]   Diercks, T., Coles, M., and Kessler, H., Curr. Opin. Chem. Biol. **2001**, *5*, 285-291.

[76]   Pellecchia, M., Sem, D. S., and Wüthrich, K., Nature Rev. Drug Discov. **2002**, *1*, 211-219.

[77]   Jahnke, W., Floersheim, P., Ostermeier, C., Zhang, X., Hemmig, R., Hurth, K., and Uzunov, D. P., Angew. Chem. Int. Ed. Engl. **2002**, *41*, 3420-3423.

[78]   Meyer, B., and Peters, T., Angew. Chem. Int. Ed. Engl. **2003**, *42*, 842-890.

[79]   Kubinyi, H., Curr. Opin. Drug Discovery Dev. **1998**, *1*, 16-27.

[80]   Metz, G., Ottleben, H., and Vetter, D., in: Protein-Ligand Interactions. From Molecular Recognition to Drug Design, Böhm, H.-J., and Schneider, G., Eds. (volume 19 of Methods and Principles in Medicinal Chemistry, Mannhold, R., Kubinyi, H., and Folkers, G., Eds.), Wiley-VCH, Weinheim, **2003**, pp. 213-236

[81]   Huc, I., and Lehn, J.-M., Proc. Natl. Acad. Sci. USA **1997**, *94*, 2106-2110.

[82]   Lehn, J.-M., and Eliseev, A. V., Science **2001**, *291*, 2331-2332.

[83]   Ramström, O., and Lehn, J.-M., Nature Rev. Drug Discov. **2002**, *1*, 26-36.

[84]   Hochgürtel, M., Kroth, H., Piecha, D., Hofmann, M. W., Nicolau, C., Krause, S., Schaaf, O., Sonnenmoser, G., and Eliseev, A. V., Proc. Natl. Acad. Sci. USA **2002**, *99*, 3382-3387.

[85]   Erlanson D. A., Braisted, A. C., Raphael, D. R., Randal, M., Stroud, R. M., Gordon, E. M., and Wells, J. A., Proc. Natl. Acad. Sci. USA **2000**, *97*, 9367-9372.

[86]   Erlanson, D. A., Lam, J. W., Wiesmann, C., Luong, T. N., Simmons, R. L., DeLano, W. L., Choong, I. C., Burdett, M. T., Flanagan, W. M., Lee, D., Gordon, E. M., and O'Brien, T., Nature Biotechnol. **2003**, *21*, 308-314.

[87]   Maly, D. J., Choong, I. C., and Ellman, J. A., Proc. Natl. Acad. Sci. USA **2000**, *97*, 2419-2424.

[88]   Swayze, E. E., Jefferson, E. A., Sannes-Lowery, K. A., Blyn, L. B., Risen, L. M., Arakawa, S., Osgood, S. A., Hofstadler, S. A., and Griffey. R. H., J. Med. Chem. **2002**, *45*, 3816-3819.

[89]   Lewis, W. G., Green, L. G., Grynszpan, F., Radic, Z., Carlier, P. R., Taylor, P., Finn M. G., and Sharpless, K. B., Angew. Chem. Int. Ed. Engl. **2002**, *41*, 1053-1057.

[90]   Wolber, G., and Langer, T., in: Rational Approaches to Drug Design (Proceedings of the 13th European Symposium on Quantitative Structure-Activity Relationships, Düsseldorf, 2000), Höltje, H.- D., and Sippl, W., Eds., Prous Science, Barcelona, **2001**, pp. 390-399.

[91]   Schneider, G., Clement-Chomienne, O., Hilfiger, L., Schneider, P., Kirsch, S., Boehm, H.-J., and Neidhart, W., Angew. Chem. Int. Ed. Engl. **2000**, *39*, 4130-4133.

[92]   Schneider, G., Lee, M. L., Stahl, M., and Schneider, P., J. Comput.-Aided Mol. Design **2000**, *14*, 487-494.

[93]   Lewell, X. Q., Judd, D. B., Watson, S. P., and Hann, M. M., J. Chem. Inf. Comput. Sci. **1998**, *38*, 511-522.

[94]   Schneider, G., Neidhart, W., Giller, T., and Schmid, G., Angew. Chem. Int. Ed. Engl. **1999**, *38*, 2894-2896.

[95]   Wang, R., Lu, Y., and Wang, S., J. Med. Chem. **2003**, *46*, 2287-2303.

[96]   Bleicher, K. H., Böhm, H.-J., Müller, K., and Alanine, A. I., Nature Rev. Drug Discov. **2003**, *2*, 369-378.

[97]   Muchmore, S. W., and Hajduk, P. J., Curr. Opin. Drug Discovery Dev. **2003**, *6*, 544-549.

[98]   Kubinyi, H., in: 3D QSAR in Drug Design. Volume II. Ligand-Protein Interactions and Molecular Similarity, Kubinyi, H., Folkers G., and Martin, Y. C., Eds., Kluwer/ESCOM, Dordrecht, **1998**, pp. 225-252; also published in Persp. Drug Design Discov. **1998**, *9-11*, 225-252.

[99]   Martin, Y. C., Kofron, J. L., and Traphagen, L. M., J. Med. Chem. **2002**, *45*, 4350-4358.

[100]   Caron, P. R., Mullican, M. D., Mashal, R. D., Wilson, K. P., Su, M. S. and Murcko, M. A., Curr. Opin. Chem. Biol. **2001**, *5*, 464-470.

[101]   Bleicher, K. H., Curr. Med. Chem. **2002**, *9*, 2077-2084.

[102]   Jacoby, E., Schuffenhauer, A., and Floersheim, P., Drug News Perspect. **2003**, *16*, 93-102.

[103]   Müller, G., Drug Discov. today **2003**, *8*, 681-691.

[104]   Kubinyi, H., and Müller, G., Eds., Chemogenomics in Drug Discovery - A Medicinal Chemistry Perspective (Volume 22 of Methods and Principles in Medicinal Chemistry, Mannhold, R., Kubinyi, H., and Folkers, G., Eds.), Wiley-VCH, Weinheim, **2004**, in print.

[105]   Wermuth, C. G., Med. Chem. Res. **2001**, *10*, 431-439.

[106]   Wermuth, C. G., J. Med. Chem. **2004**, *47*, 1303-1314.

[107]   Kubinyi, H., in: Kubinyi, H., and Müller, G., Eds., Chemogenomics in Drug Discovery - A Medicinal Chemistry Perspective (Volume 22 of Methods and Principles in Medicinal Chemistry, Mannhold, R., Kubinyi, H., and Folkers, G., Eds.), Wiley-VCH, Weinheim, **2004**, pp. 43-67.

[108]   Prentis, R. A., Lis, Y., and Walker, S. R., Br. J. Clin. Pharmac. **1988**, *25*, 387-396.

[109]   Kennedy, T., Drug Discov. today **1997**, *2*, 436-444.

[110]   van de Waterbeemd, H., and Gifford, E., Nature Rev. Drug Discov. **2003**, *2*, 192-204

[111]   Horrobin, D. F., Nature Rev. Drug Discov. **2003**, *2*, 151-154.

[112]   Ryan, J. F., Ed., The Pharmaceutical Century. Ten Decades of Drug Discovery, Supplement to ACS Publications, American Chemical Society, Washington, DC, **2000**.

[113]   Wermuth, C. G., Ed., The Practice of Medicinal Chemistry, 2nd Edition, Academic Press, London, **2003**.

A TALE OF TWO STATES: REACTIVITY OF CYTOCROME P450 ENZYMES

S. Shaik

Department of Chemistry, and The Lise Meitner-Minerva Center for Computational Quantum Chemistry, The Hebrew University, Jerusalem 91904, Israel, E-mail: sason@yfaat.ch.huji.ac.il

1 INTRODUCTION

Cytochromes P450 (P450s) are heme enzymes that are present in all aerobic species, and carry out vital oxidative processes, namely, detoxification of foreign compounds and biosynthesis of hormones and other essential molecules.[1-3] As such, the enzymes are extremely versatile and perform a variety of stereoselective transformations that are otherwise difficult to achieve, e.g., hydroxylation of strong C-H bonds, epoxidation of C=C bonds, heteroatom oxidation (sulfoxidation), and also some desaturation reactions.

The enzyme is a nano-machine that operates by means of the catalytic cycle illustrated in Figure 1.[4] The cycle begins with the resting state (**1**), in which a water molecule is bound to the heme in the distal side and to the thiolate side chain of a cysteine residue (abbreviated as SCys) in the proximal side. The entrance of the substrate (for example, an alkane, Alk-H) displaces the water molecule and triggers a transfer of an electron from a reductase protein. The resulting ferrous complex (**3**) binds molecular oxygen, which gets activated by an additional electron transfer from the reductase followed by protonation from a suitable proton relay system in the distal side of the protein, thereby generating the ferric-hydroperoxide species (**6**), also called Compound Zero (Cpd 0). The resulting Cpd 0 accepts an additional proton and splits off a water molecule to form **7**, the so-called Compound I (Cpd I) species. Cpd I possesses a high-valent oxo-iron species, $Fe^{IV}=O$, and a radical cationic state in the porphyrin, i.e., $Por^{+} \cdot Fe^{IV}=O$. Cpd I is known from studies of synthetic models, to be a powerful oxidant[3,5] and as such, it is thought that this is the ultimate oxidant species that transfers an oxygen atom to the substrate leading to the product (e.g., ferric-alcohol) complex, **8**. The cycle is restored by release of the oxidized substrate (e.g., the alcohol, Alk-OH) and complexation of a water molecule to the heme to regenerate the resting state.

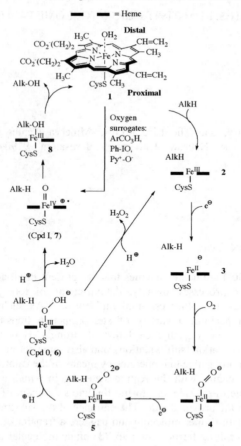

Figure 1   *The proposed catalytic cycle of cytochrome P450*

Despite the very appealing catalytic cycle, two features remain hypothetical and controversial to this day. Firstly, Cpd I of P450 is still elusive, and secondly, the reactivity patterns of P450 suggest that in addition to Cpd I, the catalytic cycle of the enzyme utilizes other species to oxidize organic molecules.[6] Theory, and specifically density functional theoretic (DFT) calculations, can be helpful in such a situation by characterizing the active species and delineating their reactivity patterns. And this is where our story of two-state reactivity (TSR) and multi-state reactivity (MSR) enters.[7] Let us first follow with a slightly more detailed description of the major problems.

2   SOME CONTROVERSIAL ISSUES IN THE CHEMISTRY OF P450 ENZYMES

2.1 How many active species does the enzyme have?

While Cpd I appears to be the consensus oxidant, its existence could so far only be inferred indirectly; e.g., by EPR/ENDOR follow-up during camphor hydroxylation,[4] or by transient electronic spectroscopy and decay kinetics[8] of the putative species, or still by analogy to the same species in the enzyme chloro peroxidase (CPO).[9] Attempts to generate Cpd I using peracetic acid or other oxygen surrogates (see Figure 1) led to one electron reduced species, PorFe^{IV}=O, so called Cpd II, and a Tyr radical, presumably formed via oxidation of Tyr by the putative Cpd I.[10] A recent characterization of Cpd I under conditions of cryogenic X-ray crystallography,[11] generated initially high hopes, but has since been cast in doubt.[4] Even the consensus technique of generating Cpd I species using iodosylbenzene has been questioned recently.[12] A strong support for the operation of Cpd I has been provided by the reactivity of synthetic Cpd I species and P450 towards the same substrates, where both sets of reactions led to similar product distribution and stereochemical scrambling information.[5] Thus, despite the lingering uncertainty, the inferences that Cpd I exists and is responsible for the normal oxidation processes of the enzyme in the cycle are quite strong, and it is assumed that Cpd I cannot be isolated in the cycle because it is very reactive.

The evidence for a second oxidant species in the cycle of P450 is indirect; it derives from reactivity studies, lifetime measurements of putative intermediates, and product distributions of P450s and their mutants. Thus, mutant enzymes, in which the proton relay that converts Cpd 0 to Cpd I is disrupted by site-directed mutagenesis,[6,13] exhibit reactivity patterns different than those of the wild type (WT) enzyme, e.g., with 2-butenes, the mutant enzyme yields more epoxidation products compared with C–H hydroxylation products.[13] Similarly, the T252A mutant of P450$_{cam}$,[14] in which Thr$_{252}$ that plays a key role in the conversion of Cpd 0 to Cpd I was replaced by Ala, does not hydroxylate camphor, but is capable of epoxidizing the double bond of camphene, albeit more sluggishly than the WT enzyme. Since the catalytic cycle of T252A is thought to terminate at Cpd 0, this and the reactivity of the mutant toward camphene were taken as evidence that Cpd 0 is involved in the epoxidation of an activated double bond. Curiously, however, the double mutant of P450$_{cam}$, T252A/D251N, in which the protonation machinery has been disrupted by mutations of both Thr$_{252}$ and Asp$_{251}$, is able to hydroxylate camphor,[15] despite the T252A mutation that should hypothetically have blocked the formation of Cpd I. This implies that either Cpd 0 has a variable reactivity in different mutants or that Cpd I may be present even in the case of the T252A mutant. Moreover, a recent mechanistic study,[16] using an oxygen surrogate provided compelling evidence for the reactivity of Cpd I as the sole oxidant. Thus, the participation of Cpd 0 is still a clouded issue, but the fact remains that the data behave, as though more than one oxidant is present in the cycle of P450.

2.2 The "Rebound Controversy"

The major controversial issue is associated with the above uncertainties in the identity of the oxidant, and it emerged primarily from mechanistic investigations of C–H hydroxylation.[5,6,13] The consensus mechanism of C–H hydroxylation by P450 enzymes is the "rebound" mechanism proposed by Groves and McClusky[17] and depicted in Figure 2a. The first step in the mechanism involves an initial hydrogen abstraction from the alkane by Cpd I. Subsequently the alkyl radical (Alk•) is partitioned between two competing processes. It can either instantly rebound to form an alcohol complex, where the alcohol is unrearranged (*U*), keeping the original stereochemical information possessed by the alkane, or it can first undergo skeletal rearrangement and then rebound to give a rearranged (*R*) alcohol product. The rebound mechanism accounts for partial loss of stereochemistry and geometrical rearrangement data, as well as for the large kinetic isotope effect (*KIE*) observed[5] when the activated C–H bond is replaced by C–D.

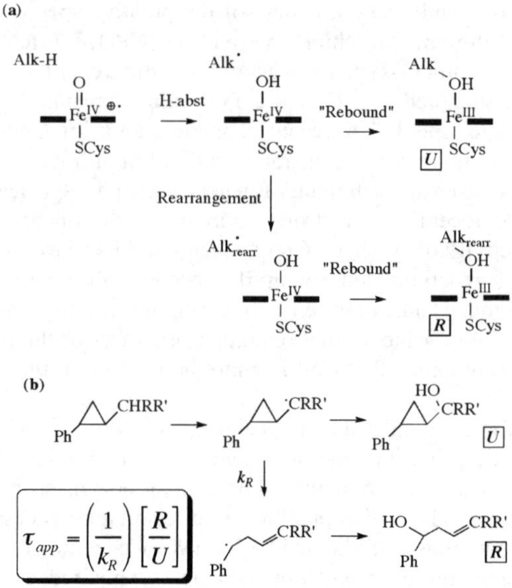

Figure 2   *The "rebound" mechanism in (a). The method of determination and expression for the apparent lifetime (τ_{app}) of the radical, using a probe substrate, in (b).*

Initial measurements of the lifetime of the radical[18] indicated a short but finite lifetime. Subsequent studies of Newcomb et al.[6] using radical clocks as the one depicted in Figure 2b, cast the rebound mechanism into doubt. Based on Figure 2b, the apparent radical lifetime (τ_{app}) is given by the free radical lifetime ($1/k_R$) times the ratio of rearranged (*R*) to unrearranged (*U*) product yields. These studies led to radical lifetimes of the order of 100 fs, which are considered much too short for a real intermediate. Such short lifetimes rule out the presence of radical as intermediates and altogether question the validity of the rebound mechanism. Furthermore, in mutant P450s, in which Cpd I was presumably absent, the ratio [*R/U*] was generally larger by comparison with the WT enzyme. This in turn suggested that the rearranged products arose from non-radical intermediates that were generated due to the presence of another oxidant species that became prominent in the mutants. This hypothesis was supported[6] by use of probe substrates, which can distinguish between radical and carbocationic rearrangement patterns, and which suggested that the major reaction intermediate is in fact a carbocation and not a radical. Therefore, Newcomb et al.[6,19] proposed that C–H hydroxylation proceeds via multiple oxidant species (Scheme 1); Cpd I that leads to concerted oxygen insertion into the C–H bond, and Cpd 0 or ferric hydrogen peroxide that transfer an OH$^+$ species and generate a protonated alcohol that subsequently undergoes rearrangements typical of a carbocationic species. Interestingly, similar Cpd 0 species in heme oxygenase and in mutants of e.g., Myoglobin are known to insert OH into the *meso* position of the heme.[20,21] However such a reactivity pattern has never been reported for P450 Cpd 0. Thus, in all rigor, there are no direct proofs for the reactivity of P450 Cpd 0 towards C–H or C=C activation.

The multiple-oxidant hypothesis:

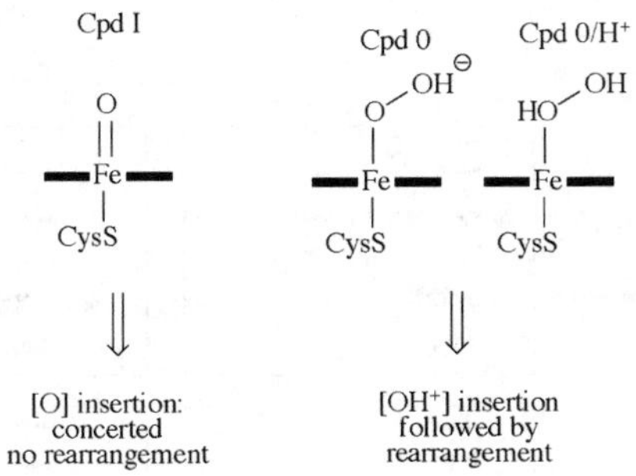

Cpd I

$$O$$
$$||$$
$$\text{—Fe—}$$
$$|$$
$$\text{CysS}$$

Cpd 0

$$O \diagdown \overset{\ominus}{OH}$$
$$|$$
$$\text{—Fe—}$$
$$|$$
$$\text{CysS}$$

Cpd 0/H⁺

$$HO \diagdown OH$$
$$|$$
$$\text{—Fe—}$$
$$|$$
$$\text{CysS}$$

$$\Downarrow$$

[O] insertion:
concerted
no rearrangement

$$\Downarrow$$

[OH⁺] insertion
followed by
rearrangement

Scheme 1

In sum: the evidence on P450 reactivity has so far been indirect; the primary oxidant species has never been observed, its identity cast in doubt, and the reactivity patterns behave as though more than one oxidant is responsible for the product distribution. The identity of this second oxidant is however still shrouded. This is an intellectually intriguing scenario that calls for the involvement of theory. Theoretical consideration since 1998[22] offered a possible reconciliation of the controversy based on the two-state reactivity (TSR) paradigm. Subsequently, the Jerusalem group used DFT to study the mechanism of C-H hydroxylation reactions between Cpd I and a variety of alkanes, by now more than 10.[23] In collaboration with the Mülheim group, the two groups explored the mechanism of camphor hydroxylation by Cpd I of $P450_{cam}$ using hybrid quantum mechanical (QM) / molecular mechanical (MM) calculations.[24] The reactivity of Cpd 0 was explored in epoxidation[25,26] and sulfoxidation reactions, without and with acid catalysis.[27] The reactivity of the putative ferric hydrogen peroxide complex, $PorFe^{III}(H_2O_2)$, is currently being investigated in Jerusalem. These extensive computational studies[28] have revealed that: (a) Cpd 0 is much less reactive than Cpd I, and cannot be invoked as an oxidant in the presence of Cpd I, (b) ferric-hydrogen peroxide is less reactive than Cpd I, (c) the reactivity of Cpd I involves at least two productive states, and can be described as TSR, or more accurately as multi-state reactivity (MSR), and (d) TSR and MSR reconcile the major mechanistic controversies.

3 ORIGINS OF TWO-STATE AND MULTI-STATE REACTIVITY

3.1 States of Cpd I

The key orbitals of Cpd I species are shown in Figure 3; these are the five d-block orbitals, with δ, π^* and σ^* characters, so denoted based on the type of antibonding interaction with the ligand orbitals, and the porphyrin based orbital, labelled as a_{2u}, which involves an antibonding interaction from the thiolate ligand.

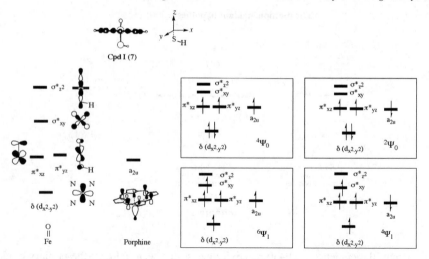

Figure 3   *Key orbitals of Cpd I and occupation diagrams in the low-lying sates.*

Based on DFT and QM(DFT)/MM calculations using the B3LYP functional and a variety of basis sets,[23,24,28] the ground state of Cpd I is a pair of doublet and quartet spin-states, which are labelled as $^{4,2}\Psi_0$ and possess three unpaired electrons in the π^* and a_{2u} orbitals that are coupled ferromagnetically (to a quartet spin) or antiferromagnetically (to a doublet spin); the superscript to the left of the state label signifies the spin multiplicity, $2S+1$. Since the π^* and a_{2u} orbitals are virtually disjointed, the coupling is weak and the quartet and doublet states are virtually degenerate, the doublet lies ca. 10-21 cm^{-1} below the quartet.[24] This state ordering, as well as the calculated spin distribution are in accord with experimentally determined features of the analogous Cpd I of chloroperoxidase.[9] Another pair of states, labelled as $^{6,4}\Psi_1$ in Figure 3, can be generated from the $\delta^1\pi^{*2}\sigma_{xy}^{*1}a_{2u}^1$ configuration that has five unpaired electrons coupled to a sextet and quartet spin states. Although the δ-σ_{xy}^* orbital gap is quite large, these states enjoy significant stabilization due to the large exchange interactions of the d-block electrons, and as a result are expected to be not much higher than the pair of ground states. These states were not considered explicitly in the original theoretical studies,[28] but recent DFT calculations[29] show that the $^{6,4}\Psi_1$ states lie only ca. 12 kcal/mol above the doublet and quartet states with the in $\delta^2\pi^{*2}a_{2u}^1$ configuration. Since this energy gap is of the order of typical barriers for C-H hydroxylation or even less, the $^{6,4}\Psi_1$ states may well become accessible along the reaction path. Clearly, therefore, *Cpd I is an oxidant, which involves at the least two states in its reactions.* More states become accessible along the oxidation pathways.

3.2 States of intermediates along the oxidation reaction path

The heme of Cpd I contains two oxidation equivalents above the resting state and the product complexes, which involve Fe^{III} and a closed-shell porphyrin. These oxidation equivalents are stored in Cpd I, in the Fe^{IV} center and the hole in the porphyrin, (i.e., Por•$^+$). Thus, any reaction of Cpd I will ultimately involve two formal electron "transfer" events from the substrate being oxidized to the heme. Since the reagent has a dense orbital manifold, this creates many possibilities of dispensing with these oxidation equivalents and hence quite a few more states will become energetically accessible along the reaction path. Figure 4 shows the changes in orbital occupancy along the reaction paths for C-H hydroxylation and C=C epoxidation by considering the d-block orbitals, the a_{2u} orbital of the porphyrin and the substrate orbitals, i.e. π_{CC} is the orbital for the π-bond, activated during epoxidation, and σ_{C-H} is the orbital for the C–H bond that is broken during hydroxylation. The orbital occupancies correspond to the lowest quartet and doublet spin states that are nascent from reactivity of the ground states, $^{4,2}\Psi_0$, of Cpd I. In each diagram, the doublet state is indicated using an inversed electron spin within parentheses.

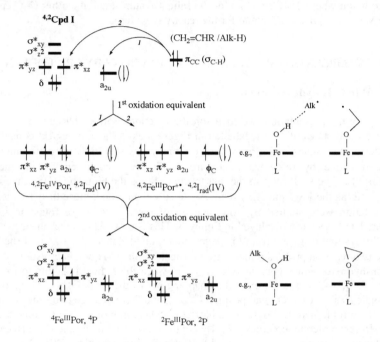

Figure 4 *Changes in orbital occupation and intermediate states during C-H hydroxylation and C=C epoxidation by Cpd I*

The initial bond formation step involves an electron "transfer" event from the substrate into Cpd I. Two of the possibilities for this "transfer" are: (1) into the porphyrin hole (a_{2u}-orbital) to generate doublet and quartet Fe^{IV}-type intermediates with a closed-shell porphyrin, i.e., the $PorFe^{IV}$ electromers, or (2) into the π_{xz}^*-orbital to generate Fe^{III}-type intermediates with a radical cationic situation on the porphyrin ring, i.e. the $Por^{+\bullet}Fe^{III}$ electromer. In both cases the substrate retains a radical center with a singly occupied orbital labelled ϕ_C. During the second bond formation, an additional electron is "transferred" from the substrate to fill, among the many options, either the a_{2u} orbital of $Por^{+\bullet}Fe^{III}$ or one of the d-orbitals of $PorFe^{IV}$. Filling the π_{xz}^* orbital generates the doublet-spin ferric product, whereas the filling of the σ_{z2}^* orbital leads to the quartet-spin ferric product.

In addition to the states included in Figure 4, there are other possible states. For example, shifting electrons from the ϕ_C orbital of the radical intermediate center into one of the d-orbitals or into a_{2u} will generate a variety of doublet and quartet spin-states in which the substrate possesses a carbocationic species. Excitation of an electron from δ to σ^*_{xy} or an initial electron shift to σ^*_{z2} will generate the corresponding sextet states. Since the orbital manifold is dense and the exchange interactions are quite strong, many of these states will be accessible and will contribute to product formation. As such, the reactivity of Cpd I will involve at least two spin states, and possibly many electromeric states, hence TSR and MSR. In the following section we show two examples, one for C-H hydroxylation where TSR accounts for the salient features, and the other C=C epoxidation where MSR is required to account for the reactivity patterns.

4 EXAMPLES OF TSR AND MSR IN OXIDATIONS BY CPD I OF P450

4.1 TSR in C-H hydroxylation

To simplify the presentation, we focus on the reaction profiles that are nascent from the pair of ground states of Cpd I, labelled in Figure 3 as $^{4,2}\Psi_0$, and neglect consideration of the other pair of states, $^{6,4}\Psi_1$, as well as the many possible varieties of the intermediates discussed above by reference to Figure 4. To keep consistency with the original designations,[23] we shall refer to the quartet state $^4\Psi_0$ as the high-spin (HS) state and to the doublet $^2\Psi_0$ as the low-spin (LS) state. As already mentioned, more than 10 cases of C-H hydroxylation were studied by now, and all of these cases were found to exhibit the computed TSR scenario displayed in Figure 5. This is a two-state mechanism with HS and LS pathways that originate from the degenerate ground states $^{2,4}\Psi_0$ of Cpd I (hence ^{4,2}Cpd I). The two reaction profiles remain close throughout the C-H bond activation phase, up to the iron-hydroxo-radical complexes ($^{4,2}I_{rad}$), in which the radical is bound to iron hydroxo by a weak OH---Alk interaction. After snapping out of the weak OH---Alk interaction, the alkyl radical assumes a rebound position ($^{4,2}I_{reb}$), and thereafter the two states bifurcate. On the HS manifold, there is a significant barrier (3-5 kcal/mol or so[23,24]) for rebound *en route* to the HS ferric-alcohol complex (^4P). By contrast, the rebound on the LS surface is barrier free, leading spontaneously to the LS ferric-alcohol complex (^2P). Thus, the DFT calculations lend support to the rebound mechanism,[17] but extend the number of states and intermediates that give rise to products.

The profiles of Figure 5 depicts only two intermediates, $^{4,2}I_{rad}$, but as we argued above there are quite a few more. The inset in Figure 5 shows a typical situation exemplified by the case of camphor hydroxylation, with four radical intermediates of the $Por^{+\bullet}Fe^{III}$ and $PorFe^{IV}$ electromer varieties.[23,24,28] It is seen that the four states are condensed within 2 kcal/mol or less, and hence, all the four species are accessible and will generate products. For each electromer type, the LS intermediate collapses to the respective product complexes without a barrier, while the HS intermediate has a significant barrier for rebound. Therefore, generally speaking, C-H hydroxylation proceeds via two competing processes on two different spin-state manifolds; one on the LS manifold is effectively concerted, with ultra short or zero radical lifetime, the other on the HS manifold is stepwise with a significant lifetime for the intermediate.

It is this feature of TSR that creates *the likeness of two different oxidants* and appears, in our view, to be the cause of the controversial conclusions which were derived from the clock experiments.[6] In a typical clock experiment (Figure 2b) the apparent lifetime (τ_{app}) of the radical intermediate, during P450 hydroxylation, is determined from the extent of skeletal rearrangement observed in the alcohol product (***R/U***) and the known lifetime of the free radical ($1/k_R$). However, the expression for the apparent lifetime arises from the assumption that there is a single intermediate that branches to give the two products. However, using the TSR scheme, it is apparent that ***R*** and ***U*** do not originate from the same intermediate; ***R*** arising only from the HS manifold and ***U*** mostly, if not only, from the LS manifold. As such, the apparent lifetime of the clock experiments, e.g., in Figure 2b, is not the real lifetime, since it reflects the relative importance of the HS and LS pathways. Furthermore, since the LS barrier is generally lower than the corresponding HS quantity (see Figure 5), the measured [***R/U***] quantity will give apparent lifetimes that are too short. In cases of C-H hydroxylation in which the yield of the HS process is rather small, most of the product will be the unrearranged variety, and the apparent lifetime will be artificially ultrashort. In reality, however, radicals exist on the HS manifold and have normal lifetimes, while the ultra short apparent lifetimes originate from the assumption that the two products arise from a single state.

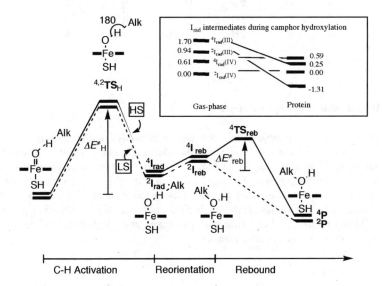

Figure 5 *Energy profiles for C-H hydroxylation. The inset shows a variety of other intermediates that exist past the C-H activation step and are close in energy (the energy scale is in kcal/mol)*

The simplified TSR scenario, *with rearrangement occurring only on the HS manifold*, creates further order in the experimental data. Analysis of the barrier for rebound on the HS manifold,[23] led to the simple prediction that as the radical becomes a better electron donor (lower ionization potential, *IP*, or oxidation potential), so should the rebound barrier decrease until it would altogether disappear. This prediction is manifested in the series of probe substrates used by Newcomb,[6] as shown in Scheme 2. Thus, probes **i** and **ii** lead to significant amounts of the rearranged product, while probe **v** that produces the radical with the lowest *IP* exhibits virtually no rearrangement (<1%). Recent DFT calculations,[30] show that indeed **ii** should give significant amount of the rearranged product, estimated as 20-30% based on the relative HS and LS barriers, while **v** should yield no rearranged products, since *its LS and HS pathways are both effectively concerted.* In addition, the TSR model predicts a product isotope effect that is different for *U* and *R*, and dependence of [*R/U*] on the strength of the Fe-S bond; the stronger the bond, in a particular P450 isozyme, the more rearranged products are expected.[23]

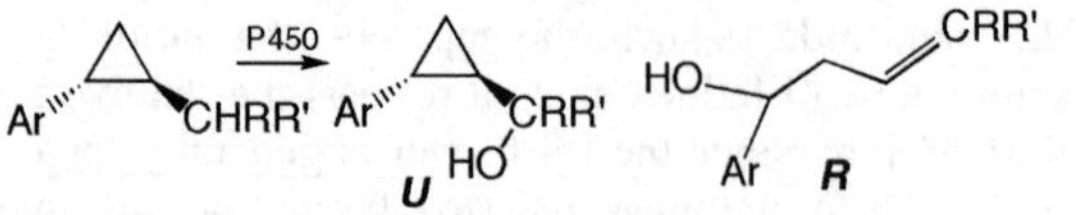

probe	Ar	R,R'	IP_{Alk}	[*U/R*]
i	pCF_3	H, H	highest	4
ii	pH	H, H	↑	4.3
iii	pCF_3	H, CH_3		12
iv	pH	H, CH_3		25
v	pH	CH_3, CH_3	lowest	> 100

Scheme 2

4.2 MSR in double bond epoxidation

As shown in Scheme 3, double bond epoxidation is normally attended by formation of aldehydes and suicidal complexes, in which the olefin alkylates the heme through one of its nitrogen atoms.[1-3] In addition, if one starts with pure *cis* or *trans* olefins one observes a mixture of *cis* and *trans* epoxides. The mechanistic details of C=C activation by P450 have been a subject of debates, albeit not as heated as in the case of C-H hydroxylation. Our studies of three different olefins (ethene, propene and styrene)[31,32] revealed a multi-state reactivity (MSR) scenario where different spin states and electromers participate in the product formation in a state specific manner.

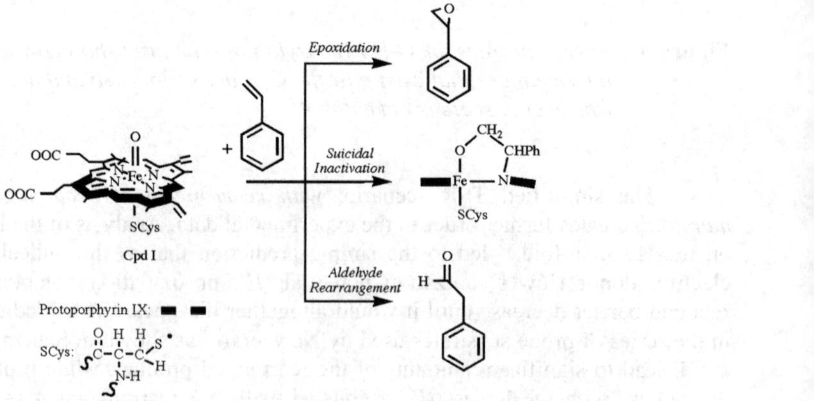

Scheme 3

A representative example where the various intermediate states lead to MSR is the epoxidation of styrene.[32] The gas-phase energy profile, which is nascent from the pair of LS and HS ground states of Cpd I, is depicted in Figure 6. It is seen that here in addition to the usual pair of LS and HS transition states, $^{2,4}TS_{rad}$, which lead to the radical intermediates, $^{2,4}I_{rad}$, there is a LS cationic transition state, $^2TS_{cat,yz}$, which leads to the cationic intermediate, $^2I_{cat,yz}$, having a $\delta^2\pi^*_{xz}{}^2\pi^*_{yz}{}^1$ configuration on the porphyrin iron-oxo moiety and a carbocationic center on the benzylic carbon of styrene. The bond activation leads here to five low-lying intermediate states (no stable solution was found for $^2I_{rad}$(III), which fell to one of the cationic states), of the radical and cationic varieties. All the LS intermediates undergo ring closure to give the LS epoxide product, ^2P, without a barrier and hence, give rise to effectively concerted LS processes.

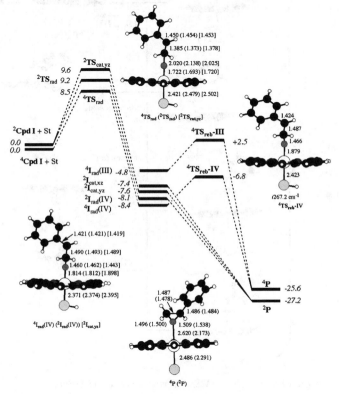

Figure 6   *Energy (in kcal/mol) profiles for C=C epoxidation of styrene.*

By contrast to the LS species, the HS intermediates encounter barriers to ring closure, and therefore give rise to stepwise mechanism with sufficiently long-lived intermediates that can participate in a variety of side product formation and in scrambling of stereochemistry (e.g., if one starts with *trans*-2-deuterio-styrene) to yield a mixture of *cis* and *trans* epoxide products.

Figure 7 shows the MSR scheme for the reaction of Cpd I with *trans*-2-deuterio-styrene. This is a state-specific scheme, where the spin-states and the electromers participate in specific reactions. The LS intermediates undergo ring closure to *trans*-

epoxides and conserve the initial stereochemistry. By contrast, the HS intermediates participate in stereochemical scrambling and in side product formation. The rotational barrier around the C-C bond of the radical intermediate is only 1.3 kcal/mol, which is smaller than the barriers for ring closure for both HS intermediates. Therefore, both $^4I_{rad}(IV)$ and $^4I_{rad}(III)$ will have sufficient lifetime to lose stereochemistry and undergo ring-closure to produce a mixture of *cis* and *trans* epoxide products.

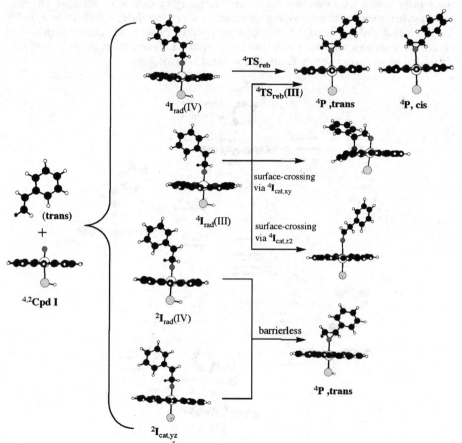

Figure 7   *Multi-state reactivity (MSR) during styrene oxidation.*

The suicidal complex and aldehyde side products were found to originate only from the HS cationic states; one state, $^4I_{cat,z2}$, with configuration $a_{2u}^2\delta^2\pi^*_{xz}{}^1\pi^*_{yz}{}^1\sigma^*_{z2}{}^1$ correlates in a barrier free fashion to the aldehyde complex, and the other, $^4I_{cat,xy}$, with the configuration $a_{2u}^2\delta^2\pi^*_{xz}{}^1\pi^*_{yz}{}^1\sigma^*_{xy}{}^1$ correlates similarly to the suicidal complex. The production of these side-product complexes requires crossover from the HS radical states to the HS cationic states. The barriers for these crossovers are of the order of 10 kcal/mol or so. Consequently, $^4I_{rad}(IV)$, which possesses a much smaller barrier for ring closure will give scrambled epoxide and will not produce suicidal complex or aldehyde. By contrast, the $^4I_{rad}$(III) intermediate possesses a sufficiently large barrier for ring closure to be able to participate also in production of the suicidal complex and aldehyde side products. This is a remarkable state specificity without precedence in organic chemistry.

4   ARE THERE OTHER ELECTROPHILIC OXIDANTS IN THE P450 CYCLE?

Cpd 0 was probed by EPR/ENDOR and appears to be the last station that is still "visible" in the cycle, before the appearance of the C-H hydroxylation product for camphor hydroxylation by P450$_{cam}$. Since the mutation in the proton relay system (T252A) of P450$_{cam}$ arrests the hydroxylation,[4] while the mutant is still capable of camphene epoxidation,[14] the search of a second oxidant focused naturally on Cpd 0. It should be noted that in the heme and nonheme chemistry, Fe^{III}OOR complexes are known to carry oxidation.[33] As such, the question is not whether Cpd 0 can ever perform oxidation by itself, but rather, whether Cpd 0 can compete with Cpd I if both are present, as suggested by the "two-oxidant hypothesis"[6,13,19] (Scheme 2)?

4.1   Computational results on the reactivity of Cpd 0: Comparison with Cpd I

Since C=C epoxidation is thought to be a marker reaction of Cpd 0, it was studied by two different groups.[25,26] The two studies concluded that the barriers for Cpd 0 are exceedingly high; the computations revealed that the lowest energy pathways involve radical intermediates and are not OH$^+$ transfer processes. Furthermore, the computed barriers for epoxidation by Cpd 0, through either one of the two oxygen atoms of the OOH moiety, were much higher than the corresponding barriers for ethene epoxidation by Cpd I, for example, >36 kcal/mol vs 14-15 kcal/mol.[25] Another study explored the reactivity of Cpd 0 in sulfoxidation of dimethyl thioether.[27] Here too, the barriers were exceedingly high; >30 kcal/mol higher than the corresponding barriers for sulfoxidation by Cpd I. Acid assistance from a potent proton source, H$_3$O$^+$(H$_2$O) was found to lower the barrier for Cpd 0. Still the resulting barrier, 27.5 kcal/mol, was 10 kcal/mol higher compared to sulfoxidation by Cpd I. One must remember that Cpd 0 of P450 carries a negative charge and this does not favor electrophilic reactivity. While the effect of the negative charge of Cpd 0 is partially offset by the acid catalysis, the sulfoxidation study shows that this does not necessarily make Cpd 0 as reactive as Cpd I. Thus, based on the present computational picture, Cpd 0 is too sluggish an oxidant to compete with Cpd I. As such, at present, the reactivity patterns of C-H hydroxylation, which were interpreted to arise from the presence of two oxidants,[6,13,19] are best understood in terms of TSR.

5   SPIN-SELECTIVE REGIOCHEMISTRY IN REACTIONS OF CPD I

TSR and MSR transpire because of the proximity of the reactive states. However, whenever this is not the case, the reactivity will be dominated by a single state, hence single state reactivity (SSR).[22] This was found to be the case for two processes.[27,28,34] In benzene hydroxylation, the lowest bond-activation TS, were found to be the LS species that arises from electrophilic and radical attacks by Cpd I on the π-system of benzene. The electrophilic-attack TS was slightly lower than the LS radical TS, and both were much lower than the corresponding HS TSs. Thus, the HS species are not really accessible for benzene hydroxylation, for reasons analyzed in the original literature.[34] Similarly, in the

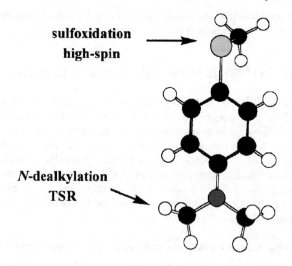

**sulfoxidation
high-spin** ⟶

N-**dealkylation
TSR** ⟶

sulfoxidation of dimethyl thioether by Cpd I, the HS TS was found to be significantly lower than the LS species, again for reasons explained in the original paper.[27] Very intriguing situations may arise in substrates that possess two different moieties, one reacting via TSR, the other via SSR; or two moieties that react via SSR but have a preference for different spin states.

For example, a common substrate used in P450 oxidations is dimethyl-(4-methylsulfanyl-phenyl)-amine, which is shown in Scheme 4. This substrate possesses two moieties that undergo oxidation by Cpd I; sulfoxidation that is predicted to occur via the high-spin state of Cpd I, and *N*-dealkylation, which involves initial C-H hydroxylation via both low- and high-spin states. Since C-H hydroxylation proceeds via TSR, while sulfoxidation by SSR (high-spin mainly), oxidation of this substrate by Cpd I is likely to generate *reactivity patterns as though belonging to two different oxidants*. Indeed, the studies of Jones et al[35] with a few P450 isozymes (e.g., P450$_{2E1}$) led to the conclusion that either two-oxidants or two different spin states are responsible for the experimental observations. Thus, a mutation of the Thr residue, known to be involved in the protonation machinery that converts Cpd 0 to Cpd I, increased sulfoxidation over *N*-dealkylation. However, substituting the C-H's of the N-methyl group by C-D's led to a significant intrinsic kinetic isotope effect, but did not affect the ratio of sulfoxidation to *N*-dealkylation. Since the mutation of Thr increases the yield of Cpd 0 at the expense of Cpd I, these patterns are consistent with sulfoxidation being mediated mostly, or only, by Cpd 0 while *N*-dealkylation exclusively by Cpd I. However, since Cpd I by itself is able to perform both sulfoxidation.[27] and *N*-dealkylation,[27] and since theory shows that Cpd 0 is a very sluggish in sulfoxidation,[27] this suggestion of co-reactivity of Cpd 0 and Cpd I is not compelling. Furthermore, Jones and coworkers[35] pointed out that since the mutation of Phe (F87A), which does not affect the conversion of Cpd 0 to Cpd I, nevertheless reduces the amount of sulfoxidation, the regioselectivity changes observed for the two different mutants may not necessarily reflect changes in the oxidizing species. In fact, Jones and coworkers[35] postulated an alternative scenario with *a regioselective reactivity of the two spin states of Cpd I*, which thereby masquerade as two oxidants. Our theoretical results point out that the postulate of Jones et al is reasonable and Cpd I may indeed oxidize dimethyl-(4-methylsulfanyl-phenyl)-amine in a spin-regioselective manner as indicated in

Scheme 4. More recent results[36] show that the C-H hydroxylation of *N,N*-dimethyl aniline is dominated by the LS state of Cpd I. Thus, the two moieties in Scheme 4 are selectively oxidized by different spin-states of Cpd I. Other substrates, which may exhibit this behaviour are the Newcomb probes depicted in Scheme 2; in these cases, C-H hydroxylation involves TSR while the phenyl hydroxylation proceeds by SSR (via the LS state).

6 OTHER POTENTIAL OXIDANT SPECIES

Two other potential electrophilic oxidants that are being considered in our group at the moment. These are protonated forms of Cpd 0; protonation on the proximal oxygen leads to the ferric hydrogen peroxide complex, $Fe^{III}(H_2O_2)$, while protonation on the distal oxygen leads to the ferric water oxide complex, $Fe^{III}(OOH_2)$. There is no hard experimental data that these species exist. Nevertheless, computationally $Fe^{III}(H_2O_2)$ appears to be stable, while $Fe^{III}(OOH_2)$ appears as an intermediate en-route to the formation of Cpd I from Cpd 0.[28] We have preliminary results for the reactivity of the ground doublet spin-state of $Fe^{III}(H_2O_2)$ in C-H hydroxylation towards the probe substrate (**ii**) in Scheme 2. The results show that the species is less reactive than Cpd I. As yet, we have no results for the reactivity of $Fe^{III}(OOH_2)$. When these studies are complete, it will be possible to draw a more conclusive judgement on the multi-oxidant hypothesis (Scheme 1).

7 CONCLUSIONS

The degeneracy of the spin states of Cpd I, its electron deficiency and dense orbital manifold lead to TSR and MSR scenarios, which describe the simultaneous reactivity of the many states. These scenarios lead to many intriguing consequences, and generate in some cases reactivity patterns as though belonging to two or more different oxidants. These states emerge from generally trustworthy computational procedures, and can be reasoned apriori also by theoretical considerations. As such, these states exist "also in the flasks" of the experimentalist. The problem is how precisely can one probe these states? An essential element, which is still missing in the TSR scheme, is some quantitative information on the rate of spin crossover between the two spin states. Understanding this factor may provide the mechanistic chemists with additional means to articulate the concept and design experiments for testing it. The advent of spin-orbit calculations can fill this want.[37]

Acknowledgements: The paper is dedicated to S. Efrima, whose untimely death is a great loss to Israeli chemistry and a tragedy to his family and friends. The research is supported by and ISF grant to SS.

References

1 P. R. Ortiz de Montellano, Ed., *Cytochrome P450: structure, mechanism and biochemistry.* New York: Plenum Press, 2nd ed, 1995.

2   Ortiz de Montellano, P. R. Ed., *Cytochrome P450: structure, mechanism and biochemistry*. New York: Kluwer Academic/Plenum Publishers, 3rd Ed., 2004.

3   M. Sono, M.P. Roach, E.D. Coulter, J.H. Dawson, *Chem. Rev.* 1996, **96**, 2841.

4   R. Davydov, T.M. Makris, V. Kofman, D.E. Werst, S.G. Sligar, B.M. Hoffman. *J. Am. Chem. Soc.* 2001, **123**, 1403.

5   J.T. Groves, *Models and Mechanisms of Cytochrome P450 Action*, Chapter 1 in reference 2.

6   M. Newcomb, P.H. Toy, *Acc. Chem. Res.* 2000, **33**, 4493.

7   S. Shaik, S.P. de Visser, D. Kumar, D. *J. Biol. Inorg. Chem.* 2004, **9**, 661.

8   D.G. Kellner, S.-C. Hung, K.E.Weiss, S.G. Sligar SG, *J. Biol. Chem.* 2002, **277**, 9641.

9   R. Rutter, L.P. Hager, H. Dhonau, M. Hendrich, M.Valentine, P. Debrunner, *Biochemistry*, 1984, **23**, 6809.

10   V. Schünemann, C. Jung, J. Terner, A.X. Trautwein, R. Weiss, *J. Inorg. Biochem.* 2002, **91**, 586.

11   I. Schlichting, J. Berendzen, K. Chu, A.M. Stock, S.A. Maves, D.E. Benson, R.M. Sweet, D. Ringe, G.A. Petsko, S.G. Sligar, *Science* 2000, **287**, 1615.

12   M.N. Bhakta, P.F. Hollenberg, K. Wimalasena, *J. Am. Chem. Soc.* 2005, **127**, 1376.

13   A.D.N. Vaz, D.F. McGinnity, M.J. Coon, *Proc. Natl. Acad. Sci. U. S. A.* 1998, **95**, 3555.

14   S. Jin, T.M. Makris, T.A. Bryson, S.G. Sligar, J.H. Dawson, *J. Am. Chem. Soc.* 2003, **125**, 3406.

15   Hishiki, T.; Shimada, H.; Nagano, S.; Egawa, T.; Kanamori, Y.; Makino, R.; Park, S. Y.; Adachi, S. I.; Shiro, Y.; Ishimiura, Y. *J. Biochem.* **2000**, *128*, 965.

16   T.S. Dowers, D.A. Rock, J.P. Jones, *J. Am. Chem. Soc.* 2004, **126**, 8868.

17   J.T. Groves, G.A. McClusky, *J. Am. Chem. Soc.* 1976, **98**, 859

18   P. R. Ortiz de Montellano, R.A. Stearns, *J. Am. Chem. Soc.* 1987, **109**, 3415.

19   M. Newcomb, D. Aebisher, R. Shen, R.E.P. Chandrasena, P.F. Hollenberg, M.J. Coon, *J. Am. Chem. Soc.* 2003, **125**, 6064.

20   P.R. Ortiz de Montellano, *Acc. Chem. Res.* 1998, **31**, 543.

21   L. Avila, H.-w. Huang, C.O. Damaso, S. Lu, P. Möenne-Loccoz, M. Rivera, M. *J. Am. Chem. Soc.* 2003, **125**, 4103.

22   S. Shaik, M. Filatov, D. Schröder, H. Schwarz, *Chem. Eur. J.* 1998, **4**, 193.

23   S. Shaik, S. Cohen, S.P. de Visser, P.K. Sharma, D. Kumar, S. Kozuch, F. Ogliaro, D. Danovich, *Eur. J. Inorg. Chem.* 2004, **35**, 207.

24   J.C. Schöneboom, S. Cohen, H. Lin, S. Shaik, W. Thiel, *J. Am. Chem. Soc.* 2004, **126**, 4017.

25   F. Ogliaro, S.P. de Visser, S. Cohen, P.K. Sharma, S. Shaik, *J. Am. Chem. Soc.* 2002, **124**, 2806.

26   T. Kamachi, Y. Shiota, K. Yoshizawa, *Bull. Chem. Coc. Jpn.* 2003, **76**, 721.

27   P.K. Sharma, S.P. de Visser, S. Shaik, *J. Am. Chem. Soc.* 2003, **125**, 8698.

28   S. Shaik, D. Kumar, S.P. de Visser, A. Altun, W. Thiel, *Chem. Rev.* 2005, **105**, 2279.

29   J.C. Schöneboom, F. Neese, W. Thiel, *J. Am. Chem. Soc.* 2005, **127**, 5840.

30   D. Kumar, S.P. de Visser, P.K. Sharma, S. Cohen, S. Shaik, *J. Am. Chem. Soc.* 2004, **126**, 1907.

31   S.P. de Visser, D. Kumar, S. Shaik, *J. Inorg. Biochem.* 2004, **98**, 1183.

32   D. Kumar, S.P. de Visser, S. Shaik, *Eur. J. Chem.* 2005, **11**, 2825.

33   W. Nam, H.J. Lee, S.-Y. Oh, C. Kim, H.G. Jang, *J. Inorg. Biochem.* 2000, **80**, 219.

34   S.P. de Visser, S. Shaik, *J. Am. Chem. Soc.* 2003, **125**, 7413.

35   T.J. Volz, D.A. Rock, J.P. Jones, *J. Am. Chem. Soc.* 2002, **124**, 9724

36   C. Li, W. Wu, D. Kumar, S. Shaik, *J. Am. Chem. Soc.* 2006, **128**, 394.

37   D. Danovich, S. Shaik, *J. Am. Chem. Soc.* 1997, **119**, 1173.

THE ROLE AND LIMITATIONS OF COMPUTATIONAL CHEMISTRY IN DRUG DISCOVERY

Mike Hann

GSK

Introduction: What we dream of being able to do *in silico* to aid drug discovery

The dream is to take a sick person (or preferably before the onset of sickness), a knowledge of their genetic make up, the symptoms and relevant bioassay data then deliver an instantaneous diagnosis and effective treatment whereby:

- The molecular structure of the drug's active ingredient was designed on a computer to interact with a target receptor of known 3D molecular structure. Its selectivity was built in on the basis of an understanding of the structure of related proteins and its Absorption, Distribution; Metabolism, Elimination and Toxicology (ADMET) profile was optimised with the knowledge of mathematical models derived from measurements on thousands of diverse molecules and specific insights from how related compounds interact with some of the key ADMET biological targets.

- The array of structures from which the drug compound was eventually selected was enhanced by computational chemists using design techniques embedded in algorithms developed to exploit the best mathematical processes and the depth of experimentally derived knowledge.

- Clinical trials will have been accelerated and made more effective by using surrogate markers of the drug's potential – the surrogate marker being selected on the basis of computational analysis of cellular signalling networks. The clinical trials will have been further enabled by computational analyses of the genetic make up of individual patients and by insights gained through the prior analysis of the compounds in bio-array assays of all known phenotypes derived from SNP analyses and splice variants.

- The patients diagnosis will also have been carried out based on computational analysis of in vivo and ex vivo test samples and comparisons to genomic and disease state databases.

Computational chemistry has made enormous progress in the last decade and is a viable tool contributing to making some of this dream a reality. However progress has perhaps been even more dramatic on the experimental side with access to increasingly informative data on multiple targets and ligands. But it still remains true that the more we can do **correctly** *in silico* the less we waste in time and money making compounds which have no chance of making it as a drug. This is increasingly so as the cost of drug discovery continues to escalate and the reality of the need for and value of drugs tailored to an individuals genetic make up emerges from new clinical trials paradigms.

We should measure what we can, predict what we can't and try and understand the differences as a way to refine our expertise.

The challenge of molecular space

While the experimental world of synthesis of arrays of compounds has made many more compounds accessible for screening it is worth remembering that the number of possible drug like compounds that could be made is enormous. Very large numbers in excess of 10^{40} compounds are often discussed and this compound space we often to refer to as the Virtual (V) world of compounds. However most companies have of the order of 10^6 compounds registered/available for High Throughput Screening (HTS). This we call the Real (R) world of available samples (Ref 1). Compared to the Virtual world we therefore have perhaps at best, a sampling rate of $1:10^{34}$. There is however a middle ground of those compounds that we can model in some way *in silico*. We can with modern computers consider compound collections of maybe 10^{12} in what we term the Tangible (T) world of compounds that we can reasonably and rationally select from for inclusion in the Real world. The sampling rate for the Tangible world is perhaps still not very good ($1:10^{28}$) but it is better than working in just the Real world.

This VTR concept allows Computational chemistry to help decide what to move from **V** to **T** to **R** for real screening (Ref 1). And importantly this construct allows us to take the learnings from each experiment to return to the V world with interpreted data for another iteration. As our computational methods and power increases then the Tangible world grows and we become more effective at sampling the Virtual world. But in order to do this we often have to sacrifice accuracy for speed. Thus while, for instance, Quantum Mechanical (QM) methods have the potential to be accurate in studying a few compounds they are not appropriate to study orders of magnitude larger numbers of compounds.

So what is realistic to consider doing with computational chemistry in this VTR era of drug discovery?

Small molecule modelling

Table 1 summarises the realms of what we can and can't effectively do in small molecule modelling. As we move from the 2D wiring diagram to the 3D property description of a molecule we get closer to representing what a receptor sees but at the same time the complexity and difficulty grows exponentially.

Table 1 *The boundaries of small molecule modelling*
What we can do in a computer:

– Draw a molecule or select from a database of 2D structures
    – Very rapidly generate millions of 2D structures by *in silico* combinatorial chemistry
    – Search for 2D substructure related molecules in a 2D sense in databases of billions of molecules
    – Generate reliably a set of 3D co-ordinates from a 2D structure
    – Generate multiple conformations of compounds (rule based, Molecular Dynamics (MD) or Monte Carlo (MC)) to generate 3D property data.

What we can sort of do:

– Store and abstract pharmacophore concepts using biological activity or protein/ligand x-ray structures for guidance
   – Search with these pharmacophores for 3D similar compounds based on simple concepts of lipophilic, Hydrogen Bond positions etc
   – Apply explicit and continuum solvation calculations in both QM and MM formalisms

What we can't or choose not to do:

– Describe effectively continuums of conformations and their properties
       ● we tend to use arithmetic averages or single conformers.
   – Describe effectively complex energetics
       ● We tend to use simpler descriptors than is appropriate in the interests of doing calculations faster. Such an example would be "yes, it is a Hydrogen Bond Acceptor" rather than describing its strength of interaction
       ● We ascribe properties to atom centres when we know that it is electrons - or lack of them - which are responsible for interactions
   – Differentiate entropy and enthalpy contributions to binding free energies.
       ● Boltzman populations for genuine free energy calculations require realistic sampling over correct energies

Protein structure prediction

In the protein domain there is some similarity to the small molecule domain in that sequence similarity can be considered to be formally equivalent to the small molecule 2D similarity problem and is similarly tractable. Unfortunately 2D to 3D predictions for proteins (i.e. homology models and *de novo* folding) do not rank alongside the small molecule equivalent (ref 2). This is because the conformational search problem is many orders of magnitude greater and the number of particles that need to be considered in energy calculations escalates similarly. It is clearly possible to do reasonable estimates of likely fold but refining this to the level of accuracy that an X-ray or NMR derived structure can provide is currently impossible although modern force fields are now making this a realistic possibility. This is not to say that homology models *per se* (especially when supported by Site Directed Mutagenesis) can not give "useful" models but the level of accuracy required for doing structure based ligand design is invariably absent. This problem can be illustrated by considering the uncertainty in the position of the OH group of a Tyrosine side chain in a homology model. Figure 1 illustrates the effect of moving the OH group from gauche to trans conformation when there is no certainty about its position. The distance is about 9A which is of the scale of the dimension of a typical drug molecule. When this uncertainty is magnified by the equivalent uncertainty for each other residue then it is clear why such models are generally ineffective for true Structure Based Design. Homology models do however remain of value for both assessing targets and providing against which ideas may be triaged.

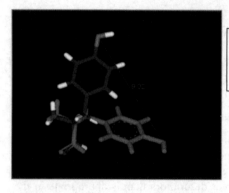

Figure 1. *Positional effect on OH group of moving from g- to t in a Tyrosine residue.*

Ligand Docking

Ligand docking and scoring is at the heart of much of what computational chemistry tries to do for drug design (ref 3,4,5). Can we really predict if and how a proposed ligand will bind to a protein? In order to assess this capability we often ask whether given a *known* protein-ligand complex can we regenerate the correct position automatically and rank its binding relative to other compounds – this is exploring how and if a compound binds.

- When a variety of software docking programmes are surveyed then it is generally found that they can predict binding modes quite well.
- Enrichment plots (e.g. Figure 2) illustrate how useful this can be in selecting likely compounds that will "fit and bind" to a target active site.
- Different algorithms are better for some protein classes but there is no general best overall technique (reference 3).

Fig 2. *Typical enrichment plot for displaying effectiveness of different ligand docking methods (adapted from ref 3b)*

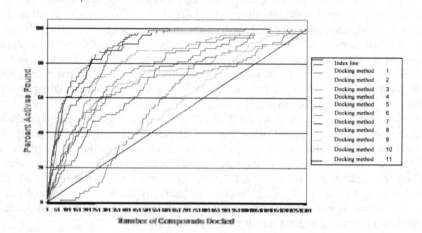

So why aren't we even more effective with the docking techniques? One issue which is often overlooked is the fact that observed crystal structures, which are routinely used to parameterise these methods, only contain information about successful poses of compounds. Interestingly there are very few examples of protein-ligand structures observed in which there are bad interactions and thus it is likely that docking interactions really need to get everything right and hence nothing wrong, if they are to be truly predictive. In protein structures it is very rare to see a bad interaction (e.g. a repulsive charge-charge interaction) indicating that either ligand or protein or both normally find mutually effective ways to avoid them. There are actually very few examples of genuine "bad" interaction in the public domain. One of the best known examples (Figure 3) could be considered to have been intentionally engineered in to test Free Energy Perturbation calculations (ref 7,8). Training on only good interactions is unlikely to give very good discriminant power to the method because the known structures do not give us information on bad interactions except that nature finds a way to avoid them!

Figure 3. *Thermolysin with "good" and "bad" interacting ligand examples*

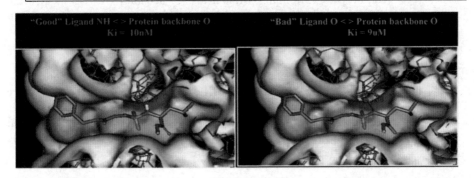

A further complication is that both the ligand and protein should be considered to be truly flexible in both relative orientations and conformation and this results in a huge search problem which is still beyond the reach of modern computers. Many techniques are being developed to address this problem. These often exploit pre-existing techniques but with different criteria applied to force faster sampling e.g. by the use of poling methods to force conformational search in MD methods (ref 8).

However we clearly still need better techniques which take account of all these aspects and ultimately get closer to true simulations in a realistic time frame and therefore address the ubiquitous problem of induced fit. Only when we have those capabilities will we be able to address problems of how signalling (e.g agonism etc.) actually happens through biomolecular interactions. This type of modelling will also need a much better understanding of kinetics rather than just thermodynamics.

Ligand Scoring

While ligand docking is tractable in the sense that we can make use of the technique in a predictive sense, the effective estimation of binding free energies is far from achievable on a routine basis. This is because of the complexity and subtlety of the interacting forces which ultimately contribute to the binding free energy that we typically observe in a biochemical/physical experiment. With extreme care and attention to detail then satisfactory estimations of free energies can be achieved but doing so in high throughput mode remains a holy grail that we are some way from achieving.

- Thus while QM/MM methods should in theory deliver what we need they are just not fast enough to do ensemble averaging to get free energies. Even Free Energy Perturbation (FEP) calculations often need extensive re-parameterisation of force fields to get useful results (ref 8).
- The faster MM/PBSA methods are probably the best compromise for *"ab initio"* free energy calculation but one protein/ligand complex typically may require significant calculation time, of the order of several hours cpu time (ref 10,11). If sampling is just done over a relaxed complex rather than using MD sampling then calculation times can drop to 15 minutes without loss of accuracy (ref 12) . Often the difficulty with all these methods for routine usage is the time required to fully understand the protein system and explore, for example, possible   protonation   states.

- There are more paramaterised methods such as Linear Interaction Approximation or Energy LIA/LIE (ref 13) but these need even more extensive parameterisation with known inhibitors so at best can be considered a semi-empirical method. While it is possible to do several compounds per cpu day once parameterised, the consensus seems to be that   these   methods   are   not   very   predictive.

- For the very fast calculation of "free energy" that is typically estimated to complement fast docking methods it is common to use even simpler interaction scores based on vdw/electrostatics/surface area effects. These are easy to calculate in a rapid mode but as we know that small structural changes cause large energy differences it is unrealistic to expect single snapshots to not miss a tight binder or to overestimate a weak binder.

- *It must always be remembered that one order of magnitude difference in an observed binding affinity is actually only a difference of 1.4 kcal/mol in free energy and this level of accuracy is just not achievable at the moment.*

Pharmacophore Methods

Pharmacophore methods have been extensively developed since Garland Marshall's early work on the Active Analogue Approach (ref 14). The power of the techniques available today is their flexibility in being able to handle differing levels of information derived from a few or many compounds and to incorporate exclusion areas derived from ligand or protein information (ref 15). Added to this is the availability of smart conformational search algorithms and the ability to incorporate directionality information or site point extensions. Finally the methods are ideally suited to coarse grain parallel computing thus enabling high throughput searching of very large databases. The fuzziness of the approach is ideally suited to refinement by real screening and the methods are very good at scaffold hopping to find new templates.

It is interesting to compare the success rate of pharmacophore selections with those derived from docking processes. While each target protein and pharmacophore comparison may give differing comparative results, one such example illustrates some further issues. Figure 4 shows a plot of the ClogP vs Molecular Weight profile of compounds selected by both a Pharmacophore method (spheres) and a docking procedure (triangles) using the information from several 3D structures of an enzyme. The colour coding of both triangles and spheres refers to the actual measured activity of the compounds selected in a real experimental assay. It can be seen that the ratio of spheres coloured blue is much higher than triangles indicating that in this case the pharmacophore method was the more successful in selecting real hits. While there are clearly actives with a wide range of logP values, the actives are more concentrated to the lower MW range and this coincides with the majority of the compounds selected by the pharmacophore method as compared to the docking method. This is a phenomenon which seems to be general about docking methods in that the scoring functions tend to just add up non-specific interactions which accumulate with increased size of compounds. It may also be that these observations are a manifestation of the hypothesis that reducing the complexity of compounds (for which MW is a surrogate) gives an increased probability of success. (ref 16)

Figure 4. *Comparison of Pharmacophore and Docking effectiveness for an enzyme*

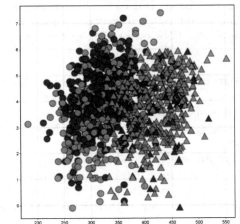

Comparing pharmacophore selections with docking selections and reality

Scatter Plot

**Sphere=
Pharmacophore
Selections**

clogp

Molecular_Weight

Blue: active
Red: weak active
Grey: inactive

**Triangle=
Docking
selections**

Entropy and experimental information

The real Achilles heel of the scoring methods that are used for high throughput docking is that they rely on approximations which satisfy the desire to do fast calculations but deviate strongly from a true interpretation of the thermodynamics of binding. Thus when binding energies (which are free energies) are used to aid in the development of algorithms, it is common practice to have algorithms which are based solely on enthalpic terms such as van der Waal and electrostatic terms and the entropic contributions are mainly ignored. Unfortunately there is a paucity of good thermodynamic data from microcalorimetry (or NMR) which can used to give the separated enthalpy and entropy contributions. With the development of new chip based screening methodologies based on microcalorimetry maybe this situation will begin to change (ref 17).

Bulk Property calculations

The bulk properties of molecules (eg lipophilicity, solubility, acidity, polar surface area, etc.) are important experimental terms which have been shown to correlate with biological activity and formed the basis of the early QSAR technologies. The use of calculated rather than experimentally derived properties is obviously an advantage in that this gives access to predictions on Virtual compounds prior to synthesis thus enabling decision making on what compounds to actually make. The bulk properties are often key components of more sophisticated models used in ADMET models which are increasingly in use to help design into molecules a good developability profile. We often need fast calculation methods to be able to address the numbers of compounds that a medicinal chemist might wish to assess for potential synthesis. But again our methods are often not able to deliver the combination of speed and accuracy. Methods based on a large number of experimental observations (data driven models) are fast and usually good for related compounds while the more ab initio methods (typically QM and implicit solvent with configurational sampling) that are required for regions of unparameterised chemical space are slow.

Notwithstanding these issues, it is now common place to have models which give at least some guidance on ADMET properties such as bioavailability, absorption, solubility, permeability, CNS penetration, protein binding and metabolism (ref 18, 19).

Moving from Data driven models to Phenomenological models

As mentioned above there are essentially two types of models that we try to make use of and these are described below.

- *Data driven models* - based on large numbers of observations and properties of compounds.
    - Typified by developing a linear or non-linear equation of the sort
        - *activity = f(property parameters)+...*
- *Phenomenological models* - based on building models with an understanding of the underlying molecular phenomena and then accurately simulating from an atomic view point.

The data driven models are essentially a mathematical representation for encoding experience (or observed data) and enabling interpolation and extrapolation but like all experience it is best used within the "field" that led to the experience. These models are also highly dependent on the quantity and quality of the experimental observations and the diversity & relevance of properties and parameters.

By contrast phenomenological models are highly dependent on the correct understanding of processes and their accurate representation. There is a need to understand the basis of the interplay between phenomena such as feedback and interactions, affinities and rates, etc. Essentially these methods involve rebuilding from the bottom up after a reductionist approach has identified the atomic and molecular basis of the components.

As more and more understanding is derived from the expanding number of protein crystal structures and other experimental data that are now available, and the accuracy and speed of calculations based on them is improved, then there is clearly the opportunity to move from data driven to phenomenological models. This has the benefit of providing models that have the power to be more predictive than the limits that can be expected for the data driven models.

Often, however, there is interplay between the two approaches with learning from atomic insights suggesting improvements in descriptors to include in data driven models. However the holy grail of becoming truly phenomenological in modelling remains compromised by our ability to simulate entropic and time related events.

It is important to understand the limitations of any type of model and often they are asymmetric in their abilities to predict. For instance, the bioavailability model that is used at GSK, is much better at predicting what compounds will not be bioavailable rather than those that will be. It is therefore important, in the use of any models, to explain carefully to potential users the inherent limitations.

Where next?

As the worldwide crystallographic community is rapidly becoming "omically productive" by using robotics and automation there is going to be a dramatic improvement in our access to high resolution protein structures from which we could potentially build phenomenological models (ref 20). But can Computational Chemistry really deliver the tools to accurately and correctly simulate so as derive the predictive models? We have seen that there are many short comings to our methods and clearly as the complexity of the systems exponentially grows as the network of interactions is rebuilt from a bottom up approach we have a long way to go to fulfil our dreams of entirely designing drugs *in silico*.

However there is an enormous impact that Computational Chemistry and Molecular Modelling is having in drug discovery. The subject matter is increasingly being taught and documented as a discrete science (ref 21) and the subject is now firmly embedded into the broad subject of Medicinal Chemistry and Drug Discovery (Figure 5).

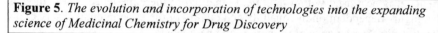

Figure 5. *The evolution and incorporation of technologies into the expanding science of Medicinal Chemistry for Drug Discovery*

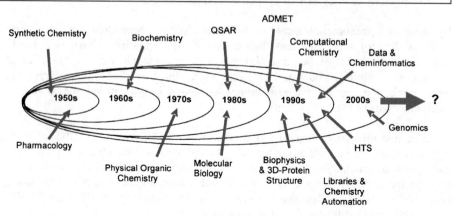

As computers became more effective in the 1980s and 1990s there was a time when Artificial Intelligence (AI) was talked about as being likely to replace attributes of the creative processes by computers. Computational and Medicinal Chemistry are highly creative subjects which require the interpretation of large amounts of data and the creation of hypothesis to be tested in the laboratory. Fred Brooks at UNC coined the term Intelligence Amplification (IA) to express what he believed to be the most effective use of computers is to support human creativity and that at any given level of technological advancement a person plus a machine can beat a machine only system (ref 22). He pointed out that although computer speed has increased dramatically over the years and continues to follow Moore's law of doubling every one to two years, this exponential growth can never overtake the chemistry problem. This problem is that chemistry phenomena are essentially combinatorial in nature in that every particle we add to a problem has to be considered in the context of every other one that is already there. While there are methods to restrict this combinatorial explosion the problems will usually outstrip the simpler exponential growth of computational capability. This is the real Achilles heal of true phenomenological modelling in that there will always be bigger problems that we want to solve than we can do with a brute force computational approach (Figure 6).

Thus the IA approach espoused by Fred Brooks is to use the computer and computer graphics to enhance the capabilities and creativity of the human mind. This is why computer graphics has been such a key interface between the computer and the chemist in the creation of the science of computational chemistry. Thanks to developments driven by the computer games industry, computer graphics for the sciences has been transformed so that real time manipulation of sophisticated representations of molecules is now routine. Graphics has become the key interface for explaining the complex models developed by computational chemists to, for example, medicinal chemists.

Figure 6. *Computational Chemistry problems are usually not simply exponential but combinatorial. Computer enhancements are exponential but are always outstripped by real chemistry problems.*

n	2^n	n!
1	2	1
2	4	2
3	8	6
4	16	24
5	32	120
6	64	720
7	128	5040
8	256	40320
9	512	362880
10	1024	3628800

In summary Computational Chemistry is clearly now a key component of modern Drug Discovery and increasingly computational chemists are cited as inventors on patents describing new drugs. However our methods are often not as good as we would want them to be. As our capabilities continue to improve, we do get closer to the holy grail as set out in the introduction but almost by definition the challenges always remain beyond our abilities as we strive to model correctly even more complex systems, more accurately and in a realistic time frame.

Gay-Lussac (1778-1850) wrote in the early 19th century that "we are perhaps not far removed from the time when we shall be able to submit the bulk of chemical phenomena to calculation". Encouragingly and at the same time regrettably this is still true!

Acknowledgements

The author would like to thank all the many colleagues within and external to GSK who have contributed to furthering the science of computational chemistry. My interpretation of our deficiencies is in no way a reflection on them! The author also thanks Andrew Leach and Drake Eggleston for helpful suggestions in the text.

References

1. Hann MM, Leach AR, Green DVS. Computational Chemistry, Molecular Complexity and Screening Set Design, in Cheminformatics in Drug Design, Edited by Oprea TI, Wile-VCH, 2004, p 43-52.
2. Venclovas C, Zemla A, Fidelis K, Moult J. **Assessment of progress over the CASP experiments.** Proteins: Structure, Function, and Genetics (2003), 53(Suppl. 6), 585-595.
3. a) Kitchen Douglas B; Decornez Helene; Furr John R; Bajorath Jurgen **Docking and scoring in virtual screening for drug discovery: methods and applications.** Nature reviews. Drug discovery (2004 Nov), 3(11), 935-49.
   b) Warren, Gregrory L.; Andrews, C Webster; Capelli, Anna-Maria; Clarke, Brian; LaLonde, Judith; Lambert, Millard; Lindvall, Mika; Nevine, Neysa; Semus, Simon; Senger, Stefan; Tedesco, Giovanna; Wall, Ian D.; Woolven, James M.; Peishoff, Catherine E.; Head, Martha S. **A Critical Assesment of Docking Programs and Scoring Functions.** Submitted to J Med Chem, 2005

4. Campbell, Stephen J.; Gold, Nicola D.; Jackson, Richard M.; Westhead, David R. **Ligand binding: functional site location, similarity and docking.** Current Opinion in Structural Biology (2003), 13(3), 389-395.

5. Taylor, R. D.; Jewsbury, P. J.; Essex, J. W.. **A review of protein-small molecule docking methods.** Journal of Computer-Aided Molecular Design (2002), 16(3), 151-166.

6. Merz, Kenneth M., Jr.; Kollman, Peter A.. **Free energy perturbation simulations of the inhibition of thermolysin: prediction of the free energy of binding of a new inhibitor.** Journal of the American Chemical Society (1989), 111(15), 5649-58.

7. Bartlett, Paul A.; Marlowe, Charles K.. **Evaluation of intrinsic binding energy from a hydrogen bonding group in an enzyme inhibitor.** Science (Washington, DC, United States) (1987), 235(4788), 569-71.

8. Ensing, Bernd; Laio, Alessandro; Parrinello, Michele; Klein, Michael L. **A new approach to efficiently explore the free energy surface, trace the reaction pathways and converge the free energy barriers.** Abstracts of Papers, 229th ACS National Meeting, San Diego, CA, United States, March 13-17, 2005 (2005),

9. Oostenbrink, Chris; Villa, Alessandra; Mark, Alan E.; van Gunsteren, Wilfred F.. **A biomolecular force field based on the free enthalpy of hydration and solvation: The GROMOS force-field parameter sets 53A5 and 53A6.** Journal of Computational Chemistry (2004), 25(13), 1656-1676.

10. Srinivasan, Jayashree; Cheatham, Thomas E., III; Cieplak, Piotr; Kollman, Peter A.; Case, David A.. **Continuum Solvent Studies of the Stability of DNA, RNA, and Phosphoramidate-DNA Helixes.** Journal of the American Chemical Society (1998), 120(37), 9401-9409.

11. Gouda, Hiroaki; Kuntz, Irwin D.; Case, David A.; Kollman, Peter A. **Free energy calculations for theophylline binding to an RNA aptamer: comparison of MM-PBSA and thermodynamic integration methods.** Biopolymers (2003), 68(1), 16-34.

12. Bernd Kuhn, Paul Gerber, Tanja Schulz-Gasch, and Martin Stahl **Validation and Use of the MM-PBSA Approach for Drug Discovery.** J. Med. Chem.; **2005;** ASAP Web Release Date: 10-May-2005;

13. Almloef, Martin; Brandsdal, Bjorn O.; Aqvist, Johan. **Binding affinity prediction with different force fields: Examination of the linear interaction energy method.** Journal of Computational Chemistry (2004), 25(10), 1242-1254.

14. DePriest, Scott A.; Mayer, Dorica; Naylor, Christopher B.; Marshall, Garland R.. **3D-QSAR of angiotensin-converting enzyme and thermolysin inhibitors: a comparison of CoMFA models based on deduced and experimentally determined active site geometries.** Journal of the American Chemical Society (1993), 115(13), 5372-84.

15. Horvath D, Mao B, Gozalbes R, Barbosa F, Rogalski S. Strengths and Limitations of Pharmacophore Based Virtual Screening in Cheminformatics in Drug Design, Edited by Oprea TI, Wile-VCH, 2004, p 117-140.

16. Hann, Michael M.; Leach, Andrew R.; Harper, Gavin. **Molecular Complexity and Its Impact on the Probability of Finding Leads for Drug Discovery.** Journal of Chemical Information and Computer Sciences (2001), 41(3), 856-864.

17. Beezer, Anthony E.; O'Neill, Michael A. A.; Urakami, Koji; Connor, Joseph A.; Tetteh, John. **Pharmaceutical microcalorimetry: recent advances in the study of solid state materials.** Thermochimica Acta (2004), 420(1-2), 19-22.

18. Baringhaus K, Matter H. Efficient Strategies for Lead Optimization by Simultaneously Addressing Affinity, Selectivty and Phamacokinetic Parameters in Cheminformatics in Drug Design, Edited by Oprea TI, Wile-VCH, 2004, p 333-379

19. van de Waterbeemd Han; Gifford Eric **ADMET in silico modelling: towards prediction paradise?.** Nature reviews. Drug discovery (2003 Mar), 2(3), 192-204.

20. See for example: The Structural Genomics Consortium. http://www.sgc.ox.ac.uk/

21. Leach, Andrew R.. **Molecular Modelling: Principles and Applications.** 2nd edition, Pearson Education EMA, 2001.

22. Hansen CD, Johnson CR. The Visualisation Handbook, Academic Press, 2004

IMPROVING CATALYTIC ANTIBODIES BY MEANS OF COMPUTATIONAL TECHNIQUES

Sergio Martí,[1] Juan Andrés,[1] Vicent Moliner[1], Estanislao Silla,[1,2] Iñaki Tuñón,[2] Juan Bertrán.[3]

[1]Departament de Ciències Experimentals; Universitat Jaume I, Box 224, Castellón (Spain) Tel. +34964728084 Fax. +34964728066 E-mail. moliner@exp.uji.es
[2]Departament de Química Física/ IcMol; Universidad de Valencia, 46100 Burjasot, (Spain) Tel. +34963544880 Fax. +34963544564 E-mail. Ignacio.Tunon@uv.es
[3]Departament de Química; Universitat Autònoma de Barcelona,08193 Bellaterra, (Spain)

1 INTRODUCTION

Catalytic Antibody (CA) design has become a major area of experimental and computational design since the first CA was made against a transition state analog (TSA) three decades ago.[1] The progress and accomplishments of this field is amply attested in a recent book edited by E. Keinan.[2] In the preface, the idea of chemists using antibodies to catalyze their non-biological reactions is compared to the biblical metaphor of swords being converted into ploughshares. In this sense, the work of scientists in the CA field remains to obtain the best tool to catalyze chemical reactions.

An antibody is part of the immune system, which can recognize and produce antibodies to a wide range of antigens. This enormous diversity arises from the ingenious reshuffling of the DNA sequences encoding components of the antibody.[3] Once a germline is selected from the pool based on its initial affinity for the antigen (also known as the hapten), additional structural diversity is generated by the affinity maturation process in which somatic mutations are introduced through changes in the variable region. This process leads to a high affinity monoclonal antibody. The most abundant of these, the immunoglobulin proteins, consists of two identical light chains and two identical heavy chains connected by disulfide bridges. Both the light and heavy chains have a constant domain and a variable one, being the antigen recognizing site composed of the variable region of both chains.

In order to understand how an antibody can be used as a catalyst the fundamentals of enzyme catalysis has to be remembered. The decrease of the activation barrier in an enzymatic process arises from the relatively larger binding energy of the transition state (TS) when compared to the reactants (Michaelis Complex, MC). This idea can be translated to antibodies utilising the immune response to a stable molecule that mimics as closely as possible the presumed structural and electronic features of the TS of a particular reaction. This is the so called Transition State Analogue (TSA). Therefore the TSA used as the hapten is a stable molecule which resembles the TS. Just as a human hand can adapt its shape to a large number of shapes, an antibody active site can change its shape to complement different ligands.[4] Due to the variability available to it, a significant reorganization of the germline active site can occur upon ligand binding. In the maturation process of a monoclonal antibody, the optimal active site conformation is fixed, increasing the affinity for a particular antigen. The result should be the evolution of the antibody binding site with maximum complementarity to the TSA and, hopefully, to the real TS.

Following these strategies, CA have been produced for a plethora of chemical reactions, ranging from hydrolytic reactions, sigmatropic or cycloaddition rearrangements, carbon-carbon bond-forming reactions, redox reactions, etc...[5] However, as a general feature, the catalytic power of CA is never as high as the one obtained in enzymes (10^6 in CA and up to 10^{17} in enzymes). Furthermore, not all antibodies that stabilize TSAs are catalysts of the reaction and what is more intriguing; there are cases where after a process of maturation (that yields an increased affinity of the CA for the TSA) a paradoxical decrease in the catalytic power is observed, with respect to the initial or germline CA.[6]

Different arguments have been invoked to explain this behaviour. The first one is that it is impossible to derive a very accurate TSA, which means that the TSA is never equal to the TS. This means that an improving of the binding of the TSA can not be directly translated in an improvement of the binding of the TS. The second one is the low affinity reached between the TSA and the CA during the maturation process, as compared between the TS and the enzymes. In terms of dissociation constants, the values reached by CA are never lower than 10^{-10}, while in enzymes they can arrive up to 10^{-23}. The third one, which can be related to the previous observation, is that antibodies appear to be less successful than enzymes in discriminating between TS and reactants. Finally, the modest efficiency observed in CA can be the consequence of the simple strategy used to derive them; while natural selection has optimize enzymes during millions of years on the basis of catalytic activity (the difference between the binding energy of TS and MC) the immune system has optimized antibodies for a period of weeks on the basis of binding affinity to the TSA (and hopefully to the TS) with no attention to the structure of the reactant state.

It has been proposed at least three strategies to improve the catalytic efficiency of the CA:[7] the search for a better hapten (TSA), a more extensive screening of the immune response and the use of engineering, based on site-directed mutagenesis. We show in this paper how the methods and techniques provided by Computational Chemistry can be applied to better understand and improve the site-directed mutagenesis on CAs. As a benchmark, an oxy-Cope rearrangement catalyzed by AZ-28 and related antibodies has been selected,[8] a system which is extremely interesting because no enzyme is known for this reaction. The monoclonal AZ-28 catalyzes the unimolecular rearrangement of a substituted hexadiene to a enol that spontaneously proceeds to the corresponding aldehyde (see Scheme 1). As can be observed in the scheme, the TSA presents a chair-like structure in the substituted cyclohexanol, similar to the structure of the TS. Nevertheless, there is also a difference which is the hybridization of the C2 and C5 atoms of the central ring; while these two carbon atoms in TSA are sp3, the hybridization of the equivalent atoms in the TS is sp2.

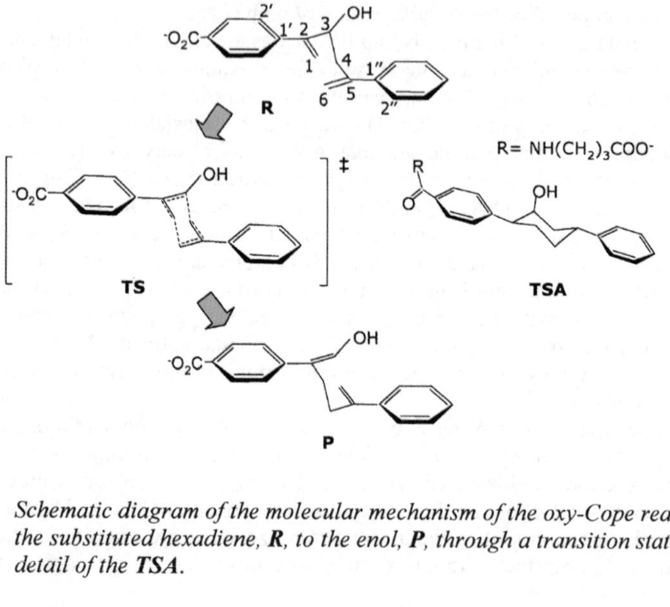

Scheme 1   *Schematic diagram of the molecular mechanism of the oxy-Cope reaction from the substituted hexadiene, **R**, to the enol, **P**, through a transition state, **TS**, and detail of the **TSA**.*

From the germline antibody, generated against this TSA, a matured AZ-28 form was obtained after six different mutations; four of them in the heavy chain and two in the light one. Only one of these mutations, SerL34Asn, is located in the active site. The result of this mutation is that in the case of the AZ-28, it was found that the matured AZ-28 presents a 40-fold higher affinity for the TSA but affords a 30-fold lower rate enhancement than its germline precursor.[6] The aforementioned mutation on the active site seems to be largely responsible for the decrease of the catalytic efficiency of the antibody. This paradoxical behaviour of the AZ-28 has been previously addressed in three theoretical papers.[9, 10, 11] In the paper of Kollman and co-workers, binding free energies were calculated using the MM-PB/SA approximation for AZ28-TS and AZ28-R complexes, both for the germline and the matured CA forms. The energetic analysis was done for only a single MD trajectory of the desired antibody-hapten complex. Unbound antibody and hapten snapshots are taken from snapshots of that trajectory. Due to the fact that classical MD calculations are run for TS complexes a reparametrization of the TS force field was required. In the work of Houk and co-workers, a more complete study of the reaction in gas phase at B3LYP level is carried out. Flexible ligand docking in rigid germline and matured antibodies enable exploration of the stereoselectivity of the reaction. Finally, in our previous paper, [11] a quite different strategy through free energy profiles using an appropriate distinguished reaction coordinate is employed. This procedure leads to a more accurate estimation of the free energy variation along the reaction coordinate taking also into account the flexibility of the CA.

2   METHODS

The starting point for the PMF calculations was obtained from an overlap of the gas phase TS on the TSA-CA X-ray coordinates, maintaining the position of the C2 and C5 atoms. Then, 500 ps of dynamics calculations were run applying an harmonic constrain on the distances of the breaking and forming bonds. In this way, the TS is fitted into the active site allowing the relaxation of not only the chemical system but the antibody.

Reaction profiles were obtained as a Potential of Mean Force (PMF) along a specific reaction coordinate, obtained as the antysimmetric combination of the C1-C6 and C3-C4 distances.[12] Selection of this reaction coordinate is based on the exploration of the PES including the environment by means of the IRCs traced down to the corresponding reactants and products valleys from transition structures located and characterised in the CA active site. Simulations were carried out with in the DYNAMO program[13] based on the original idea of the hybrid Quantum Mechanical / Molecular Mechanics (QM/MM) methods.[14] All the degrees of freedom were sampled by means of a series of molecular dynamic simulations. Umbrella sampling[15] was used to place the chemical system at different values of the reaction coordinate that cannot be sampled just by thermal fluctuations and then Molecular Dynamics simulations were run. This is done by adding an adequate parabolic energy function centered at the value of the reaction coordinate that is being explored. The fluctuations of the reaction coordinate are finally pieced together by means of the weighted histogram analysis method, WHAM,[16] obtaining the full distribution function and thus the PMF profile.

The system consisted of a cubic box of 79.5 Å side with the CA and 15492 water molecules (50167 atoms in total) all treated using the OPLS-AA force field,[17] plus the substrate (39 atoms) that was treated by AM1 semiempirical hamiltonian. Because of its size, only those CA atoms lying inside a sphere of 30 Å of radius from the substrate centre of mass were allowed to move (~11500 atoms, starting point structure dependent), keeping frozen the position of the rest. For the system in solution, the PMF was obtained following a similar protocol. In this case, the QM molecule was placed in the centre of a cubic box of TIP3P water molecules of 31.4 Å side. Periodic boundary conditions and a switched cut off radius of 12 Å were used.

3 RESULTS

In Figure 1 the PMF profiles of the oxy-Cope rearrangement obtained in solution and in the active site of the CA, both in the germline and the matured counterpart, are presented. The differences of the free energy barriers between the reaction in solution and in the matured CA, as well as between this one and its germline counterpart, are 4.6 kcal·mol⁻¹ and 1.9 kcal·mol⁻¹, respectively. These values are in very good agreement with the experimental data of 5.0 kcal·mol⁻¹ and 2.2 kcal·mol⁻¹, respectively.[6] The difference of free energy barrier between the reaction in solution and in the active site of the CA can be related with geometrical features of the MC. In this sense, in solution the substrate is less constrained than in the active site of the CA and the value of the reaction coordinate is larger than the values presented for the MC in the CA. The MC, both in the germline or in the matured CA, is closer to the TS than the reactants in solution, in agreement with previous predictions for other enzyme reactions.[18]

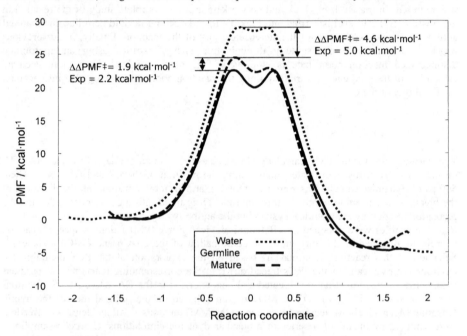

Figure 1. *Free energy profiles (in terms of PMF) of the oxy-Cope rearrangement in solution (dotted line) and for the AZ-28 catalytic antibody in its germline (solid line) and maturedd form (dashed line) for the R to P transformation (see Scheme 1).*

The calculated free energy barrier in solution (23.7 kcal·mol⁻¹) is 2.5 kcal·mol⁻¹ higher than the experimental value.[6] We have explored the potential energy surface of the reaction by means of gas phase calculations with density functional theory methods (B3LYP/6-31G*) as implemented in the Gaussian03 package of programs.[19] Our results enable the exploration of two mechanisms, depending on the orientation of the hydroxyl substituent (inwards or outwards the cyclohexane ring), both being stepwise. Gas phase calculations were also carried out by Houk and co-workers[10] with B3LYP methods showing two possible mechanisms; a concerted one obtained with restricted RB3LYP and a stepwise one with a biradical intermediate at unrestricted UB3LYP level. Both mechanisms appeared to be closely coupled.

The energetics of the closed-shell TS and the rate limiting TS obtained with the unrestricted treatment are very close. We have carried out gas phase AM1 calculations in order to test the suitability of the semiempirical method to be used in the QM/MM calculations, obtaining similar results as the ones at B3LYP level, although slightly higher energy barriers for the rate limiting steps. For instance, the difference obtained between the two methods in gas phase appears to be around 2 kcal·mol^{-1} for the hydroxyl substituent outwards conformation (which is the relevant structure in solution and in the CA). This difference between the B3LYP and the AM1 method would explain the differences observed in the computed free energy barrier in solution compared with the experimental data. As a consequence, we can be confident of the relative free energy values so obtained and hence analyzing, from the results in both CAs and free solution, why the matured CA presents higher barriers than the germline.

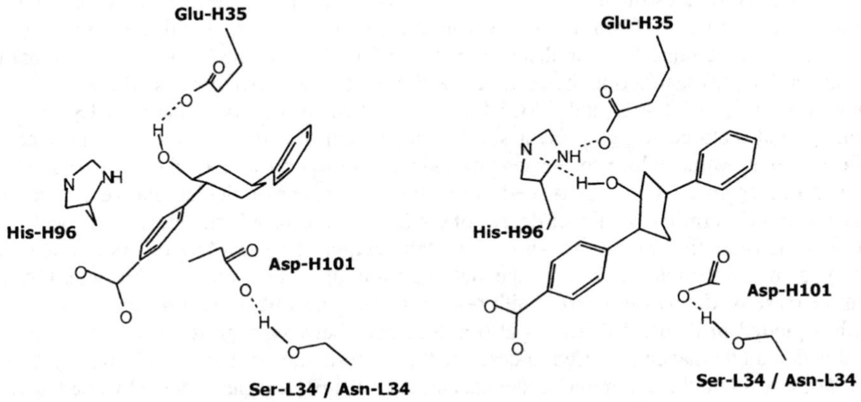

Scheme 2 *Schematic diagram of the two different pattern of interactions observed between the hapten and the CA. Ser-L34 stands for the germline CA and Asn-L34 for the matured CA.*

To achieve this goal, MD dynamics trajectories of 1ns long were carried out in the TS and MC with harmonic restrains applied to the reaction coordinate, as well as in the TSA-CA complex. The averaged results obtained in the TSA, TS and MC trajectories reveal two different patterns of interactions, which are schematically represented in Scheme 2. It can be observed that there is an important difference regarding the interaction between the hydroxyl group of the hapten and the residues of the active site. Thus while in some structures (left hand side of Scheme 2) the hydroxyl group interacts with the glutamate residue of the heavy chain (GluH35), in others (the right hand side of Scheme 2) the interaction is established with a histidine residue of the same chain (HisH96). In the case of the TSA, both germline and matured present a pattern of interactions similar to the first one. The effect of the maturation is that the distance between the heavy and light chains is diminished making a narrower cavity. This effect can be monitored by means of the distance between the SerL34/AsnL34 and the AspH101, which changes from 4.3 to 2.2 Å. This is also reflected in the averaged distance between the hydroxyl group and the glutamate (H···O distance), which is diminished from 2.3 to 2.0 Å, showing a stronger interaction between the CA and the hapten in the matured than in the germline. When analyzing the TS-CA complexes, a dramatic change in the pattern of interactions is observed with respect to the TSA-CA one, probably due to the different hybridization of the C2 and C5 atoms. As mentioned in the previous section, these carbon atoms present sp^2 and sp^3 hybridizations in TSs and TSAs, respectively, defining an important change in the orientation of the reactive ring. This change results in the hydroxyl substituent being unable to interact with the same residues in the TS as in the TSA. A rotation around the C2-C3 bond makes the interaction with the GluH35 no longer exist but a new hydrogen bond between the hydroxyl group and HisH96 is established, which may enhance the rate of the process by increasing the electron density on the hydroxyl oxygen. Also in this case the maturation process led to a slight diminution of the H-bond distance between the hapten and the CA from 2.9 to 2.8 Å. Regarding the reactant states, while in the germline CA the pattern of interactions is equal to the TS-CA, the matured CA pattern of interactions is similar to the TSA presenting the interaction between the hydroxyl and the GluH35. The shorter intermolecular distances between the matured CA and the substrate, due to the smaller size of the active site cavity, cause the reactant state to be trapped in a more stable conformation, resembling the TSA rather than the TS, and thus making the reaction more unfavourable. These comments are reflected in the internal coordinates of the hapten that are listed in Table 1. The reactant complex in the matured state is more different to the TS than in the germline case, as reflected in the reaction coordinate and the bond forming distance (r.c. and C1···C6 columns, respectively). Regarding the dihedral angles that define the relative position of the two phenyl rings with respect to the central ring (C5-C2-C1'-C2' and C2-C5-C1''-C2''), it can be observed that both TS structures, germline and matured forms, present similar values. On the contrary, the reactant states are quite different: values close to 90 degrees (which correspond to a coplanar orientation) are present for reactant state of the matured form while a more perpendicular relative orientation is observed in the germline. A higher degree of coplanarity means a better electron delocalization between the rings thus provoking an energy stabilization of the conformer. As this feature is observed only in the reactant state of the matured form, the overall effect in this latter CA form is an increase of the barrier energy with respect to the germline, in accordance with the PMF profiles plotted in Figure 1. There is also a difference between both reactant states regarding the dihedral angle that can be used to describe the conformation of the central ring (C2-C3-C4-C5). Thus, while in the matured CA the reactant complex presents an averaged value far from a chair-like conformation, in the germline it renders a value similar to the TS that defines a chair-like conformation. These differences on the orientation of the central ring are also related with the pattern of interactions with the CA mentioned above. Considering that both TS conformations display a very similar C2-C3-C4-C5 angle, the use of this dihedral angle dictates, again, that the difference between the germline and the matured CA is in the reactant complex of the latter, similar to the TSA and not to the TS.

Table 1. *Selected internal parameters (distances in Å and angles in degrees) of the TS and reactants obtained in the matured and germline AZ-28 antibody.*

		r.c.	C1···C6	C5-C2-C1'-C2'	C2-C5-C1''-C2''	C2-C3-C4-C5
germline	TSA	--	--	17 ± 11	20 ± 17	32 ± 31
	TS	-0.20	1.78	56 ± 13	64 ± 12	-44 ± 7
	R	-1.47	3.03	65 ± 15	70 ± 16	-70 ± 16
matured	TSA	--	--	16 ± 11	34 ± 23	49 ± 15
	TS	-0.14	1.71	52 ± 8	57 ± 11	-48 ± 6
	R	-1.56	3.10	91 ± 17	82 ± 13	65 ± 13

Taking into account that the increase in the energy barrier from the germline to the matured can be attributed from our previous analysis to the different patterns of interactions observed in the reactant complex of the matured form, theoretical simulations on this complex are carried out in order to improve the catalytic efficiency of the CA. In particular, we tested the back mutation of AsnL34Ser which seems to contribute decisively to the narrowing of the cavity. This mutation was carried out in ten different snapshots of the MD trajectory of the reactant complex of the matured hapten-CA complex observing a spontaneously opening of the cavity after ~0.5 ns in all the structures. This effect is accompanied by the change of the interaction pattern between the hapten and the CA; the H-bond between the hydroxyl substituent of the hapten and the GluH35 is broken and a new interaction is established with the HisH96. A representative example of this behaviour is presented in Figure 2, where the SerL34-AspH101 is monitored as a measure of the opening of the cavity, together with the distance between the hydroxyl group and the GluH35. As can be observed, the opening of the cavity and the breaking of the H-bond follow a coupled evolution and the interaction pattern changes to a new one, closer to the TS.

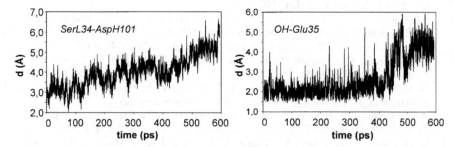

Figure 2.- *Time evolution of some relevant distances corresponding to a MD trajectory of the reactant complex of the mutated CA after introducing the mutation AsnL34 by SerL34*

From the new reactant complex obtained in the mutated CA a PMF profile was computed following the same strategy as explained before.[11] This new PMF gives an activation free energy which is 0.7 kcal·mol⁻¹ lower than that of the matured CA, in agreement with the estimated value by Schultz and coworkers.[6] This test shows how a theoretical simulation of a directed mutation can be used to predict an improvement of the catalytic efficiency of the CA.

4   CONCLUSION

Mutational engineering to further improve enzymes is often not very successful probably because Nature has already been doing the job for millions of years. However, in CAs, somatic mutations take place in a period of weeks and they aim to improve the TSA hapten binding and because they have no knowledge of the real TS they do not "set out" to improve the actual catalytic mechanism that the TSA mimics. Thus there is a real possibility for chemists to improve CA efficiency. Site directed mutagenesis is therefore an attractive tool to carry out this job. Furthermore, these mutations may be guided by computational techniques as suggested in the present work. The target of these mutation is the relative stabilization of the TS with respect to the Michaelis Complex, and not only the absolute stability of the former, emulating Nature's evolution. From the knowledge reached in the present computational study, another way of improving the catalytic power of CA is to build TSA that emulate better the TS of the reaction to be catalyzed. In this case, haptens with better proton donor groups in the central ring and presenting sp2 hybridization on carbons C2 and C5 are expected to produce better catalytic antibodies and, as consequence, the TS will be better stabilized by the antibody matured against this new molecule.[7, 20]

Acknowledgements

We are indebted to DGI for project DGI BQU2003-04168-C03, BANCAIXA for project P1A99-03 and Generalitat Valenciana for projects GV04B-021, GV04B-131 and GRUPOS04/08, which supported this research, and the Servei d'Informàtica of the Universitat Jaume I for providing us with computer capabilities. We thank professor Ian H. Williams for valuable discussions.

References

1   V. Raso, B. D. Stollar, *Biochemistry,* 1975, **14**, 584.
2   E. Keinan, *Catalytic Antibodies*, Wiley-VCH Verlag, Weinheim, 2005.
3   G.J. Nossal, *Nature* 2003, **421**, 440.
4   J. Yin and P. G. Schultz, "Inmunological Evolution of Catalysis" in *Catalytic Antibodies*, ed. E. Keinan, Wiley-VCH Verlag, Weinheim, 2005, Chapter **1**, pp 1-29.
5   P. G. Schultz, J. Yin, R. A. Lerner, *Angew. Chem. Int. Ed.* 2002, **41**, 4427.
6   H. D. Ulrich, E. Mundorff, B. D. Santarsiero, E. M. Driggers, R. C. Stevens, P. G. Schultz, *Nature,* 1997, **389**, 271.
7   D. Hilvert, "Critical Analysis of Antibody Catalysis" in *Catalytic Antibodies*, ed. E. Keinan, Wiley-VCH Verlag, Weinheim, 2005, Chapter 2, pp 30-71.
8   A. P. Marchand, R. E. Lehr, *Pericyclic Reactions*, Academic Press, New York, **1977**.
9   T. Asada, H. Gouda and P. A. Kollman, *J. Am. Chem. Soc.* 2002, **124**, 12535.
10   K. A. Black, A. G. Leach, M. Yashar, S. Kalani and K. N. Houk, *J. Am. Chem. Soc.* 2004, **126**, 9695.
11   S. Martí, J. Andrés, V. Moliner, E. Silla, I. Tuñón and J. Bertrán, *Angew. Chem. Int. Ed.* 2005, **44**, 904.
12   M. J. Field, *A practical Introduction to the Simulation of Molecular Systems*, Cambridge University Press, 1999.
13   M. J. Field, M. Albe, C. Bret, F. Proust-de Martin, A. Thomas, *J. Comput. Chem.* 2000, **21**, 1088.
14   A. Warshel, *Proc. Natl. Acad. Sci. U.S.A.* 1978, **75**, 5250.
15   G. M. Torrie; J. P. Valleau, *J. Comput. Phys.* 1977, **23**, 187.
16   W. D. Cornell, P. Cieplak, C. I. Bayly, I. R. Gould, K. M. Merz, Jr, D. M. Ferguson, D. C. Spellmeyer, T. Fox, J. W. Caldwell and P. A. Kollman, *J. Am. Chem. Soc.* 1995, **117**, 5179.
17   W. L. Jorgensen, D. S. Maxwell, J. Tirado-Rives, *J. Am. Chem. Soc.* 1996, **118**, 11225-11236.
18   S. Martí, M. Roca, J. Andrés, V. Moliner, E. Silla, I. Tuñón, and J. Bertrán, *Chem. Soc. Rev.* 2004, **33**, 98.
19   Gaussian 03, Revision A.1, M. J. Frisch, G. W. Trucks, H. B. Schlegel, G. E. Scuseria, M. A. Robb, J. R. Cheeseman, V. G. Zakrzewski, J. A., Jr. Montgomery, R. E. Stratmann, J. C. Burant, S. Dapprich, J. M. Millam, A. D. Daniels, K. N. Kudin, M. C. Strain, O. Farkas, J. Tomasi, V. Barone, M. Cossi, R. Cammi, B. Mennucci, C. Pomelli, C. Adamo, S. Clifford, J. Ochterski, G. A. Petersson, P. Y. Ayala, Q. Cui, K. Morokuma, D. K. Malick, A. D. Rabuck, K. Raghavachari, J. B. Foresman, J. Cioslowski, J. V. Ortiz, B. B. Stefanov, G. Liu, A. Liashenko, P. Piskorz, I. Komaromi, R. Gomperts, R. L. Martin, D. J. Fox, T. Keith, M. A. Al-Laham, C. Y. Peng, A. Nanayakkara, C. Gonzalez, M. Challacombe, P. M. W. Gill, B. Johnson, W. Chen, M. W. Wong, J. L. Andres, C. Gonzalez, M. Head-Gordon, E. S. Replogle, J. A. Pople, Gaussian, Inc.; Pittsburgh PA, **2003**.
20   I. H. Williams. Personal communication

THE "THEORETICAL" CHEMISTRY OF ALZHEIMER'S DISEASE: THE RADICAL MODEL

Patrick Brunelle, Duilio F. Raffa, Gail A. Rickard, Rodolfo Gómez-Balderas, David A. Armstrong and Arvi Rauk*

Department of Chemistry
University of Calgary
Calgary, AB T2N 1N4
Canada

1 INTRODUCTION

Protein oxidative damage caused by reactive oxygen species (ROS)[1,2,3] is believed be the underlying cause of numerous physiological conditions ranging from cataract formation, ischemia/reperfusion injury, cardiovascular problems, inflammatory events, neurological diseases, and apoptosis. The accumulation of free radical damage may be the ultimate cause of aging.[4,5,6] Formation of ROS occurs by a variety of mechanisms, by external radiation or toxic chemicals, detoxification of pollutants in the liver,[7,8,9] or as side products of normal respiration.[10] Our research has shown that the most stable radical species that can be generated in proteins are radicals on the peptide backbone ($^{\alpha}$C-centered radicals).[11] In a series of studies on the structure and reactivity of $^{\alpha}$C-centered radicals,[11,12,13,14] we have demonstrated theoretically the bond dissociation capacity of the $^{\alpha}$C-H bond in all amino acid residues and the possibility that damage to this site may be caused by thiyl radicals (of cysteinyl residues and glutathione).[15] The role of secondary structure in conferring protection to the $^{\alpha}$C-site has also been clarified to some extent. The theoretical studies show that α-helical and β-sheet structures effectively protect the $^{\alpha}$C-site of all residues toward damage by ROS or thiyl radicals, *with the sole exception of glycyl residues in antiparallel β-sheets*.[14] These early results have been elaborated into a theoretical model, the *Radical Model*, for Alzheimer's disease (AD) centered on the chemical action of Aβ peptide.[16,17]

Aβ is cleaved from the interior of a much larger membrane-spanning protein, the amyloid precurser protein (APP), by two "secretases"; at the N-terminal end by β-secretase which has been identified as BACE1 (also known as Asp 2 or memapsin 2),[18] and at the C-terminal end, by γ-secretase, which has been identified as presenilin1.[18] The sequence of Aβ(1-42) is:

1				5					10					15
Asp	Ala	Glu	Phe	Arg	His	Asp	Ser	Gly	Tyr	Glu	Val	His	His	Gln

16				20					25					30
Lys	Leu	Val	Phe	Phe	Ala	Glu	Asp	Val	Gly	Ser	Asn	Lys	Gly	Ala

31				35					40		42	
Ile	Ile	Gly	Leu	Met	Val	Gly	Gly	Val	Val	Ile	Ala	

The first 16 residues of Aβ are largely hydrophilic and contain His6, His13, and His14, which have been reported to form a binding site for Cu(II) with attomolar affinity (10^-^{17}).[19] The remaining 24-26 residues comprise a largely hydrophobic domain.

A strong case may be made that Aβ is the causative agent for AD.[20,21] In recent years, more details have emerged regarding the mode of toxicity to neurons *in vivo*.[22,23,24,25] The salient points for the active involvement of Aβ in AD are:

1) AD brain is characterized by extensive oxidative stress[26,27]

2) Free radicals have been detected in post mortem AD brain[28]

3) Free radicals initiate lipid peroxidation

4) Lipid peroxidation is correlated with brain degeneration[29] and is ameliorated by antioxidants,[30,31,32] including glutathione[33]

5) Aβ, in combination with metal ions and oxygen, generates reactive oxygen species,[34,35] is toxic to neurons,[36,37] and causes lipid peroxidation and protein oxidation[38]

6) Aβ toxicity is ameliorated by antioxidants,[30,39,40] including vitamin E,[30,41] and glutathione[33]

7) Aβ is itself damaged by Cu(II)- and Fe(III)-catalyzed oxidation[42]

8) Aβ can form pores in membranes[43,44,45]

9) Aβ changes neuronal and endothelial cell membrane permeability[46,47,48,49,50]

All of the first 7 points are explained in some detail by the *Radical Model*[16] described below. Point 4) requires special comment since it forms the endpoint of the *Radical Model*. Neuronal membranes are enriched in bis(vinylic) methylene groups which are particularly susceptible to free radical initiation of lipid peroxidation. Lipid peroxidation disrupts cell membrane structure, reduces membrane fluidity, increases permeability to Ca^{2+} and other ions, and generates α,β-unsaturated carbonyl biproducts such as 4-hydroxynonenal, which are themselves highly toxic. It is the beginning of a complex cascade of events that results in cell death. Cell death may be the result of accumulated damage, and/or apoptosis.[22]

Point 8) may be accommodated within the *Radical Model*, which otherwise deals with mechanisms for radical generation. Recent literature is consistent with the view that the formation of ion-permeable pores by Aβ may be a mechanism for toxicity to neurons that is independent of lipid peroxidation. The Met residue of Aβ has been deemed essential for toxicity, but mutations in which the Met side chain is replaced by a hydrocarbon chain still show residual toxicity.[36,51] Indeed, a recent report indicates that M35V is even more toxic, ostensibly because of more favourable membrane interactions.[52] Pore formation may provide a mechanism for transporting the hydrophobic Gly radical into the interior of the membrane where it can initiate lipid peroxidation. However, this feature has yet to be demonstrated.

Point 9) is an important one. There is ample evidence that Aβ causes endothelial cell damage, most likely through production of superoxide[53], and involvement of homocysteine.[54,55,56,57] A hypothesis has been put forth under the acronym, ABSENT, that the real cause of AD is a combined assault by Aβ on cerebral vasculature and neurons.[58] Taken together, the above points suggest that oxidative stress is a key feature of AD, that oxidative stress is caused by free radicals, that Aβ can generate free radicals in redox chemistry with metal ions, and that free radicals can damage neuronal and endothelial membranes to the point of causing cell death. In addition, a form of Aβ, which may be the same as the radical-generating form, causes physical damage to cell membranes through pore formation. Therefore, a search for the underlying chemistry of Aβ and its interaction with cell membranes may reap tangible benefits towards finding a cause and prevention of AD.

Knowledge of the structure of Aβ would be invaluable for understanding the chemistry. However, Aβ does not crystallize and its solution structure is not readily accessible. A "collapsed coil" structure lacking α-helical or β-sheet content is found by nmr spectroscopy.[59] Two independent determinations by nmr spectroscopy were carried out under conditions that promote α-helix formation (SDS micelles and 40% trifluoroethanol). Not surprisingly, these showed substantial α-helical secondary structure.[60,61] CD spectra consistently show a significant amount of β-sheet structure in water, as do *a priori* structure predictors such as ROSETTA.[62] The aqueous solution structure of Aβ has previously been investigated by "directed-pathway" molecular dynamics simulations.[63] That study found that α-helical, β-sheet, and "collapsed coil" conformations were similar in energy. The primary structure of Aβ in plaques is in the form of fibrils assembled from β-sheets. The soluble form of the naturally occurring protein undergoes a conformational change to β-sheet prior to its removal from circulation as plaques or fibrils.[64,65]

Aβ(1-40) and Aβ(1-42) directly promote free radical oxidative stress.[66] Aβ(1-42) is more toxic than Aβ(1-40), which differs by the absence of two hydrophobic residues, Ile41 and Ala42. Aβ(1-42) has a greater propensity for β-sheet aggregation and can seed precipitation of the more soluble Aβ(1-40).[67] The $i,i+4$ sequence of four Gly residues, Gly25, Gly29, Gly33, and Gly37 is helix-destabilizing. Precipitation into insoluble fibrils (composed of β-sheets) is promoted by metal ions, primarily Zn(II), but also Cu(II) and Fe(III).[72,68] It is likely that Aβ is most toxic in protofibril form.[69,70,71] It is likely that in AD, as in other amyloid diseases like Parkinson's (PD), Huntington's (HD), and dementia with Lewy bodies (DLB), that deposition of the plaques may actually be an inactivation mechanism.[72] The plaques in AD are enriched in Aβ(1-42), and Cu(II), as well as Zn(II), Fe(III),[73] and other metal ions,[74,75] suggesting that metals are present in Aβ in protofibril form. The single methionine residue, Met35, is not essential for neurotoxicity,[52] but if present, the toxicity appears to be correlated with the ability of the lone Met residue of Aβ to be oxidized by Cu(II).[76,76,77]

This work describes high level computational studies of model systems with the aim of establishing the viability of key steps associated with the *Radical Model*, specifically Step 1 (Figure 1)

2 METHODS

Quantum chemistry calculations have been carried out using the Gaussian 98[78] and Gaussian 03[79] suites of programs. Molden 4.0[80] and Molekel 4.0[81] visualization programs were extensively employed. Geometry optimizations were performed without geometry or symmetry constraints at the B3LYP/6-31G(d) level of theory. For each optimized structure a frequency analysis at the same level of theory was used to verify that it corresponded to a stationary point on the potential energy surface. Frequencies scaled by 0.9806[82] were used to compute the zero-point vibrational energy, no scaling factor was used to calculate the thermal correction ($H_{298}° - H_0°$) or the vibrational entropies. Where necessary, the contributions of structural conformers to the gas phase entropies have been taken into account, assuming that the gas is a mix of low lying conformers with the entropy of mixing approximated as $R\ln(n)$, where n is an estimate of the number of conformers.

More accurate ground state energies were calculated at the CBS-RAD[83] or B3LYP/6-311+G(2df,2p) levels of theory in order to obtain reliable energy changes for the reaction pathways under study. Finally, the solvation energies (ΔG_{solv}), employed for the calculation of free energies of reaction in aqueous solution ($\Delta G_{(aq)}$), were obtained using the continuum COSMO[84] procedure as implemented in Gaussian 03 (SCRF = CPCM).[85,86] For the definition of the solvent cavity, the atomic radii were systematically adjusted to fit the molecular isodensity surface of 0.001 electrons bohr^{-3}.[87,88] To improve the computation of the thermodynamic parameters of our reactions, the experimental values of $\Delta G_{solv}(H_2O) = -26.4$ kJ mol^{-1} [89] and $\Delta G_{solv}(H^+) = -1107$ kJ mol^{-1} [90] have been employed.

3 RESULTS AND DISCUSSION

The *Radical Model* is shown in Figure 1, in which the key steps are identified. The ultimate toxicity is attributed to cell-membrane damage by lipid peroxidation (right hand side of Figure 1) probably combined with pore formation or interference with existing ion channels. The Radical Model identifies the radical species that initiate lipid peroxidation. The initial and final products and the energetic feasibility of many of the proposed steps have been demonstrated by high-level theoretical computations (see Computational Methods). **Step 1)** has proved to be one of the most problematic and is the subject of the present study. We begin by briefly describing the theoretical and experimental support for the other steps. A complex between Aβ and Cu(II) results in the reduction of Cu(II) to Cu(I) and the oxidation of the S of Met35 to the radical cation in **Step 1)** (see below). In the majority of cases, the Met side chain is restored in **Step 2)** by the action of endogenous reducing agents such as glutathione (GSH) or vitamin E by hydrogen atom transfer and deprotonation of the resulting sulfonium ion. H-atom transfer is energetically unfavourable in each case because the bond dissociation enthalpy (BDE) of the sulfonium ion, BDE(>S+-H) = 331 kJ mol^{-1}, is less than that of S-H bond of Cys (BDE(S-H) = 366 kJ mol^{-1}, or the O-H bond of a typical phenol, 360 kJ mol^{-1}. However, rapid deprotonation of the resultant sulfonium ion (**Step 5**), p$K_a \approx -5$),[91] will prevent the reverse step. Alternatively, due to the high reduction potential, E°(>S$^+$/>S) = 1.6 V,[92] of dialkyl sulfides, the oxidized Met may also be repaired by direct electron transfer from

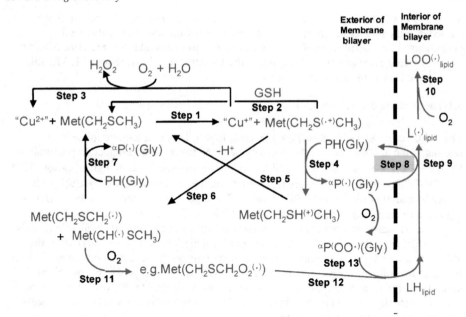

Figure 1 *Radical Model: see the text for an explanation*

endogenous reducing agents such as ascorbate (vitamin C). In **Step 3)**, the Cu(I) is reoxidized to Cu(II) by molecular oxygen in a complex process that has characteristics of superoxide dismutase (SOD). H_2O_2 is produced but superoxide ($O_2^{(\cdot-)}$) is not detected as an intermediate. It is a possibility that the hydrogen peroxide, or hydroxyl radicals derived from it, are the damaging species that cause oxidative stress. We consider this to be unlikely because damage caused by these species is probably very short range (close to the Cu binding site) because of the high reactivity of OH˙. The cytotoxicity of Aβ is thought to be maximum while in soluble oligomeric form. While the structures and state of aggregation of oligomeric Aβ are not known, it is reasonable to assume that it will have some characteristics in common with its structure in Aβ fibrils. These have been shown to consist of in-sync parallel β-sheets in which individual Aβ monomers have a hairpin fold at residues 24-26, probably stabilized by a salt bridge between Asp23 and Lys28. In the monomeric or small oligomeric form, such a turn may also stabilize an internal antiparallel β-sheet which brings the Met35 residue into close proximity of the Cu-binding region at His13 and His14, thus permitting opportunistic oxidation of the Met residue by bound Cu(II). In the *Radical Model*, it is postulated that while in the toxic oligomeric form with β-sheet secondary structure, the Met35 radical cation may also be reduced by the $^\alpha$C-H bond of a glycine residue, thereby generating a Gly peptide backbone radical **(Step 4)** in an endothermic step. However, the resulting sulfonium ion will be rapidly deprotonated **(Step 5)**, completing the Met35 part of the cycle. The *Radical Model* admits the prior deprotonation of the oxidized Met **(Step 6)**, $pK_a \approx 0$,[93] yielding a C-centered radical that can also generate a Gly radical **(Step 7)**. It is a key postulate of the *Radical Model* that the radical propagation steps from Cu(II) to

Met35$(^{+\cdot})$ to Gly takes place outside a cell membrane but that the Gly radical may be carried into the membrane (**Step 8**) during pore formation, and there generate a pentadienyl-like radical (**Step 9**), the first event of lipid peroxidation (**Step10**). Alternate steps involving prior addition of oxygen to the C-centered radicals (**Steps 11, 12,** and **13**)[93,94] can lead to the same outcome.

3.1 The Radical Model – Step 1

The oxidation of Met35 by Cu(II)/Aβ (Step 1, Figure 1) is problematic for several reasons. First and foremost is the disparity between the measured reduction potential of Cu(II)/Aβ, 0.7 V versus S.H.E.,[95] and that of a typical dialkyl sulfide radical cation, 1.6 V (for CH_3SCH_3). If these values are representative, the oxidation of Met by Cu(II)/Aβ would be endothermic by about 90 kJ mol^{-1}. The nature of the binding of the Cu(II) to Aβ is poorly understood, as is the structure of Aβ itself, with or without bound Cu(II). There is no reliable assessment of the binding affinities of biologically available ligands (bioligands, from the side chains of amino acid residues) for Cu(II) or Cu(I), or of the effect of various ligand combinations and orientations on the reduction potential of Cu(II) bound to these ligands. Specifically, the role of Met as a ligand is not known. In the absence of structural information, no assessment is available of the probability that the Met may actually be a ligand within the coordination sphere of the Cu(II). We address each of these questions in the following sections.

3.1.1 Model studies of Cu(II) and Cu(I) coordination complexes with bioligands. The structures of the four-coordinate square planar Cu(II) complexes with bioligands that are available in Aβ have been modelled by $Cu(II)(NH_3)_i(H_2O)_{3-i}X$, $i = 0, 1, 2$, or 3, and X is a model for a possible Cu-binding ligand from Aβ: $X = S(CH_3)_2$, representing Met, CH_3NH_2 for Lys, or the N-terminus, MeImid (4-methylimidazole for His), $C_6H_5O^-$ for Tyr, and $CH_3CO_2^-$ for Glu, Asp, or the C-terminus.[96] The NH_3 molecules are surrogate representatives of other nitrogen-donating ligands and approximate the effect of coordination of additional CH_3NH_2 or MeImid groups. The water molecules play the role of explicit waters of hydration but also may represent the hydroxyl functionality of Ser or Thr, or other oxygen-donation groups, including amide carbonyl groups. All of the complexes possess four-coordination with slightly distorted square planar geometry. Examples of complexes with $X = S(CH_3)_2$ are shown in Figure 2a. A *trans* effect is evident. Structures with ammonia and/or MeImid groups disposed *anti* to each other are more stable than the respective *syn* isomers. For instance, structure **1b** is more stable than structure **1c** (Figure 1a) by 17 kJ mol^{-1}. The binding affinities of the ligands as measured by their ability to displace a single water from $Cu(II)(H_2O)_4$ at pH = 7, are in the sequence: MeImid > $CH_3CO_2^-$ > CH_3NH_2 ≈ NH_3 > $C_6H_5O^-$ > $S(CH_3)_2$. All of the ligands except $S(CH_3)_2$ will displace water from $Cu(II)(H_2O)_4$ at pH = 7. That MeImid is the best ligand is not surprising given that 98% of Cu(II) in blood serum is found attached to His. The carboxylate is a close second by virtue of its ability to use the non-coordinated O atom to make a short hydrogen bond to a *syn*-coordinated water molecule. However, the strength of the interaction depends on what else is attached to the Cu(II). For instance, with respect to the ability to displace water from a complex that already has

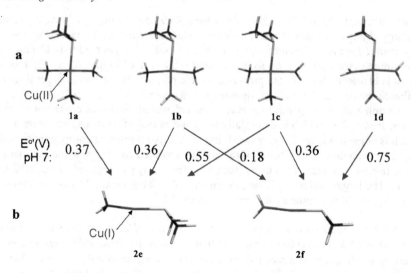

a

Cu(II)

1a   1b   1c   1d

E°'(V)
pH 7:   0.37   0.36   0.55   0.18   0.36   0.75

b

Cu(I)

2e   2f

c $H_2O(NH_3)_2Cu(II):X + 2H^+ + e^- \longrightarrow H_2OCu(I):X + 2NH_4^+$
     X: = $(CH_3)_2S$

Figure 2 *Model complexes of copper with dimethyl sulfide as a ligand: a Cu(II) complexes; b Cu(I) complexes; c Half reaction for the reduction of Cu(II) to Cu(I). The numbers on the arrows are predicted reduction potentials at pH = 7 relative to the standard hydrogen electrode (S.H.E.)*

three N ligands, namely from $Cu(II)(NH_3)_3(H_2O)$, the order is: MeImid > $CH_3NH_2 \approx$ NH_3 > $CH_3CO_2^-$ > $S(CH_3)_2$ > $C_6H_5O^-$. This study suggests that in the observed Cu(II) coordination motif, 3N1O, in agreement with conclusions from experiment, the oxygen-bearing ligand is unlikely to be Tyr[97,98,99] or water.[99]

The complexes of Cu(I) with the same ligand set have also been investigated.[100] The complexes $(CH_3)_2SCu(I)(NH_3)$ **(2e)**, and $(CH_3)_2SCu(I)(H_2O)$ **(2f)**, shown in Figure 2b, are typical. At the present level of theory, all of the most stable structures are dicoordinated and quasi-linear in geometry. Measured by the ability to displace water from $Cu(I)(H_2O)_2$ at pH = 7, the ligand binding order is: MeImid > $S(CH_3)_2 \approx CH_3NH_2 \approx$ NH_3 > $C_6H_5O^-$ > $CH_3CO_2^-$. The affinities for displacing water from $Cu(I)(H_2O)_2$ are somewhat greater than from $(NH_3)Cu(I)(H_2O)$. Thus MeImid is the best ligand for Cu(I), as it is for Cu(II). However, in the case of Cu(I), it is followed closely by $S(CH_3)_2$ which was among the poorest ligands for Cu(II). In the case of Cu(I), neither of the last two ligands, $CH_3CO_2^-$ or $C_6H_5O^-$ will displace water from either $Cu(I)(H_2O)_2$ or $(NH_3)Cu(I)(H_2O)$ at pH = 7.

3.1.2 Reduction Potentials of Model Cu(II) Complexes. The effect of structure and ligand composition on the reduction potentials of Cu(II) complexes has been discussed in detail.[100] Reduction potentials, E°, relative to S.H.E. are computed from the free energy change of half reactions such as shown in Figure 2c by application of the relation, $E°("Cu(II)"/"Cu(I)") = -\Delta G_{(aq)}/F$, where $F = 96.485$ kJ mol^{-1} V^{-1}. $\Delta G_{(aq)}$ is the free energy change for the half reaction for one-electron reduction to which the free energy change for the S.H.E. half reaction, -418 kJ mol^{-1}, has been added. If n protons are consumed in the half reaction for reduction (in the case of Figure 2c, $n = 2$), the potential is pH-dependent. In the present case, the pK_a of NH$_4^+$ is much larger than 7 and the value of the potential at pH = 7, denoted E$^{o'}$, is given by E$^{o'}$ = E$°$ + n*0.414. As seen in the example reductions given in Figure 2, the calculated potentials cover a wide range of values, from 0.2 V, which is essentially the same as that of aqueous cupric ion, to 0.75 V, which is comparable to that measured for Cu/Aβ. An important component for raising the reduction potential is the ability of the complex to shed two ligands upon reduction, since the resultant increase in entropy makes the free energy change of the reaction more negative. The largest values of E$^{o'}$ are obtained with $(CH_3)_2S$ (i.e. Met) as the ligand since dialkyl sulfides stabilize Cu(I) relative to Cu(II).

    3.1.3 Complexation of copper to two imidazoles. We have also examined the binding of Cu(II) and Cu(I) to a compound that contains two imidazole rings and an intervening amide group as a model for the H13-His14 region of Aβ.[101] Typical binding motifs are shown in Figure 3. The binding of aqueous Cu(II) to the ligand to give complex **3a** is exergonic by 139 kJ mol^{-1} and that forming **3b** is only slightly less so, 131 kJ mol^{-1}. Both **3a** and **3b** have the two imidazole rings coordinated through their Nπ atoms and the rings are in the favourable *anti* orientation. One coordination site of the square planar complex is occupied by the backbone amide carbonyl (**3a**) or the deprotonated amide N atom (**3b**). The fourth site in the model systems is occupied by a water, but in Aβ itself, may be occupied by any of the ligands discussed above, most probably an N atom from the N-terminus[98,99] but also possibly a third imidazole from

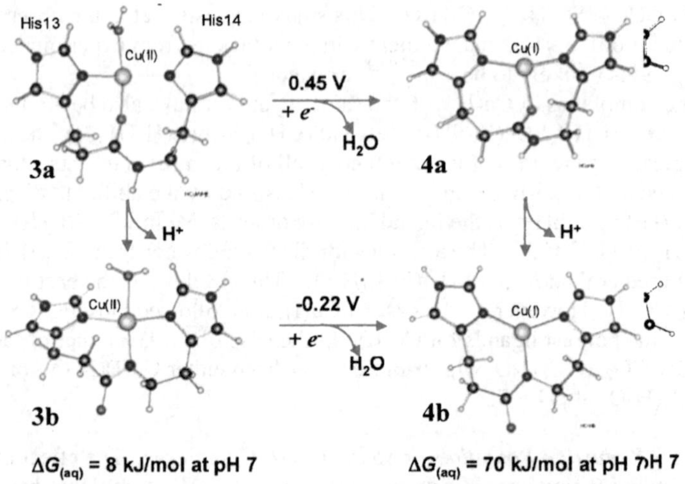

Figure 3 *Complexes of Cu(II) and Cu(I) bound to a model of the His13-His14 region of Aβ. Values in V are reduction potentials at pH = 7 relative to S.H.E.*

His6 or a Lys side chain. Less probable, according to the model studies, is the involvement of the Met or Tyr side chains, although it might be argued that oxidation of the Met would require that it be in a position to coordinate, at least transiently, to the Cu(II).

Reduction of the Cu(II) to Cu(I) yields complexes **4a** and **4b** (Figure 3).[102] As in the model systems shown in Figure 2, reduction of the copper is accompanied by a reduction of the number of coordinating ligands, in this case from four to three. The two imidazoles remain coordinated to the Cu(I) and the backbone involvement cannot be avoided because of the adjacency of the two model His residues. Whichever group was occupying the fourth coordination site of the precursor Cu(II) complex is released. Coordination of the deprotonated amide N atom to Cu(I) (**4b**) is disfavoured by 70 kJ mol^{-1} relative to amide O coordination (**4a**). This difference manifests itself as an approximately 0.7 V difference in the calculated E° values.

The predicted reduction potentials are given in Figure 3: E°(**3a/4a**) = 0.45 V and E°(**3b/4b**) = -0.22 V. The results clearly indicate that the species (of Cu/Aβ) for which the high reduction potential, 0.7 V, was measured cannot have a deprotonated amide group coordinated to the copper, since such a bonding motif yields very low reduction potentials. The higher value of E°(**3a/4a**) is still lower than the measured value. In Figure 2 it was seen that involvement of the Met in the reduction process may result in higher E° values provided that the Met remains bound to the reduced copper. In the more realistic model shown in Figure 3, the Met would have had to have been the fourth ligand of the Cu(II) complex (in place of the water), but it is likely that it is ejected from the Cu(I) in favour of the more strongly coordinating His side chains. We explore below the possibility that the Met was initially involved but departs, possibly in oxidized form as the Met S radical cation, Met(S$^{+\cdot}$).

3.1.4 Coordination of Cu(II) to the Met Residue. The possible coordination complexes between *N*-formylmethioninamide and copper have been investigated in detail as a model system for the possible coordination of Cu(II) to the Met35 residue of Aβ.[103] The most stable complex **5** is shown in Figure 4. Despite the seemingly

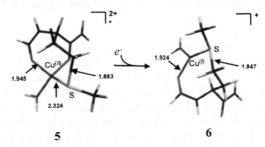

5   **6**

Figure 4 *Structures of Cu(II)* **5** *and Cu(I)* **6** *complexes with a Met residue in a peptide environment. Distances are in Å.*

favourable three-point chelation involving the S atom and the oxygen atoms of both flanking amide groups, **5** is only bound by 19 kJ mol^{-1} with respect to dissociation into aqueous Cu(II) and the Met model. The corresponding Cu(I) complex **6** is also shown in

Figure 4. In **6**, the number of ligands is reduced to three not by release of the bound water molecule but by release of one of the carbonyl oxygen atoms. The stability of **6** with respect to dissociation, 16 kJ mol^{-1}, is similar to that of **5**. In Aβ, the water molecule would be replaced by a more strongly binding residue, such as His, and the overall binding energy of copper to the Met region may be higher. The predicted reduction potential, $E°(5/6) = 0.41$ V, is comparable to that found in the case of copper bound solely in the His13-His14 region of Aβ.

3.2 Oxidation of Met by Cu(II) complexes

We have examined the possibility that a Met residue may depart the copper coordination sphere in oxidized form, thereby reducing the copper. An example involving model complexes of the type discussed above, specifically structure **1b** in Figure 2, is shown in Figure 5a. The voltage change at pH = 7 for the reaction is

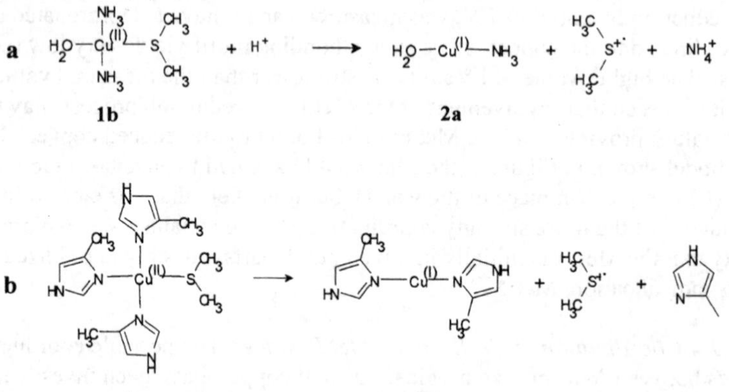

Figure 5 *Equations showing oxidation of a dialkyl sulfide (dimethyl sulfide) by Cu(II) from a coordination complex a Cu(II) in 2N1O complex. b Cu(II) in a 3N complex.*

calculated to be −1.11 V, corresponding to an endothermicity of about 110 kJ mol^{-1}. Figure 5b depicts a hypothetical case in which a Cu(II) complex consisting of three His residues (modelled by 4-methylimidazole, MeImid) and a bound Met (modelled by (CH$_3$)$_2$S) dissociates to yield a *bis*(MeImid)Cu(I) complex, a free MeImid, and the oxidized sulfide. This reaction is predicted to be endothermic by about 65 kJ mol^{-1} (ΔE° = -0.66 V). The relatively low endothermicity is due to the instability of the oxidized form, **7**, which is predicted to dissociate in water. It is possible that the tertiary structure of Cu/Aβ or its oligomeric form, may hold the Met residue in a position to coordinate the S to the Cu(II), but such an eventuality remains to be demonstrated.

3.3 Stabilization of one-electron oxidized Met by electron pair donors in an aqueous environment

Stabilization of the S radical cation of oxidized Met, Met(S$^{+\cdot}$), by electron pair donors has been invoked as a mechanism for reducing the high oxidation potential ($\approx$1.6 V for a typical dialkyl sulfide) thereby facilitating the redox chemistry of **Step 1** of the *Radical Model* (Figure 1). More specifically, it has been proposed that complexation of the S radical cation to an accessible amide carbonyl O atom or to a C-terminal carboxylate group, is the key enabling chemical step in the oxidation/reduction and is responsible for the neurotoxicity.

3.3.1 Three-electron Bonding. The mechanism of stabilization by three-electron bonding (or σ,σ^* bonding) is understood by means of the orbital interaction diagram shown in Figure 6. The singly occupied non-bonding molecular orbital n_S (essentially, the 3p orbital) of a dialkylsulfide radical cation (left hand side) interacts with a doubly-occupied highest MO of the donor, shown on the right hand side

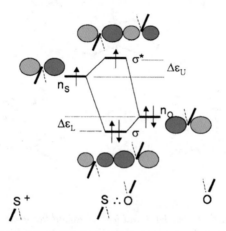

Figure 6 *Orbital interaction diagram for three-electron bonding. The σ bondiing combination is lowered by an amount $\Delta\varepsilon_L$ that depends on the energy separation of n_S and the electron-pair donor orbital, and their σ-type overlap. The σ^* combination is raised by a similar amount if the overlap is small, leading to stabilization.*

as the non-bonded orbital n_O (essentially a 2p orbital) of a water molecule. The two orbitals interact to give a doubly occupied σ orbital, and a singly occupied σ^* orbital. If the overlap is not very large (as is the case with 3p-2p overlap), the lowering of the σ combination, $\Delta\varepsilon_L$, is about the same as the raising of the σ^* combination, $\Delta\varepsilon_U$, and because two electrons go down and one goes up, there is a net stabilization. The so-called three-electron bond has an effective bond order of about ½. The stabilization may be increased, i.e., $|\Delta\varepsilon_L|$ and $|\Delta\varepsilon_U|$ increased, if the interacting orbitals are closer in energy, namely if the energy of the donor orbital were higher. This would be the case if the donor were a negatively charged O atom as in a carboxylate group, or if it were a N atom, as in an amino group.

3.3.2 Reduction Potential of Mid-chain Met Radical Cation, Met(S⁺˙). We have examined the effect on the reduction potential of Met(S⁺˙) of various biologically accessible electron pair donors.[104] Oxidation of *N*-acetylmethioninamide **9** (a model for Met in a peptide environment) yields a mixture of three-electron-bonded structures, four of which are shown in Figure 7. The most stable of these is **9O**$^{(i-1)}$ in which the oxidized

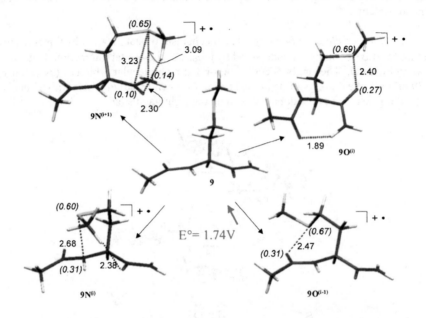

Figure 7 *Model of a mid-chain Met **9** and four oxidized three-electron bonded forms. Distances are in Å. Numbers in italics are spin densities. Reduction potential is relative to S.H.E.*

sulfide makes a three-electron bond to the carbonyl oxygen atom of what would be the *i*-1 residue in the peptide chain. Reduction of **9O**$^{(i-1)}$ has a calculated potential, $E°(9O^{(i-1)}/9) = 1.74$ V, as shown in Figure 7. This value is even higher than for an isolated (i.e., aquated) dialkyl sulfide radical cation, indicating that the three-electron bonded structure, while a minimum in the gaseous phase, is not stable in aqueous solution. The three-electron bond between Met(S⁺˙) and an amide carbonyl donor does not form in aqueous solution. Therefore there is no possibility of facilitating the oxidation of Met by the intervention of amide carbonyl oxygen atoms.

3.3.3 Reduction Potential of C-Terminal Met Radical Cation. The eleven-residue fragment of Aβ, Aβ(25-35), is even more toxic than Aβ(1-42). The increased toxicity has been attributed to the fact that the Met residue is at the C-terminus and the oxidized form can be stabilized by very strong three-electron bonding, as shown in structure **10a** in Figure 8. Structures **10** and **10a** are models for the reduced and S-oxidized forms, respectively, of a C-terminal Met. The predicted reduction potential, E°(**10a/10**) = 1.99 V, is even higher than for a mid-chain Met (Figure 7). While it is true

Figure 8 *a Model C-Terminal Met **10** and **b** its oxidized form **10a**. Distances are in Å. Numbers in italics are spin densities. Reduction potential is relative to S.H.E.*

that the three-electron S∴O bond of **10a** is quite strong in the gaseous phase, it is destabilized in water by preferential solvation of the charged reduced form compared to the neutral oxidized form. Therefore, whatever the explanation for the increased toxicity of Aβ(25-35), it is unlikely to be carboxylate accelerated oxidation of the Met.

3.3.4 Reduction Potential of N-Terminal Met Radical Cation. The possibility that stabilization of the S radical cation may be effected by S∴N bond formation was examined for the case of a N-terminal Met, modelled by structures **11**, **11**[(+)], and **11a**, as illustrated in Figure 9. At pH = 7, the N-terminus is protonated and the proton must be removed before stabilization by S∴N bond formation can occur. The oxidized form, **11a**, has a short S∴N bond and a large amount of spin transfer is evident (Figure 9). Reduction of **11a** is pH-dependent; the value of the reduction potential at pH=7 is E°(**11a, H⁺/11**[(+)]) = 1.11 V.[104] Thus S∴N bond formation rather than S∴O bond formation is predicted be effective in facilitating the oxidation of a Met residue. Of course, Aβ does not have an N-terminal Met, but it does have an N-terminus and two Lys residues at positions 16 and 28. If any one of these can approach the Met during the oxidation process, its amino group may assist.

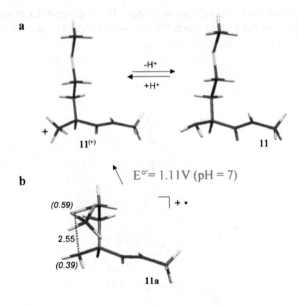

Figure 9 *a Model N-Terminal Met 11, 11⁺ and b its oxidized form 11a. Distances are in Å. Numbers in italics are spin densities. Reduction potential is at pH = 7, relative to S.H.E.*

4 CONCLUSIONS

The theoretical and computational analysis described above supports the 3N1O coordination motif for the cupric ion bound to Aβ. Two of the N atoms and the O ligand originate from His13 and His14 and the carbonyl O atom of the intervening amide group. Involvement of the deprotonated backbone amide group at neutral or higher pH is supported, but it cannot be the species with the high reduction potential. The third N ligand is probably reversibly bound and may be any of His6, the N terminal amino group, or an amino group from one of the Lys residues. The S atom of Met35 is unlikely to be bound strongly, but must be accessible to the copper coordination sphere. One-electron oxidation of the Met sulfur atom by the Cu(II)/Aβ complex requires stabilization of the S radical cation by three-electron bonding to an amino group. Such bonding to an amide carbonyl group or to a carboxylate group does not lead to stabilization in an aqueous environment.

MECHANISTIC MODELING IN DRUG DISCOVERY: MMP-3 AND THE HERG CHANNEL AS EXAMPLES

Jian Li, Ramkumar Rajamani,[1] Brett A. Tounge, and Charles H. Reynolds

Johnson & Johnson Pharmaceutical Research and Development, P.O. Box 776, Welsh and McKean Roads, Spring House, Pennsylvania 19477, USA
[1] Current address: Bristol-Myers Squibb Pharmaceutical Research Institute, Wallingford, Connecticut, USA

1 INTRODUCTION

Computer modelling has assumed a large role in modern drug discovery. Computational approaches are used to design drug molecules with improved potency, selectivity, bioavailability and reduced toxicity.[1] Since drug discovery is focused on developing commercially viable drugs, most computer modelling is very applied. Often largely empirical or highly fitted models are used to answer fairly narrow questions surrounding ligand binding or other molecular properties. In some cases, however, it is necessary to delve deeper into the fundamental mechanisms involved in protein targets of interest. Examples include examining the protonation states of titratable residues, modelling dynamical properties of proteins, and studying reaction mechanisms that are relevant to protein or ligand function.[2-4]

The methods available to computational chemists have never been more capable. It is now possible to carry out very detailed simulations on large macromolecular complexes, including in some cases detailed electronic structure calculations. These new capabilities and the relentless increase in computer power now widely available have made it possible to address two long-standing issues: the flexibility of proteins, and chemical reactions in macromolecular systems where bonds are being made or broken. We have investigated two protein targets that are emblematic of these two issues. The first is the metalloproteinase, MMP-3 enzyme. MMP-3 has a critical zinc in the active site that presents special problems for efforts to model this enzyme using classical force fields. One is forced to make a decision with regard to the electronic state of the zinc and the nature of the interaction between zinc and any putative ligand. This is a case where examining the active site with a method capable of modelling the zinc's reactivity would be invaluable. The second target is of interest not as a traditional medicinal chemistry target, but as a potential source of cardiac toxicity liability to be avoided. This is the hERG potassium channel. In the case of this ion channel there is experimental evidence that it adopts at least two configurations: open and closed. Determining the structure experimentally is unlikely because of the well know difficulties associated with obtaining crystal structures of membrane bound proteins. While there are homologous bacterial ion channel structures available that can be used for homology modelling, one is still left with the question of multiple possible channel states and how these states might affect binding. In both examples we have used physically rigorous chemical models to examine the basic mechanisms responsible for ligand binding.

2 METHOD AND RESULTS

2.1 MMP-3

2.1.1 Linear Scaling Quantum Mechanical Methods. Although classical force field methods are widely used in many aspects of mechanistic modeling of ligand-protein binding, they are limited in their ability to model chemical or biochemical process involving bond breaking/forming, proton transfer, charge transfer, or metal ion coordination. These systems are better treated by quantum mechanical (QM) methods. However, conventional QM calculations scale as N^3 to N^5 to the number of basis functions, which makes these methods prohibitively expensive for proteins even using semi-empirical approaches. To overcome this bottleneck, linear scaling QM methods have been developed that scale more favourably with molecular size. One commonly used approach is the divide-and-conquer (D&C) scheme.[5] In this scheme, a large molecular system is partitioned into a set of small subsystems. Another approach is based on solving the SCF equation by using localized molecular orbital (LMO).[6] This approach has been implemented in the semi-empirical QM software package MOZYME with a PM5 Hamiltonian.[7] The MOZYME implementation of the linear scaling QM method is very efficient, allowing geometry optimization to be carried out for modest sized proteins.

2.1.2 Nature of the Zinc-Ligand Bond in MMP-3. The MMP-3 active site contains a Zn ion coordinated with three imidazole rings from His201, His205, His211, and a water molecule. When an inhibitor binds to MMP-3, this water molecule is displaced by a zinc binding group (ZBG) in the inhibitor such as hydroxamate or carboxylate. In the second coordination shell, Glu202 is about 4.5 Å from Zn ion and plays a critical role in shuttling protons in the active site. This Glu is essential for the enzyme's catalytic function. The positive charge on the Zn ion alters the electrostatic environment of the active site and influences the protonation states of Glu202 and the ZBG group. In addition the protonation state of the ZBG plays a key factor in the potency of the inhibitors. Therefore, a clear picture of the ZBG protonation state is essential to the design of potent inhibitors.

We carried out MOZYME PM5 calculations on MMP-3 complexed with two inhibitors, inhibitor **1** with a carboxylate ZBG and inhibitor **2** with a hydroxamate ZBG.[8,9] The crystal structures of these two complexes have been reported (PDB codes 1hy7 and 1g05). We optimized the geometry of the inhibitors and all protein side chains while keeping the protein backbone fixed.

The MOZYME calculations were initiated with a neutral inhibitor and the titratable residues in MMP-3 assigned charges consistent with physiological pH. Glu202 is deprotonated. After geometry optimization the proton in the hydroxamate or caboxylate of inhibitors **1** and **2** are spontaneously transferred to Glu202 resulting in a neutral Glu202 overall negatively charged inhibitor. The MOZYME optimized geometry of the MMP-3 active site complexed with inhibitor **1** is given in Figure 1. It clearly shows that the carboxylate group is coordinated to the Zn ion in a bi-dentate mode and that the acid group in Glu202 is protonated. The same result is obtained for the MMP-3 inhibitor **2** complex. These results indicate that binding to the Zn ion decreases the pKa of the ZBG. A result that is consistent with experimental observation of the pKa reduction for zinc-bound water in zinc peptidases.[10] This is also consistent with results reported for another zinc contained enzyme, TNF-α converting enzyme (TACE), in which density functional calculations for the active site model revealed the same protonation states for the Zn-bound inhibitor and the proximate Glu406 residue.[11]

1	**2**

Charge transfer is another issue for the Zn-bound inhibitors in zinc-containing enzymes that cannot be examined using classical methods. Based on Mulliken charge analysis, it can be shown that the protein transfers 0.3 electrons to the carboxylate ZBG in inhibitor **1** and 0.2 electrons to the hydroxamate ZBG in inhibitor **2**. The charge transfer occurs between the Zn ion and oxygen atoms of the ZBG hydroxyl group. Overall, this charge transfer stabilizes the ligand in the active site and facilitates binding.

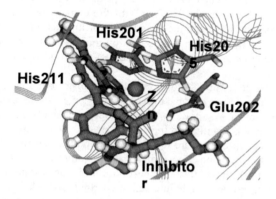

Figure 1 *MOZYME optimized binding mode of inhibitor 1 in MMP-3 active site.*

2.2 hERG Ion Channel

2.2.1 hERG K⁺ Channel. The human ether-a-go-go related gene (hERG) encodes a voltage-gated K^+ channel that plays a role in the normal repolarization phase of cardiac action potentials. It has been shown that disruption of this channel due to ligand binding can result in drug-induced long QT syndrome (LQTS) where normal repolarization is delayed.[12,13] This can result in an arrhythmia that under certain circumstances leads to sudden death. Since many drugs are known to interact with this channel, screening against hERG has become a common practice early in the drug discovery process.[14]

Electrophysiology studies have shown that the channel must be in the active open state for most ligands to bind. This is consistent with mutagenesis data that indicate that most ligands bind to the channel in a cavity under the ion selectivity filter that is only accessible in the open state.[15-18] However, once a ligand enters the pocket the channel can close to varying degrees depending of the nature of the bound ligand. Thus understanding this process of opening and closing the pore is fundamental to understanding how ligands bind to hERG.

2.2.2 Modeling the hERG Channel. Two crystal structures of bacterial K^+ channels are available. One, KcsA (pdb code 1K4C), represents the closed state of the channel while the other, MthK (pdb code 1LNQ), represents the open state.[19,20] By aligning these crystal structures we were able to identify the motion associated with pore opening and closing. This movement is characterized by a tilt of the pore forming S6 helix at the glycine hinge and can be captured by defining a rotation axis perpendicular to the S6 helical axis.[21] By varying the degree of this rotation we are able to produce any number of intermediate states of the channel. For each of these states we used PRIME to re-predict the sidechains.[22] In addition, each state was subjected to a short protocol of heating (0.4 ps), equilibration (0.6 ps), and dynamics (5 ps) to relieve any strain introduced by the helix translation.

2.2.3 Binding Affinity. As a first step in building a binding affinity model, we chose two snapshots of the channel (Figure 2). One state was a 10° tilt away from the KcsA structure, the other was a 19° tilt. With these states in hand, we used docking (GLIDE) to find appropriate ligand poses and a Linear Interaction Energy (LIE) variant to extract the changes in the van der Waals (Δvdw) and electrostatic (Δele) terms (bound verse free ligand) to build a binding affinity model of a set of hERG binders.[23,24]

The intermediate states proved to be crucial in terms of building a predictive binding affinity model. Single state models provided very poor fits to experimental binding affinities. However, even a simple two state model drastically improved the resultant models. This difference is highlight in the case of astemizole (Figure 2). The docked pose and computed interaction energy are drastically different in the two channel states. For the partially open state we find a pose that makes relatively week contacts with the S6 helices (−47.7 kcal/mol). In contrast, the best pose in the open state fills the entire cavity resulting in a much more favourable interaction energy (−164.9 kcal/mol). In the case of cisapride, the complex for the open state predicts a weak binder (interaction energy −64.2 kcal/mol), but in the partially open state the interaction energy is −87.1 kcal/mol (experimental pIC50 = 8.19). These results show that selecting the correct state for the channel has a profound effect on the ligand orientation and predicted potency.

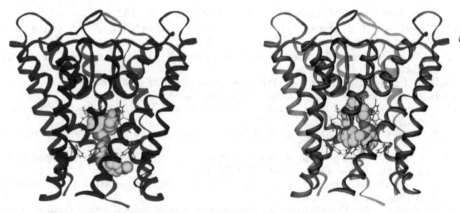

Figure 2 *Astemizole (yellow CPK representation) docked to the partially open (blue) and open (red) state of the hERG K^+ channel. The tight binding of astemizole (pIC50 = 8.83) is only capture in the open state model where the interaction energy is –164.9 kcal/mol. The interaction energy for the partially open state is only –47.7 kcal/mol.*

3 CONCLUSION

We have described two examples where mechanistic studies have provided significant insights into the inner workings of protein systems that are important pharmaceutical targets. In each case this more mechanistic understanding of the protein and interactions with its ligand is useful in better understanding the molecular interactions and properties important for effective ligand binding.

References

1   W. L. Jorgensen, *Science,* 2004, **303**, 1813.
2   M. Garcia-Viloca, J. Gao, M. Karplus, and D. G. Truhlar, *Science,* 2004, **303**, 186.
3   K. K. Lee, C. A. Fitch, J. T. Lecomte, and B. E. Garcia-Moreno, *Biochemistry,* 2002, **41**, 5656.
4   R. Rajamani and C. H. Reynolds, *J. Med. Chem.,* 2004, **47**, 5159.
5   W. Yang, *Phys. Rev. Lett.,* 1991, **66**, 430.
6   J. J. P. Stewart, *Int. J. Quant. Chem.* 1996, **58**, 5674.
7   J. J. P. Stewart, *J. Mol. Mod.* 2004, **10**, 6.
8   M. G. Natchus, R. G. Bookland, M. J. Laufersweiler, S. Pikul, N. G. Almstead, B. De, M. J. Janusz, L. C. Hsieh, F. Gu, M. E. Pokross, V. S. Patel, S. M. Garver, S. X. Peng, T. M. Branch, S. L. King, T. R. Baker, D. J. Foltz and G. E. Mieling, *J. Med. Chem.* 2002, **44**, 1060.
9   M. G. Natchus, M. Cheng, C. T. Wahl, S. Pikul, N. G. Almstead, R. S. Bradley, Y. O. Taiwo, G. E. Mieling, M. Dunaway, C. E. Snider, J. M. McIver, B. L. Barnett, S. J. McPhail, M. B. Anastasio and B. De, *Bioorg. Med. Chem. Lett.* 1998, **8**, 2077.

10 W. N. Lipscomb and N. Strater, *Chem. Rev.* 1996, **96**, 2375.

11 J. B. Cross, J. S. Duca, J. J. Kaminski and V. S. Madison, *J. Am. Chem. Soc.,* 2002, **124**, 11004.

12 J. I. Vandenberg, B. D. Walker, and T. J. Campbell, *Trends Pharmacol. Sci.*, 2001, **22**, 240.

13 R. A. Pearlstein, R. Vaz, and D. Rampe, *J. Med. Chem.*, 2003, **46**, 2017.

14 B. Fermini, and A. A. Fossa, *Nat. Rev.*, 2003, **2**, 439.

15 J. S. Mitcheson, J. Chen, M. Lin, C. Culberson, and M. C. Sanguinetti, *Proc. Natl. Acad. Sci. U.S.A.,* 2000, **97**, 12329.

16 K. Kamiya, J. S. Mitcheson, K. Yasui, I. Kodama, and M. C. Sanguinetti, *Mol. Pharmacol.*, 2001, **60**, 244.

17 J. A. Sanchez-Chapula, R. A. Navarro-Polanco, C. Culberson, J. Chen, and M. C. Sanguinetti, *J. Biol. Chem.,* 2002, **277**, 23587.

18 J. A. Sanchez-Chapula, T. Ferrer, R. A. Navarro-Polanco, and M. C. Sanguinetti, *Mol. Pharmacol.,* 2003, **63**, 1051.

19 Y. Zhou, J. H. Morais-Cabral, A. Kaufman, and R. MacKinnon, *Nature,* 2001, **414**, 43.

20 Y. Jiang, A. Lee, J. Chen, V. Ruta, M. Cadene, B. T. Chait, and R. MacKinnon, *Nature,* 2003, **423**, 33.

21 Y. Jiang, A. Lee, J. Chen, M. Cadene, B. T. Chait, and R. MacKinnon, *Nature,* **2002**, *417*, 523.

22 Prime; 1.0 ed.; Schrodinger, Inc.

23 T. A. Halgren, R. B. Murphy, R. A. Friesner, H. S. Beard, L. L. Frye, W. T. Pollard, and J. L. Banks, *J. Med. Chem.,* 2004, **47**, 1750

24 R. Rajamani, B. A. Tounge, J. Li, and C. H. Reynolds, *Bioorg. Med. Chem. Lett.,* 2005, **15**, 1737.

Subject Index